73 Structure and Bonding

W0042920

Editors:
M. J. Clarke, Chestnut Hill
J. B. Goodenough, Oxford · J. A. Ibers, Evanston
C. K. Jørgensen, Genève · D. M. P. Mingos, Oxford
J. B. Neilands, Berkeley · G. A. Palmer, Houston
D. Reinen, Marburg · P. J. Sadler, London
R. Weiss, Strasbourg · R. J. P. Williams, Oxford

Noble Gas and High Temperature Chemistry

With contributions by
D. Cremer   G. Frenking   K. Hilpert
C. K. Jørgensen   G. B. Kauffman

With 38 Figures and 43 Tables

Springer-Verlag
Berlin Heidelberg GmbH

ISBN 978-3-662-15049-8   ISBN 978-3-540-46880-6 (eBook)
DOI 10.1007/978-3-540-46880-6

Library of Congress Cataloging-in-Publication Data
Noble gas and high temperature chemistry/with contributions by D. Cremer . . . [et al.].
p.   cm. –(Structure and bonding; 73)

1. Gases, Rare. 2. High temperature chemistry.
3. Cosmochemistry. I. Cremer, D. (Dietes), 1944-. II. Series.
QD461.S92 vol. 73 [QD162] 541.2′2 s–dc20 [546′.75] 90-9434 CIP

© Springer-Verlag Berlin Heidelberg 1990
Originally published by Springer-Verlag Berlin Heidelberg New York in 1990.
Softcover reprint of the hardcover 1st edition 1990

Typesetting: Macmillan India Ltd, Bangalore-25;
2151/3140-5 4 3 2 1 0 – Printed on acid-free paper

Table of Contents

Historical, Spectroscopic and Chemical Comparison of Noble Gases

Christian K. Jørgensen[1] and Gernot Frenking[2]

[1] Section de Chimie, Université de Genève, CH 1211 Geneva 4, Switzerland
[2] Fachbereich Chemie, Philipps-Universität Marburg, Hans-Meerwein Strasse, D 3550 Marburg, West Germany

The closed-shell configuration of noble gas atoms Ng does not prevent formation of compounds, either as even, positive oxidation states of xenon, isosteric with iodine complexes (and to a smaller extent by krypton and radon) or functioning as Lewis bases. In condensed matter, Ar, Kr, and Xe form distinct $NgCr(CO)_5$ and $ArCr(NN)_5$ complexes. Gaseous noble gas molecular ions, especially HeX^+ and ArX^+, numerous organo-helium cations, and some neon-containing cations are calculated to be quite stable, and several of them are indeed detected in mass-spectra. The history of Ng chemistry and its relations with the Periodic Table, atomic spectra, and ionization energies, are discussed.

Structure and Bonding 73
© Springer-Verlag Berlin Heidelberg 1990

1 Noble Gas History and Atomic Spectra

When developing atomic spectra as a method of analysis 1859–1861, Bunsen and Kirchhoff discovered caesium and rubidium. In the following few years, thallium, indium, gallium (and, in a way, also europium, though its oxide was not prepared until 1901) were at first detected by spectral lines [1] of samples containing quite low concentrations. A major result of spectral analysis was that the *Fraunhofer lines* (thoroughly studied first in 1815, though first perceived by Wollaston in 1802) of the Sun were shown, to a large extent, to be absorption lines corresponding to some of the emission lines known from arc spectra (in the laboratory) mainly of iron atoms, but also of H, Na, Mg, Al, Si, Ca, Ca$^+$, Ti, V, Cr, Mn, Co, Ni, Cu, Zn, Sr, Sr$^+$, Zr and a few heavier elements. This choice of detected elements [2] is not a fair representation of the relative concentrations in the solar photosphere. It is recognized today that nearly all stars (with a few remarkable exceptions [2–4] containing exorbitant concentrations of either manganese, gallium, europium, dysprosium, osmium, platinum, mercury, bismuth or uranium) have a roughly standard composition, 1 ton (10^6 g) containing about 750 kg H, 230 kg He (our major subject), < 1 g Li, Be and B (all readily undergoing thermonuclear reactions), 13 kg C + N + O, and about 4 kg of all the higher atomic numbers, Z, the most abundant (0.6 to 1 kg each) being neon, magnesium, silicon and iron. Elements with Z above 30 represent only 0.7 g/t in the Sun. When Fraunhofer lines of other stars were carefully studied at the end of the last century, it turned out that A-type stars (such as Sirius) having the continuous spectrum representing an absolute temperature T about twice that (5800 K) characterizing G-type stars (such as our Sun), the Balmer series of the hydrogen atoms has the first three lines far stronger than even the two strongest Ca$^+$ lines (within the visible and near ultraviolet region cut off at 300 nm by the ozone layer in the upper terrestrial atmosphere). The hotter B-type stars show absorption line of He, N$^+$, O$^+$, Mg$^+$, Si$^+$, Fe$^+$ and other spark lines (the traditional name for lines belonging to gaseous M$^+$, M^{2+}, ...) and the (rare) stars having T reaching 40000 K show absorption lines of He$^+$.

Helium was first detected [1, 3] during the solar eclipse of August 18, 1868. When the Moon has just covered the solar disc, a very thin (a few hundred km) luminous layer called the chromosphere can be seen for a moment, emitting bright (and not dark) Fraunhofer lines. A very brilliant yellow line (at a slightly shorter wavelength than the two strong Fraunhofer lines of sodium) was ascribed by Janssen to a new element named helium after the Greek name for the Sun. Rather than restricting himself to the scarce and unreliable meteorological conditions of total solar eclipses, Lockyer, in October 1868, invented an ingenious heliograph with a black dish exactly screening out the brilliance of the solar disc, and detected the yellow line, which could not be identified with an atomic emission line seen in the laboratory, as well as a few other, weaker emission lines of the chromosphere, conceivably also due to helium. Because of the low number of visible lines, it seems to have been generally believed that helium is a metallic element of the same type as lithium or indium. Two other elements [3] were proposed, following a similar line of thought, "nebulium" giving emission lines in certain gaseous nebulae (now known to be mainly due to O^{2+}, but also N^+) and "coronium" representing the quite numerous emission lines of the irregularly shaped (radius of order 10^6 km) weakly luminous corona, observable outside the Moon disc at total solar eclipses. The strongest, green, emission line is identified with Fe^{13+}, and other lines to similar, quite high, ionic charges of iron and nickel. Only one of these three supposed elements gained recognition by chemists. In 1895, Ramsay [3] isolated a gas from uranium minerals, clearly showing the yellow, chromospheric emission line (in a Geissler tube, like our neon tube used in advertizing displays). Two months later, 5 more lines coinciding with the unidentified lines of the solar chromosphere were found by Cleve, who isolated a (much purer) sample from the mineral cleveite (UO_{2+x} also containing thorium, lanthanides, and lead originating from the radioactivity of uranium). The latter radioactivity was discovered in 1896 and mg quantities of radium compounds isolated in 1898, but the picture of spontaneous transmutation [5] of trans-bismuth elements and concomitant stopping and neutralization of α-particles to helium atoms was not clarified until 1905.

As discussed further below, helium is really distinct from the other noble gases neon, argon, krypton, xenon and radon (originally the name of the specific isotope ^{222}Rn, like ^{230}Th was called ionium; the element $Z = 86$ was called emanation, or, by a few other authors, niton). Seen from the point of view of an atomic spectroscopist, the neutral atoms have the point in common that the ground state is separated to a great extent (around 0.8 to 0.7 times the ionization energy I_1) from the first excited state. Actually, the reason why the emission lines are readily studied (in Geissler tubes) and in the near infrared and the visible regions (and not just the strong "resonance lines" in the far UV, involving the ground state as lower state) is that the first excited state in all six noble gases is metastable. Its long duration in large glass vessels at low gas pressure, submitted to an electric discharge above 80 V, reflects the quantum-mechanical selection rule that it can neither irradiate by an electric dipole nor magnetic dipole (since it changes the quantum number J by two units), nor by electric quadrupole processes (since it has the opposite parity). In the five heavier noble gases, the metastable level having $J = 2$ belongs to an electron configuration terminating $(np)^5(n+1)s^1$, and in the He atom the $^3S_1$ state belongs to 1s2s (having the same parity as the ground state $^1S_0$ belonging to $1s^2$) which happens to be unable to undergo a magnetic dipole transition. However, we can already see the major difference between the heavier noble gas atoms (ground configuration terminating with six np electrons forming a closed shell) and the helium atom exemplifying the most striking distinction between the spectroscopic [3, 6, 7] and the chemical versions of the Periodic Table. In the neon atom, the first excited configuration $1s^2 2s^2 2p^5 3s$ has the levels $^3P_2$, $^3P_1$, $^3P_0$ and $^1P_1$ found in the same order in the subsequent four noble gases, but with strongly increasing (relativistic) effects of mixing of $^3P_1$ and $^1P_1$. By the same token, the ten J-levels of $(np)^5(n+1)p^1$ and twelve J-levels of $(np)^5(n+1)d^1$ show biunique relations. It is striking [8] that the $3s \rightarrow 3p$ transitions in the sodium atom, $4s \rightarrow 4p$ in potassium, . . . have slightly higher wave-numbers than the average of the 30 allowed transitions between the four J-levels of $(np)^5(n+1)s^1$ and the ten J-levels of $(np)^5(n+1)p^1$. Said in other words, the alkali-metal atom energy difference is perturbed by the absence of both one (np) electron and one unit of Z of the nuclei. That this situation gives an overall small destabilization for the lower Z value is well-known from comparisons of the beginning and the end of a given transition group.

Spectroscopically speaking, the helium atom is an alkaline-earth. This is not the impression one gets from the wavelength in nm of the spin-allowed resonance transition from the ground state $^1S_0$ to $^1P_1$ of $(ns)^1(np)^1$:

$$He(58, 2058); \ Be(235); \ Mg(285); \ Ca(423); \ Sr(461); \ Ba(554); \ Ra(483);$$

$$Zn(214); \ Cd(229); \ Hg(185); \ Yb(399). \tag{1}$$

The two numbers in the case of He are the wavelengths for the transitions $1s^2 \rightarrow 1s2p$ and $1s2s \rightarrow 1s2p$, in the far-ultraviolet and in the infrared, respectively.

A much more convincing argument is that all the readily accessible excited configurations terminate $(ns)^1(n'l)^1$ and hence, that all atoms in the series (1), and presumably also nobelium ($Z = 102$) show the same manifold of low-lying excited states. The exceptions start in the ytterbium atom even with the first level of $4f^{13}5d6s^2$ 23188 cm^{-1} above the ground state (corresponding to 431 nm) whereas $5d^96s^26p$ of the mercury atom does not start before 68887 cm^{-1} above the ground state (the transition to the $^1P_1$ level occurs at 127 nm). It may be noted [3, 9, 10] that the ground state of all the spectroscopic alkaline-earths belong to a configuration entirely constructed from closed shells, the loosest bound being the two electrons in 1s, 2s, 3s, ...

The first noble gas to be isolated in a laboratory was argon in 1894. Though it is not everyday one gains the highest prize of the lottery by refining a specific density by one more digit, Rayleigh wanted to scrutinize the surprising result that "synthetic" nitrogen (reacting NO_2^- with NH_4^+) has a density 0.5% lower than N_2 remaining, when O_2, humidity and CO_2 are removed from air. This observation also had an obscure precursor [1]. Cavendish found in 1785 that exhaustive electric sparking of air to which excess oxygen was added made the nitrogen fix in potash solution (presumably as nitrate and nitrite; this process was industrially developed by Birkeland and Eyde in countries with cheap electricity, but later mainly replaced by the catalyst of Haber combining N_2 and $3H_2$ to ammonia); however, 1 part in 120 remained unreactive. Until as late as 1894, very few reactions of N_2 were known, but metallic magnesium was available, which could form the nitride Mg_3N_2 at red heat. Indeed, Ramsay and Rayleigh found that almost 1 volume percent of atmospheric air is unreactive to incandescent magnesium, and constitutes a new element, the "lazy" (greek) argon. It was startling to find a component of air nearly as abundant as water vapour (and much more abundant than H_2O at low T) when 0.03 volume percent of CO_2 had been known for a century. The first suggestion by Dewar (the inventor of the vessel) and by Mendeleev was an allotropic oligomer N_3 but very unreactive, quite in contrast to ozone O_3 (actually somewhat worrying if N_2 was less stable, much like the ephemeric legend of "polywater" where the Navy would be put out of business, if someone dropped a piece of this substance which turned out to be dirty silicic acid). But during 1898, Ramsay and Travers used the techniques of fractioned distillation of liquid air (and atomic spectra from Geissler tubes) to find in ppm volume (cm^3/m^3) 5 helium, 18 neon ("new"), 1.14 krypton ("hidden") and 0.09 xenon ("strange"), and Mendeleev accepted an eighth column in his Periodic Table, joining the halogens with the alkali-metals. The reason why argon is 500 to 10000 times more frequent in the atmospheric air than the four other noble gases is the presence of 0.0012% of the radioactive isotope ^{40}K in potassium (2.6 weight percent of the Earth's crust) continuously supplying ^{40}Ar. By contrast, ^{36}Ar and ^{38}Ar have very low abundances, only 30 and 6 times that of krypton. The gravitational field readily retains Ar, Kr and Xe, but allows He (and 0.5 ppm of H_2) to escape quite rapidly. Hence, our supply of helium (mainly from several sources of natural hydrocarbon gas) also derives from radioactivity, but this time from the α-particles emitted by elements [5] having atomic weights above 209.

2 Early Attempts to Form Noble Gas Compounds

In 1898, chemists had five elements distinctively far less reactive than N_2, the least reactive gas previously known. A quite febrile activity broke out, with fluorine, distilling metallic sodium, powerful electric discharges, etc. A monument over these attempts is the novel by H. G. Wells "The War of the Worlds", where the Martians attacking London (and later being irreversibly defeated by our common cold) use a toxic brown argon compound (residues giving a green spectral line). The first volume [11] of the 8th Edition of Gmelin, published in 1926, is quite reluctant to accept any noble gas compounds, with exception of the clathrate solids written as hexahydrates. It is now known to be $8Xe, 46H_2O$, and $8Kr, 46H_2O$ derived from the crystal structure of an ice modification found in a certain range of high hydrostatic pressure. The equilibrium pressure of the xenon hydrate reaches 1 atm at $-8\,°C$ and of the krypton hydrate at $-74\,°C$. If kept under 20 atm pressure, the solid xenon hydrate can still be crystallized at $+25\,°C$. On the other hand, the argon hydrate reaches an equilibrium pressure of 1 atm at $-123\,°C$ and 105 atm at $0\,°C$. Since chemists can never know what may happen tomorrow, [5], it is quite apt that Oddo [12] suggested after 1902 that xenon and krypton might be particularly suited to form compounds. The reason American and British writers transcribe lack of reactivity as "inert gas" is intimately connected with the octet rule of the Lewis paradigm [13] from 1916, which has met severe criticism [14, 15], especially for compounds of atoms having Z above 10. In German, the less absolute noun "Edelgas" is used, as for noble metals (difficult, but not impossible to make react) and sometimes the generic symbol "E". In this review, we use "Ng" for "noble gas" (like "Ln" for lanthanides [3, 7] with Z from 57 through 71), because "E" frequently is used in a more extended sense "element". Seen against the background that Pauling, in 1931, prolonged the active life of the Lewis paradigm for almost half a century by incorporating it in the hybridization model, it is fascinating that Pauling [16], in 1933, predicted the existence of several xenon and a few krypton compounds. The argument was the almost invariantly tetrahedral coordination number $N = 4$ for phosphates, arsenates, sulfates and selenates, increasing to $N = 6$ for antimonates, tellurates and the para-periodates $IO_{6-n}(OH)_n^{-5+n}$ (except meta-periodate IO_4^-), and then extrapolating to XeF_6 and XeO_6^{4-} being feasible. Actually, perxenates are globally the most stable xenon compounds, also in aqueous solution at high pH. These increases of N from 3 for nitrate may be mainly a question [14, 15, 17] of relative atomic size, but it cannot be denied that there must be a contribution from covalent bonding to unusually low N values, such as 2 or 4 for the large central atoms of gold(I), mercury(II) and thallium(III).

The effect of Pauling's valid extrapolation in 1933 was unfortunately attenuated by an unsuccessful attempt [18] to fluorinate xenon. The more tragicomic aspect of this, apparently negative, result is that a sizeable decrease of total gas pressure was ascribed to parasitic fluorination of the wall materials. Among several reasons, why nobody had the desire to repeat this experiment, was a tacit assumption that fluorination of a monatomic gas cannot need an activation energy. Although the dissociation energy of 38 kcal/mol of F_2 is unusually low, an activation needing the full dissociation of this diatomic molecule would need quite a long reaction time, or some heating. The respect for the Lewis paradigm impeded creative thought, and in addition, xenon was a very expensive gas. We know today [19] that a high yield of crystalline XeF_2 can be obtained by the photochemical action of sunshine on $Xe + F_2$ mixtures in a few days. In a technical report [20] in 1958, one of us [21] proposed a similar stereochemistry of xenon(II) and mercury(II) and the possibility that cyanogen NCCN and xenon would combine to a linear molecule NCXeCN. We are not aware of this compound being reported, but the mass spectrum [22] of such a mixture shows the peaks due to $Xe(CN)_2^+$. Laszlo and Schrobilgen [72] recently reviewed the diversified attempts to form Ng compounds between 1895 and 1935, and the intricate question whether the preparation of xenon compounds in 1962 was a case of a multiple discovery.

3 Preparative Xenon Chemistry Since 1962, and Related Krypton and Radon Compounds

The breakthrough in 1962 was an experiment by Bartlett [23] obtaining a yellow precipitate by mixing xenon with the dark red–brown gas PtF_6. The species O_2^+ was previously known from molecular spectra of electric discharges (and noted for its shorter equilibrium internuclear distance than O_2; it has only one π-antibonding electron, while O_2 has two) and a similar experiment precipitates $O_2^+ PtF_6^-$. Also colorless $O_2^+ BF_4^-$ is known. Since the ionization energy of the xenon atom is marginally smaller than of the oxygen molecule, the yellow solid was supposed [23] to be $Xe^+ PtF_6^-$, though magnetic and near infrared spectroscopic characteristics of Xe^+ had not been observed. It seems rather [24] to be a mixture of (mainly) $XeF^+ PtF_6^-$ with (smaller amounts of) $XeF^+ F_5 PtFPtF_5^-$ and/or $FXeFXeF^+ PtF_6^-$. Inorganic chemistry is sometimes too complicated to be easily rationalized by beautiful, but less complicated, ideas. Anyhow, the preparation of this yellow salt initiated hectic activity in many laboratories [25–28] and within 7 months, the three binary fluorides had been prepared as colorless crystals containing linear FXeF molecules, or quadratic XeF_4 molecules, or XeF_6 showing a complicated structure of tetragonal-pyramidal XeF_5^+ groups bridged by F^-. As the vapor, XeF_2 and XeF_4 retain essentially the same structure, whereas XeF_6 is the only gaseous hexafluoride molecule known not to remain regular octahedral, but shows a rapid fluxional behavior, as if the force constant for one of the bending normal modes vanishes. Though O_3 (or even O_2) should be thermodynamically capable of oxidizing xenon at pH 14 or 15 (and hence extract the rare element from air), no catalyst nor suitable conditions of reactions are known to us. All $M_4 XeO_6$ perxenates and the highly unstable, tetrahedral XeO_4 and the brisantly explosive white solid XeO_3 are made by cautious hydrolysis (and frequently disproportionation) from the binary fluorides. There are a large number of other xenon compounds known [27, 28] such as $XeCl_2$, $ClXeO_3^-$, $F_5 TeOXeOSeF_5$, ... and many salts of XeF_5^+. The molecules and complex ions are all colorless (or in a few cases pale straw-yellow) except for the salts of the green cation Xe_2^+ representing a fractional oxidation number. The photo-electron spectra of solid perxenates [29] and gaseous binary fluorides [30] yield interesting information about the electronic structure, also of the inner shells.

Looking back on 25 active years of xenon chemistry, it is striking how similar [31, 32] is the stereochemistry of the isoelectronic couples iodine(I) and xenon(II); mainly quadratic iodine(III) and xenon(IV); mainly pyramidal iodine(V) and xenon(VI); and mainly octahedral (rarely tetrahedral) iodine(VII) and xenon(VIII). Corresponding analogs include $ClICl^-$, ICl_4^-, IF_5, IO_3^- and $IO_{6-n}(OH)_n^{-5+n}$. Though the difficult study (the half-life of ^{222}Rn is 3.8 days) of radon fluorides generally has been interpreted [28] as the formation of salts of RnF^+, it is not easy to exclude salts of RnF_5^+. It is conceivable [10] that an isoelectronic series of oxidation states may cross a noble gas, for instance Bi(I), Po(II), At(III), Rn(IV), Fr(V), Ra(VI).

One might have expected oxo-anions [16] of Kr(VI) or Kr(VIII) (BrO_4^- was prepared in 1968), but the only well-characterized krypton compounds are the rather unstable linear molecule FKrF and the Kr(II) salts such as $KrF^+ SbF_6^-$ and $KrF^+ AuF_6^-$ containing an octahedral $5d^6$ gold(V) complex anion. There are good reasons to believe from quantum mechanical calculations (see Sect. 4.3 of

Ref. 51) that ArF^+ might form stable salts, either of BeF_4^{2-} or BF_4^- providing strong Madelung potentials [9, 32] or of SbF_6^- and AuF_6^- known from KrF^+ salts. One is well advised [10] to keep one's eyes open for the possible surprise of salts formed by F_2^+ having lost one of the four anti-bonding π-electrons occurring in the fluorine molecule. Since perchlorates are somewhat more stable than perbromates, it cannot be excluded that tetrahedral ArO_4 (with intrinsic small molar volume) may be formed by compression of argon–oxygen mixtures at very high pressures, and either remain as a metastable phase, or immediately dissociate when the pressure is removed. It may be significant to compare [15] it with the behavior of solid caesium iodide and of xenon at enormous hydrostatic pressures. A comparison with chromate and permanganate suggests that ArO_4 is the only plausible Ar(VIII) compound.

4 Noble Gas Ligands

The noble gas chemistry since 1962 has essentially shown a great number of Xe(VIII), Xe(VI), Xe(IV), Xe(II), and a few Kr(II) compounds prepared. The increasing propensity toward positive oxidation states dramatically growing along the series argon, krypton and xenon, is remarkably comparable to the behavior of chlorine, bromine and iodine, though it must be admitted that Br(VII) and I(VII) are more oxidizing than expected. On the other hand, three Br_2 disproportionate reversibly and rapidly to five Br^- and one BrO_3^- in alkaline aqueous solution. One of the main purposes of this review is to point out that noble gas chemistry has two major sides: one being the oxidation state being an even, positive integer (not at all likely for helium and neon, and only having a calculational basis for argon); the other aspect of noble gas chemistry, which has attracted relatively less attention, is the noble gases acting as Lewis bases coordinated to highly positive centers. The latter possibility had already been explored [33] in 1935 when a marginally decreased freezing-point of various mixtures of boron trifluoride and liquid argon was reported. However, it is almost certain [34] that these observations are due to the uncertain establishment of equilibrium, and it was also shown that liquid krypton and xenon had no perceptible affinity to the otherwise so strong anti-base [35] (Lewis acid; electron-pair acceptor) BF_3. In a proposal from 1969 by Kaufman and Sachs [36] of gaseous HeLiH needing 2 kcal/mol for dissociating the He–Li bond (further discussed in Sect. 6.1 of Ref. 51) evidence is also mentioned for BH_3 (formed from diborane $H_2BH_2BH_2$) being bound to xenon atoms in a cool matrix. Around 1975, two studies gave rather direct evidence for Lewis basicity of neon, argon, krypton and xenon (but not of helium). Perutz and Turner [37] photolyzed colorless $Cr(CO)_6$ dispersed in various solid matrices kept at 20 K to $Cr(CO)_5Ng$ or $Cr(CO)_5L$ where Ng is a noble gas atom and L a polyatomic ligand. These brightly colored species have one strong band in the visible spectrum, and it does not shift gradually as known from usual solvent effects. Thus, the matrix $Ne_{0.98}Xe_{0.02}$ has the higher intensity of the band belonging to the red species in pure xenon, and a somewhat weaker band of the blue species known from solid neon (which almost certainly [38, 39] is a $Cr(CO)_5$ without a definite neon atom attached). It is possible to use the AOM (angular overlap model) treatment of these $3d^6$ systems in tetragonal holohedrized symmetry [32, 40, 41] and establish a spectrochemical series [42] for the sixth ligand:

$$(\text{nothing, or Ne}) < SF_6 < CF_4 < Ar < Kr < CH_4 \sim Xe$$

$$< I^- < NH_3 \ll CO \tag{2}$$

where krypton provides slightly less sub-shell splitting between the 3d-like σ-antibonding and the three 3d-like π-backbonding orbitals than half the value provided by iodide in $Cr(CO)_5I^-$, and xenon slightly more than half the iodide effect. It may seen unfamiliar that closed-shell molecules have a similar spectrochemical position as argon, but the CrFS, CrFC and CrHC asymmetric bridges are not really so different from MXM' bridges familiar [9, 17, 41, 43] from many crystalline and vitreous solids. Anyhow, the coordination chemistry of $3d^6$ chromium(0) must be quite different from higher oxidation states [9, 17, 40, 41] such as $3d^3$ chromium(III) and $3d^6$ cobalt(III) as can already be seen from the fact that the sub-shell splitting is only 1.17 times larger in $Cr(CO)_5NH_3$ than in $Cr(CO)_5I^-$ even at room temperature, whereas the ratio would be 1.6 for positive oxidation states [9]. A related situation occurs in the six $Cr(N_2)_nAr_{6-n}$ species characterized by DeVore [44] in cool argon matrices at 10 K, using vibrational (infrared) spectra. Yellow $Cr(N_2)_6$ has all 13 nuclei on cartesian axes, and at least for n = 5 and 4, there seems to be one or two argon atoms on the remaining coordination positions of the octahedron. It may be worth noting [35] that though most strong anti-bases (electron-pair acceptors) tend to be quite strong oxidants (electron acceptors), the clear-cut counter examples boron(III) and aluminum(III) show that one cannot consider the two categories equivalent, even if one is in the mood for not considering the (rather subtle) arguments [9, 10, 45] for oxidation states (written with Roman numerals).

The noble gas adducts of chromium(0) $NgCr(CO)_5$ and in the iron(I) and manganese(0) species [46] $KrFe(CO)_5^+$ and $KrMn(CO)_5$ and molybdenum(0) complex (calculated [47] bond energy 0.23 eV) $KrMo(CO)_5$ may not be extremely weak; they are just difficult to protect against competition with more strongly bound ligands (even CrHC bridged alkanes) and are essentially metastable species under cryogenic conditions. This is not *so* different from solid yellow CrF_6 decomposing [48] by heating above $-100\,°C$. Before this compound was prepared, it could not be excluded that it were unreactive like SF_6 or, at least, might be handled at room temperature like [49] solid and gaseous UF_6. The important feature allowing study [37, 42] of $NgCr(CO)_5$ is the near-ultraviolet photolytic dissociation of colorless $Cr(CO)_6$ to form the highly reactive blue $Cr(CO)_5$. Mass-spectrometric evidence [50] for a strong adduct of six argon atoms with a gaseous Co^+ ion is most readily interpreted [42, 50] as an octahedral cobalt(I) complex $CoAr_6^+$ having a triplet ($^3A_2$) ground state like octahedral, isoelectronic $3d^8$ nickel(II).

As discussed in our more elaborate review [51] in this volume, the very large majority of helium and argon-containing species known (from mass-spectra or from precise quantum chemistry) to be stable toward dissociation (either to two fragments, or the less severe condition [15] of atomization to species with one nucleus) are *cations*, either diatomic HeX^+ and ArX^+, or organo-helium cations such as $HeCCH^+$, ... This is also true for the (relatively smaller) number of neon-containing species not dissociating exothermally in the gaseous state (and quite different from the large amount of xenon compounds stable in solids and neutral gaseous molecules, not dramatically less stable than isoelectronic iodine compounds having positive oxidation states).

It was previously pointed out [32] that *one* positive charge in NgX^+ enhances probability of detection, since dissociation only can take place to Ng^0 and X^+, or Ng^+ and X^0, favored according to whether I_1 of Ng is higher or lower than of X atoms (cf. Table 1 of Ref. [4] in this volume). If I_2 of X^+ is lower than $I_1 = 24.587$ eV of helium, HeX^{+2} does not have a repulsive potential curve at large internuclear distance (the other HeX^{+2} dissociate to He^+ and X^+ although they may stay together in a shallow potential minimum). This lack of tendency to lose He^+ occurs for all Z from 12 through 17 (Mg to Cl) and from $Z = 20$ to 35 (Ca to Br) and just marginally for $HeKr^{+2}$ making HeX^{+2} far more stable than ArX^{+2} for Z below 20. The situation is more extreme in most lanthanides [3] forming $HeLn^{+3}$ not repulsively dissociating to He^+ and Ln^{+2} with exception of europium ($Z = 63$) and ytterbium ($Z = 70$) having I_3 above I_1 of helium. Hence, gaseous Eu^{+3} and Yb^{+3} are probably too oxidizing to form helium adducts, needing at least 0.11 or 0.44 eV explicit chemical bonding effects, in addition to the compensation of the electrostatic repulsion between Ln^{+2} and He^+ at whatever internuclear distance, where a stable species might occur.

However, these clear-cut arguments for the asymptotic behavior at large internuclear distances in gaseous NgX^{+n} cannot be of help in all circumstances of chemical bonding in diatomic or polyatomic cations containing Ng, and it may be worthwhile turning to the more familiar (and far less quantitative) arguments about electronegativity of Ng and X in the next Section.

5 Electronegativities of Noble Gas Elements

Since the time of Berzelius, there has been a persistent argument used by inorganic chemists that, in many cases, valency as a positive integer should be replaced by negative and positive oxidation numbers adding up to zero in a molecule or solid, and to the ionic charge of a polyatomic cation or anion [9]. The same year, 1916, as the Lewis paradigm was formulated [13, 14], emphasizing valency as two-electron bonds, Kossel [52] proposed a description much closer related to the ideas of Berzelius, and which in more recent times [9, 10, 45] bifurcated into *oxidation numbers* (also defined for M–M contacts in F_5SSF_5, dithionate $O_3SSO_3^{-2}$, and calomel ClHgHgCl and the mercurous ion, being $+5$ in the two former, and $+1$ in the two latter examples) used as a tool for writing redox equations, and *oxidation states* (in a more spectroscopic context, such as number of d-like, 4f or 5f electrons in a ground state). The fact that MX_3 exists for X = chlorine and M(III) = boron, aluminum, phosphorus, chromium, yttrium, rhodium, indium, antimony, iodine, lanthanum, gold, thallium, bismuth and americium does not reflect the comparable extent of electrovalent bonding, expected by most chemists to attenuate along the series La, Y, Am, Al, In, Cr, Bi, Rh, Tl, Sb, B, Au, P, I.

Pauling [53] introduced the *electronegativity* as expressing the decisive factor behind the relative extent of electrovalent bonding, called "the power of an atom in a molecule to attract charge (electronic density) to itself". Originally, he compared the dissociation energy of M–X bonds with the arithmetic mean value for M–M and X–X bonds. Pauling optimized a set of electronegativities in such a way that closest feasible agreement was obtained between the excess dissociation energy of M–X and 1 eV multiplied by the square of $\chi(X) - \chi(M)$. Later, he recognized that this excess dissociation energy rather should be compared with the geometric (i.e. square-root of product) of the (M–M) and X–X dissociation energy. Such values are here called χ_P. In the case of vanishing Ng_2 dissociation energy, very readily dissociating bonds, such as Kr–F, indicate $\chi_P(Kr)$ very close to $\chi_P(F) = 3.9$, and $\chi_P(Xe)$ presumably somewhat lower, but above 3, in the original treatment involving arithmetic mean values; and the rather meaningless $\chi_P(Ng)$ close to half the F–F dissociation energy (or 0.8 eV) for all xenon and krypton fluorides in the "ameliorated" version, using geometric (vanishing) mean values. It may seem far more fruitful to extrapolate the original intuition of Pauling with equidistant χ_P from 2.0 for boron to 4.0 for fluorine, to 4.5 for neon, and the smaller interval from 1.5 for aluminum to 3.0 to chlorine to 3.4 for argon. It is noted that the dispersion of χ_P values decreases dramatically from the 2p to the 5p and 6p groups, as already seen from the opposite trend from Al to Tl compared with the (expected decrease) from P to Bi in the last line of the previous paragraph above. Accepting $\chi_P = 2.8$ for bromine and 2.5 for iodine almost precludes values above 3.2 for krypton and 2.7 for xenon, and would also tend toward 3.4 for argon.

Attempting to derive a quantitative parameter from isolated atoms, Mulliken [54] defined $\chi_M = (I_1 + I_0)/2$ as the arithmetic mean value of the first ionization energy I_1 (the energy difference between the ground state of gaseous M^+ and the ground state of the gaseous atom M^0) and the first electron affinity I_0 (the energy difference between the ground states of gaseous M^0 and M^-). χ_M has certainly a consistent definition, but in spite of many authors [55] considering it "the theoretical basis of χ" it seems rather alien to chemical thought at several points. Gaseous monoatomic entities differ from well-established oxidation states [9, 10] such as nitrogen($-III$), oxygen($-II$) and sulfur($-II$) by *never* having an electron affinity for a second electron, and by quite small I_0 values for most elements [56] except known cases of vanishing I_0 in He, Be, N, Ne, Mg, Ar, (very likely Mn), Kr, Cd, Xe, and Hg atoms. It is a vexed question whether I_0 in any sense can be negative. When performing Born-Haber calculations on crystalline MgO (with a huge Madelung potential) the implied I_{-1} close to -8 eV corresponds to an oxide ion forced to be isoelectronic ($1s^2 2s^2 2p^6$) with a neon atom. If not confined in a solid, O^{2-} would loose the tenth electron, and states as $1s^2 2s^2 2p^5 6h$ (like $1s^2 2s^2 2p^5 117h$ in neon compared to Ne^+) have energies arbitrarily close to the ground state of O^- (i.e. vanishing I_{-1}). Results of electron scattering on mercury atoms show evidence of a resonant Hg^- 0.45 eV above the ground state [57]; but such resonances in the continuum are not stationary states of the Schrödinger equation. This does not prevent that in spite of almost identical electric polarizabilities [58] of the crystalline- and the five gaseous Ng, the solids seem to have an electron affinity 0.56 eV in xenon, 0.46 eV in krypton and 0.35 eV in argon. These values are likely to be higher in the liquids, allowing for some cavitation around the free electron (like liquid ammonia). Anyhow, χ_M is larger for the halogen atoms (Cl 8.29 eV, Br 7.59 eV, I 6.76 eV) than for the subsequent noble gases (Ar 7.88 eV, Kr 7.00 eV, Xe 6.06 eV) to be compared with 7.18 eV for hydrogen and 7.54 eV for oxygen. These values are to a chemist what Einstein said about the thirteenth stroke of a clock; not only is it not plausible, but it throws doubts on the 12 previous strokes.

The strategy for making χ_M more appealing to chemists has mainly involved the increase of χ_M in valence-states pre-adapted to chemical bonding [55, 59] as has also showed up in the recent controversy [60–62] about the predicted chemical behavior of species with central charge $(Z_0 + \frac{1}{3})$ or $(Z_0 + \frac{2}{3})$ containing unsaturated quarks [5]. As a further instance of the tendency to swallow camels and screen mosquitos [3] may be mentioned two, somewhat more impressive, problems. Since I_k of M^{+k-1} tend to be proportional to $k(k+1)$ (as long as a closed electron shell is not crossed) quite redox-unreactive cations can have huge χ_M values, such as 40.5 eV for Li^+, 26.2 eV for Na^+ and 74.2 eV for Al^{3+} (and such values are actually discussed [63] in a different context). Even $\chi = 22.07$ eV for gaseous N^+ is an important ingredient in our description of the betaine molecule $(CH_3)_3NCH_2CO_2$. A transition-group chemist is firmly convinced that χ depends on the oxidation state (as also admitted [64] by Sanderson), e.g. that chromium(VI) and manganese(VII) are quite similar to S(VI) and Cl(VII); and Mg(II) and Al(III) quite analogous to Mn(II) and Cr(III). When discussing χ of atoms in organic molecules, it is more fashionable to speak about "group electronegativities", the substituents $-CH_3$; $-CH_2F$; $-CHF_2$; CF_3 behaving as if their electronegativity smoothly increases from, say, 2.5 to 3.5. The equilibration of χ which is the basis for the system [64] of Sanderson electronegativities χ_S is qualitatively very plausible, but has some problems in the lanthanides [3, 9] and also brings up the question whether 44 fluorine atoms out of 89 atoms in the aliphatic compound $C(CH_2CH_2CF_2CF_2CF_2CF_2CF_3)_4$ significantly increase χ of the 13 inner-most atoms, compared to the $C(C_7H_{15})_4$ molecule. It can be argued in textbooks that the difference between sulfate and sulfide is a similar inductive effect of the four ligands.

The concept of χ equalization as a chemical potential for electrons was introduced by Iczkowski and Margrave [65] and developed to the treatment of *differential ionization energies* [9, 66, 67] including the strong effect of a Madelung potential. There is no doubt that this treatment neglects significant effects of covalent bonding, but it may be the closest one can come to the idea of χ_M deriving chemistry from atomic spectra. The octahedral molecule SF_6, tetrahedral ClO_4^- and d-group compounds have charge separations typically corresponding to central atom fractional charges in the interval between $+1$ and $+3$. Xenon compounds have much less charge distribution, the fractional charge of xenon is $+0.56$ in XeF_2, $+0.92$ in XeF_4 and $+0.90$ in XeO_4 really making xenon and chlorine rather comparable (the atomic I_z has three parameters and cannot be directly compared to χ_M being its value [63] for vanishing z).

As a warning what the Born-Haber model can imply, if the spatial extension of anions is neglected, may be cited a recent prediction [68] of neonides, CsNe (as ionic crystals) achieving 1 atm dissociation pressure at $300\,°C$, NaNe at $-41\,°C$, and CsHe at $-147\,°C$. The major problem here is the dramatic underestimate of the ionic radius of Ne^- to $2.21\,Å$ in spite of its highly anti-bonding 3s-like electron. Any M would be unstable at sufficiently high pressure relative to M^+M^- (or $M^{4+}M^{4-}$) if we are allowed the full Madelung potential [15]. In the simplest terms [32] any stable electrostatic system is x times more stable if all distances are divided by x. Implosion of matter is prevented by the kinetic energy operator in quantum mechanics.

Among the χ_M for noble gas atoms cited in literature but transcribed to a χ_P-like scale, may be mentioned [69] 4.53 (He); 3.98 (Ne); 2.91 (Ar); 2.58 (Kr); 2.24 (Xe); and 1.98 (Rn), surprisingly enough with lower values for Ar, Kr, and Xe than of the preceeding halogen. By the way, "halogen" means "salt-former". The suggestion of Noyes [69] to call the noble gases "aerogens" would be more suitable to a mixture of $Ba(N_3)_2$ and BaO_2 brought to detonation. Two other sets of χ values are evaluated by Bing-Man Fung [70], one set from a three-parameter Iczkowski-Margrave function, corresponds (on χ_P-like scale) to 4.7 (Ne); 3.7 (Ar); 3.4 (Kr) and 3.0 (Xe), and another set derived from the Gordy treatment involving normalized covalent radii, and in this sense [64] closely related to χ_S 4.5 (Ne); 3.5 (Ar); 3.0 (Kr); and 2.7 (Xe). Probably, these values look familiar to most readers (like we all know that a unicorn has two eyes and four legs) but the more nagging question is why it is so obvious to chemists that an element *has* to have *one* χ. On the other hand, χ is not exclusively a scholastic frivolity; one of the most striking manifestations is the *optical electronegativity* χ_{opt} introduced from electron transfer spectra [71] (especially for octahedral hexa-halide complexes MX_6^{+z-6} where $\chi_{opt} = \chi_P(X)$ can be obtained, on the acceptance of the somewhat enigmatic constant $3.7\,eV$, and a moderate increase of $\chi_{opt}(M)$ as a function of increasing oxidation state z) and from photo-electron spectra of solids and of gaseous molecules [3] as well.

As far as noble gases go, the fact that the differential ionization energy [9] makes a large jump at the closed shell, creating problems for other treatments [63] is clearly related to the $\chi(Ng)$ in literature implicitly being intended for donating electronic density to halogen and oxygen atoms. The affinity for increased electronic density is disproportionately far smaller for Ng, and probably similar to magnesium atoms.

6 References

1. Weeks ME (1968) The discovery of the elements, J. Chem. Educ. Publ., Easton PA
2. Trimble V (1975) Rev. Mod. Phys. 47: 877
3. Jørgensen CK (1988) In: Gschneidner KA, Eyring L (eds) Handbook on the physics and chemistry of rare earths vol 11, North-Holland, Amsterdam, pp. 197–292
4. Jørgensen CK (this volume)
5. Jørgensen CK, Kauffman GB (this volume)
6. Jørgensen CK (1983) Radiochim. Acta 32: 1
7. Jørgensen CK (1987) Inorg. Chim. Acta 139: 1
8. Jørgensen CK (1975) Structure and Bonding 22: 49
9. Jørgensen CK (1969) Oxidation numbers and oxidation states, Springer, Berlin Heidelberg New York
10. Jørgensen CK (1986) Z. Anorg. Allg. Chem. 540: 91
11. Gmelins Handbuch der anorganischen Chemie (1926) System-Nummer 1, Edelgase, Verlag Chemie, Berlin Wertheim
12. Paoloni L (1983) J. Chem. Educ. 60: 758
13. Lewis GN (1916) J. Am. Chem. Soc. 38: 762
14. Jørgensen CK (1984) Topics Current Chem. 124: 1
15. Jørgensen CK (1989) Topics Current Chem. 150: 1
16. Pauling L (1933) J. Am. Chem. Soc. 55: 1895
17. Jørgensen CK (1983) Rev. Chim. Minér. 20: 533; (1986) 23: 614
18. Yost DM, Kaye AL (1933) J. Am. Chem. Soc. 55: 3890
19. Streng LV, Streng AG (1965) Inorg. Chem. 4: 1370
20. Jørgensen CK (1958) Absorption spectra of complexes of heavy metals. ASTIA document 157158, National Technical Information, Springfield VA
21. Jørgensen CK (1963) Inorganic complexes, Academic, London
22. Melton CE, Rudolph PS (1960) J. Chem. Phys. 33: 1594
23. Bartlett N: Proc. Chem. Soc. (London) 1962: 218
24. Sladky FO, Bulliner PA, Bartlett N: J. Chem. Soc. (A) 1969: 2179
25. Hyman HH (1963) Noble gas compounds, University of Chicago Press
26. Gmelins Handbuch der anorganischen Chemie (1970) Supplement to 8th edn vol. 1, Edelgasverbindungen, Verlag Chemie, Weinheim
27. Seppelt K, Lentz D (1982) Progress Inorg. Chem. 29: 167
28. Selig H, Holloway JH (1984) Topics Current Chem. 124: 33
29. Jørgensen CK (1975) Chem. Phys. Lett. 36: 432
30. Jolly WL, Perry WB (1974) Inorg. Chem. 13: 2686
31. Jørgensen CK (1967) In: Gutmann V (ed) Halogen chemistry vol 1, Academic, London, p 265
32. Jørgensen CK (1971) Modern aspects of ligand field theory, North-Holland, Amsterdam
33. Booth HS, Willson KS (1935) J. Am. Chem. Soc. 57: 2273
34. Wiberg E, Karbe K (1948) Z. Anorg. Allg. Chem. 256: 307
35. Jørgensen CK (1974) Chimia 28: 605
36. Kaufman JJ, Sachs LM (1969) J. Chem. Phys. 51: 2992
37. Perutz RN, Turner JJ (1975) J. Am. Chem. Soc. 97: 4791
38. Kelly JM, Long C, Bonneau R (1983) J. Phys. Chem. 87: 3344
39. Breckenridge WH, Stewart GM (1986) J. Am. Chem. Soc. 108: 364
40. Glerup J, Mønsted O, Schäffer CE (1976) Inorg. Chem. 15: 1399
41. Reisfeld R, Jørgensen CK (1988) Structure and Bonding 69: 63
42. Jørgensen CK (1988) Chem. Phys. Lett. 153: 185

43. Brookhart M, Green MLH, Wong Luet-Lok (1988) Progress Inorg. Chem. 36: 1
44. DeVore TC (1976) Inorg. Chem. 15: 1315
45. Jørgensen CK, Kauffman GB (1987) Chimia 41: 150
46. Fairhurst SA, Morton JR, Perutz RN, Preston KF (1984) Organomettallics 3: 1389
47. Rossi A, Kochanski E, Veillard A (1984) Chem. Phys. Lett. 66: 13
48. Glemser O, Roesky H, Hellberg KH (1963) Angew. Chem. 75: 346 (Int. Ed. 2: 266)
49. Jørgensen CK, Reisfeld R (1982) Structure and Bonding 50: 121
50. Lessen D, Brucat PJ (1988) Chem. Phys. Lett. 149: 10
51. Frenking G, Cremer D (this volume)
52. Kossel W (1916) Ann. Physik 49: 229
53. Pauling L (1932) J. Am. Chem. Soc. 54: 3570
54. Mulliken RS (1934) J. Chem. Phys. 2: 782
55. Bratsch SG (1988) J. Chem. Educ. 65: 34, 223
56. Zollweg RJ (1969) J. Chem. Phys. 50: 4251
57. Heddle DWO (1975) J. Phys. B8: L33
58. Lyons LE, Sceats MG (1970) Chem. Phys. Lett. 6: 217
59. Mulliken RS (1935) J. Chem. Phys. 3: 573, 586
60. Liebman JF, Huheey JE (1987) Phys. Rev. D36: 1559
61. Lackner KS, Zweig G (1987) Phys. Rev. D36: 1562
62. Jørgensen CK (1989) Chemistry of systems containing quarks, In: Liebman JF, Greenberg A (eds) From atoms to polymers, VCH, New York (Molecular structure and energetics)
63. Pearson RG (1987) J. Chem. Educ. 64: 561
64. Sanderson RT (1988) J. Chem. Educ. 65: 112, 227
65. Iczkowski RP, Margrave JL (1961) J. Am. Chem. Soc. 83: 3547
66. Jørgensen CK, Horner SM, Hatfield WE, Tyree SY (1967) Int. J. Quantum Chem. 1: 191
67. Jørgensen CK (1988) Quimica Nova (São Paulo) 11: 10
68. Purser GH (1988) J. Chem. Educ. 65: 119
69. Noyes RM (1963) J. Am. Chem. Soc. 85: 2202
70. Fung B-M (1965) J. Phys. Chem. 69: 596
71. Jørgensen CK (1970) Progress Inorg. Chem. 12: 101
72. Laszlo P, Schrobilgen GJ (1988) Angew. Chem. 100: 495; Angew. Chem. Int. Ed. Engl. 27: 479

The Chemistry of the Noble Gas Elements Helium, Neon, and Argon – Experimental Facts and Theoretical Predictions

Gernot Frenking[1] and Dieter Cremer[2]

[1] Fachbereich Chemie, Universität Marburg, Hans-Meerwein-Straße, D-3550 Marburg, FRG
[2] Theoretical Chemistry, University of Göteborg, Kemigården 3, S-41296 Göteborg, Sweden

The results of 65 years of experimental and theoretical research in light noble gas chemistry is reviewed, with particular emphasis on recent quantum chemical studies on the structures, stabilities and bonding of molecules containing He, Ne, or Ar. The scattered experimental results reported mainly for cations are interpreted using a chemical bonding model which is based on donor-acceptor interactions. A systematic view of bonding in Ng compounds (Ng = He, Ne, Ar) is presented that allows the prediction of new compounds which are theoretically predicted to be stable or metastable. The nature of the Ng,X interactions is studied with the help of the analysis of the electron density distribution and its associated Laplace field. Covalent noble gas bonds are found for many cations and dications, while closed-shell interactions are responsible for the unusually stable van der Waals compounds NgBeO.

1 Introduction

For many chemists the title of this article may sound strange and even obscure. After all, are there any compounds containing either helium, neon or argon? A quick look into a recent textbook of inorganic chemistry confirms the suspicious reader. There are no revolutionary research reports on chemical reactions involving He, Ne or Ar that the reader may have missed. Of course, most standard textbooks devote a chapter to "noble gas chemistry". However, these chapters are exclusively restricted to the chemistry of the heavier noble gas elements krypton, xenon (this in particular), and radon. In this sense, noble gas chemistry started with Neil Bartlett's landmark experiment in 1962 [1, 2]. The lighter homologues helium, neon and argon are only mentioned as chemical elements for which all attempts to isolate a chemical compound failed. So what is the topic of this article?

The answer to this question depends on the chemists judgement whether or not a particular atom or molecule is the subject of chemical research. In orthodox chemistry, one would say: "As long as a compound cannot be filled in a bottle, it is not a subject of chemical research. A molecule has to have stable chemical bonds and a well-defined molecular structure in order to become the topic of a research report."

Of course, the position of orthodox chemistry has been abandoned as chemistry has moved more into the border areas to investigate molecules with low stability, fluxional, or nonclassical geometries. But even in modern chemistry, it is common chemical thinking that there are two classes of noble gas elements, namely those that chemically react (Kr, Xe, Rn) and those that do not react at all, the "truly" noble gas elements He, Ne, Ar. The former possess quite an interesting chemistry, the study of which has deepened chemists' understanding of chemical bonding, while the latter should be left to physicists and engineers to do work on super conductors and to prepare light bulbs.

This somewhat provocative view is not correct. The gas phase chemistry of the light noble gas elements has been known ever since HeH^+ was observed by Hogness and Lunn in 1925 [3]. Nevertheless, it has failed to qualify as a truly chemical research field because it did not meet the criteria of orthodox chemistry for a long time. Until today, all experimentally observed species containing He, Ne, or Ar, are only stable in the gas phase under unusual conditions, or they are weakly bound neutral complexes that do not qualify as molecules in the chemical sense. As a result, research of species containing light noble gas elements is exclusively a domain of physics or chemical physics, a view which is confirmed by the list of references in this review.

Physicist consider different aspects and properties of molecular species to those chemists do. This explains why the chemical aspects of light noble gas elements, in particular their reactive behavior, have not been the subject so far of a review article. Chemists ask whether He, Ne or Ar can form stable covalent bonds with other elements; they are interested in the resulting electronic structure, the stability, and the shape of the molecules containing light noble gas elements. They lose interest, however, if there are just van der Waals or dispersion forces that lead to a weakly stable associate, or if the molecule in question can only be observed under extreme conditions.

This report is aimed to (re)activate chemical interest in light noble gas elements by demonstrating that these elements should undergo interesting chemical reactions, provided the right reaction partners are found. We give an overview on the results of roughly 65 years of experimental and theoretical research on molecular species containing the light noble gas elements. We discuss in more detail recent quantum chemical investigations that predict stable molecules containing either He, Ne or Ar [4–15]. These predictions are extraordinary in so far as they are based on nothing but electronic structure calculations and *they are predictions before the (experimental) fact and not after the fact.*

The modern techniques of quantum chemistry, in particular the ab initio methods, and the availability of supercomputers which perform several million operations in a second have given a variety of tools that help to predict experimental events with chemical accuracy [16]. The power of these methods has long been overlooked by orthodox chemists, since the methods of quantum chemistry have often only been used to predict a molecular property or the existence of a certain species *after* the fact. This is valuable for theoretical chemistry itself, but not for experimental chemistry in search of new compounds and new reactions. The value of theoretical predictions beyond the realm of theory is obvious only in new areas. In such a situation, theory can guide experimentalists by predicting the existence of yet unknown compounds, describing their electronic structure and properties, in particular their chemical behavior, and suggesting experiments that eventually may lead to the observation of these compounds. All this has been done in the theoretical investigations that represent the kernel of this review [4–15].

This review has four aims:

(a) To summarize the experimental and theoretical research on the chemistry of He, Ne, Ar, during the past roughly 65 years.
(b) To give an overview of the results and their interpretation laid down in five major theoretical studies and several communications, where the chemistry of the light noble gas elements is presented with an attempt to unify the various results, utilizing techniques of modern quantum chemistry.
(c) To (re)awaken the interest of experimentalists in a research area that can be explored using the theoretical predictions as a guide line.
(d) To demonstrate the predictive power of quantum chemical calculations, in particular those based on ab initio methods.

Special consideration will be given to a description of the electronic structure and the nature of bonding in noble gas (Ng) containing compounds. Because it is the peculiar electronic structure of the Ng atoms which cause their resistance to forming a chemical bond, we review briefly some of the atomic properties of Ng in Sect. 2. Then, in Sect. 3, various ways of analyzing chemical bonding are discussed, where special emphasis is given to a description of bonding based on the electron density distribution calculated for Ng containing molecules. In Sect. 4, we will systematically discuss diatomic ions NgX^{n+} with Ng = He, Ne, Ar, and X being any chemical element. In Sect. 5 we will extend our discussion to polyatomic ions of He, Ne, and Ar. On the basis of these results, we will discuss in Sect. 6 the attempts to synthesize neutral Ng compounds and the prospect for future experiments.

2 Atomic Properties of Light Noble Gas Elements

The light noble gases helium, neon, and argon are known to resist chemical bonding with other atoms or molecules. What is the origin of the peculiar resistance? The formation of a chemical bond is characterized by sharing of electrons by the atoms that are bonded. The strengths of such interactions depends on some key properties of the corresponding atoms. The energy necessary to share an electron with another atom or to accept an additional electron can be estimated from the ionization potential I_n and the electron affinity A_n, where n denotes the number of electrons removed from or accepted by a neutral atom. Another useful property, that can be used to describe the bonding ability of an atom, is the static electric dipole polarizability α which reflects the degree to which the electronic structure of an atom can be deformed by a potential binding partner.

In Table 1, first and second ionization potentials, I_1 and I_2, and dipole polarizabilities α of the noble gas elements and some first row elements are listed. The electron affinities of the noble gas elements are not shown, because they are essentially zero. This indicates that none of the noble gas elements accepts an electron from any other neutral atom or molecule.

The atomic data given in Table 1 show that in going from He to Xe there is a regular decrease in I_1 and I_2, and an increase in α. The first ionization potentials of He and Ne are higher than that of any other element. The highest I_1 for a non-noble gas element is found for fluorine (Table 1), but $I_1(F)$ is still approx. 3 eV and 6 eV below the I_1 value of Ne and He, respectively. Accordingly, no other element should be capable of withdrawing electrons from either He or Ne, thereby forcing them into chemical bonding. This is in line with all experimental observations reported so far.

Argon, however, is different from He and Ne in so far, as its first ionization potential I_1 is *smaller* than that of fluorine (Table 1). Also, there is a very large change in the dipole polarizability when going from Ne to Ar. This suggests that Ar should be closer to Kr in its chemical behavior than to Ne. As a matter of fact, there are many theoretical and experimental observations that support this assumption. Whether this implies that Ar can form chemical bonds in the same

Table 1. First and second ionization potentials[a] I_1 and I_2 (in eV) and static electric dipole polarizabilities[b] α (in Å^3) for noble gas elements and first-row atoms

	I_1	I_2	α
He	24.587	54.416	0.205
Li	5.392	75.638	24.3
Be	9.322	18.211	5.60
B	8.298	25.154	3.03
C	11.260	24.383	1.76
N	14.534	29.601	1.10
O	13.618	35.116	0.802
F	17.422	34.970	0.557
Ne	21.564	40.962	0.395
Ar	15.759	27.629	1.64
Kr	13.999	24.359	2.48
Xe	12.130	21.21	4.04

[a] Ref. 85; [b] Ref. 171

way as Kr does will be discussed in detail later on. Here, it is sufficient to stress that when discussing light noble gas elements one should distinguish between He and Ne on the one side and Ar on the other side.

3 Ways of Describing the Electronic Structure and the Nature of Bonding in Ng Compounds

While the atomic properties shown in Table 1 reflect the reluctance of the light noble gas elements to undergo chemical reactions with other elements, they do not exclude the possibility of chemical bonding totally. For example, the data in Table 1 reveal that the second ionization potential of carbon is of the same magnitude as the first ionization potential of He. *Hence, carbon dications should be sufficiently strong electron acceptors to withdraw electrons from He (or Ne and Ar) and to establish a chemical bond with these noble gas elements. This assumption was the basis of the quantum chemical investigations reviewed in this article. They lead to the prediction of strongly bound helium (neon, argon) containing organic cations* [7, 11–13].

It is essential for the quantum chemical investigation of potential compounds containing light noble gas elements that a clear distinction is made between bonding and nonbonding situations. For example, if there are interactions between He and C^{2+} the question has to be answered whether these interactions lead to chemical bonding.

3.1 The Problem of Defining and Describing a Bond Ng–X

Chemists tend to describe chemical bonds either by comparing appropriate energies such as bond energies and dissociation energies, by analyzing interatomic distances and by describing the forces exerted on the nuclei of a molecule. However, none of these methods leads to a clear distinction between bonding and nonbonding situations or helps to differentiate between covalent and noncovalent bonds. Dissociation energies depend on both the stability of the target molecule and that of the dissociation products. Therefore, they cannot provide an accurate measure for the bond strength and the character of a bond. Bond energies, on the other hand, are arbitrarily defined. Even if this were not the case, it would be problematic, if not impossible, to define an energy value that indicates the change from nonbonding to bonding situations. For example, an interaction energy of 30 kcal/mol and more does not necessarily indicate covalent bonding [4]. It can be due to electrostatic forces, e.g. charge-dipole, dipole-dipole or other multipole interactions between two atoms that do not undergo covalent bonding. We will give several examples for such cases in Sects. 4.5 and 5.5.

It is similarly problematic to define bonding on the basis of measured or calculated atomic distances. In order to identify a certain interatomic distance as typical of covalent bonding, suitable reference distances have to be found. Normally, this is not very difficult in cases where there is no doubt about the nature of bonding. However, for nonclassical structures or in the case of exceptional long bonds or bonds between "unusual" atoms the problem of the reference bond cannot be solved in an unique way. The same difficulties arise when other molecular properties are used to define and to describe chemical bonding [17].

Theoreticians often resort to an analysis of the molecular orbitals. However, orbitals are not observable entities, they have only a mathematical meaning. Furthermore, they can be transformed by any unitary transformation into a new set of orbitals without changing energy or other properties of the molecule. For example, they can be presented as delocalized or localized MOs. A unique definition of chemical bonding, however, should be independent of the form of the MOs. The chemical bond should preferentially be described with the help of a molecular quantity that is observable.

In many textbooks of chemistry, the electrostatic picture of chemical bonding is presented. Electron density accumulates in the region between bonded atoms. Since electron-nuclear attraction dominates the forces stabilizing the molecule, the negative charge in the bonding region is thought to act like a glue that keeps the atomic nuclei together thereby establishing a chemical bond.

However, the electrostatic force description is flawed in several ways. Electrons should prefer those locations in the molecule where the potential energy is most negative and electron-nuclear attraction is a maximum. Hence, they should be found in the vicinity of the nuclei rather than the internuclear region. Nevertheless, those electrons that are responsible for chemical bonding prefer the internuclear region. The reason why they behave different can only be described by quantum mechanics and not by a classical electrostatic argument. Accordingly, any description of chemical bonding based on electrostatic force arguments must be flawed as has been pointed out by several authors [18].

Because of the reasons outlined above we have refrained from using any of the traditional ways of describing chemical bonding in molecules containing light noble gas elements. Instead we have used a three-step procedure to elucidate the nature of the electronic interactions that He, Ne, and Ar can undergo with other elements. First, we have qualitatively assessed the degree of interactions between a light noble gas element and a potential binding partner with the aid of a simple donor-acceptor model. Then, we have calculated the potential NgX system and have analyzed the computed electron density distribution $\rho(\mathbf{r})$ in order to describe chemical bonding in a more quantitative way with the aid of the properties of $\rho(\mathbf{r})$. For this purpose we have utilized a model of the chemical bond suggested by Cremer and Kraka [19].

In the last step, we have complemented the information gained in the first steps by an analysis of the Laplace field $\nabla^2 \rho(\mathbf{r})$ associated with $\rho(\mathbf{r})$. It has been shown that the Laplaceian of $\rho(\mathbf{r})$ indicates where electrons concentrate in the molecule which is useful for a description of electronic structure and chemical bonding [20].

This procedure has been very successful when used to describe compounds containing noble gas elements [4, 5, 7, 13, 15]. Therefore, we will briefly outline some of the details of the three steps of describing chemical bonding.

3.2 The Donor-Acceptor Model

Since the electron affinity of noble gas atoms is zero (see above), chemical bonding can only be established if the very weak electron donor Ng is combined with a strong electron acceptor X. While the donor ability of Ng is reflected by the ionization potential I_1, the acceptor ability of X can be assessed from its electron affinity. This, in turn, depends on the energy of the lowest unoccupied MO (LUMO) of X. Furthermore, the nature of the LUMO, σ or π, symmetric or asymmetric, influences the acceptor ability of X with regard to Ng electrons.

Ab initio calculations provide a means of describing both the energy and nature of the LUMO of X and thereby its acceptor ability. All necessary data can be calculated to establish a donor-acceptor model for any combination of Ng and X and to predict the nature of Ng,X interactions in a potential molecule or complex. While these predictions are only qualitative, they are particularly useful when viewed together for a series of NgX molecules. For example, it is easy to see from the data in Table 1 that the donor ability of Ng increases from He to Ar and that for a given X the largest chance of finding a compound NgX should exist for Ng = Ar. Also, the strength of the binding interactions in NgX for a given Ng will be stronger when the acceptor ability of X increases.

On the basis of the donor-acceptor model the first calculational search for bonded systems between He and doubly charged cations was performed [7, 11, 12]. Also, the analysis of NgX systems for varying X has been facilitated by the use of the donor-acceptor model. For example, the trend in interatomic distances and bond strength for first-row diatomic ions NgX^+ and HeX^{2+} (X = Li–Ne) in their electronic ground and excited states could be rationalized [4, 5] by a qualitative discussion of the HOMO of Ng and LUMO of X^{n+}. The donor-acceptor model turned out to be very helpful not only for an understanding of the binding features of light noble gas compounds, but also for the search for new and strongly bound molecules [4–15, 21].

While a qualitative understanding of the structures and stabilities of Ng molecules was achieved by the donor-acceptor model, the quantitative analysis of the binding properties was provided by the analysis of the electron density distribution.

3.3 Analysis of the Electron Density Distribution

The electron denstiy distribution $\rho(r)$ of an atom or molecule is an observable property that can be measured by a combination of X-ray and neutron diffraction experiments [22]. Also, it is easy to calculate $\rho(r)$ once the MOs and the wave function of a molecule have been determined. The distribution $\rho(r)$ is invariant with regard to any unitary transformation of the MOs. It has been shown by Hohenberg and Kohn that the energy of a molecule in its (non-degenerate) ground state is a unique functional of $\rho(r)$ [23]. In other words, the physical and chemical properties of a molecule can be related to $\rho(r)$. Thus, $\rho(r)$ represents the best starting point for an analysis of chemical bonding.

Bader and coworkers have shown that properties of the electron density distribution $\rho(r)$ can be used to partition the molecular space into subspaces in a unique way [24]. This has been used by Cremer and Kraka (CK) to establish a model of the chemical bond that it easy to use, that allows a simple distinction between bonding and nonbonding situations, and that helps to characterize covalent bonds [19]. Therefore, the CK model has been applied when quantitatively describing chemical bonding in molecules containing light noble gas elements. In the following, we will briefly review the essentials of the Bader analysis and the CK bond model [25].

In Fig. 1, a perspective drawing of the calculated electron density distribution $\rho(r)$ of the N_2 molecule is shown with regard to a plane containing the N nuclei. At the positions of the nuclei, $\rho(r)$ adopts a maximal value, while in the off-nucleus direction $\rho(r)$ decreases exponentially. In the region between the two N atoms, however, the behavior of $\rho(r)$ is different. Along the internuclear connection line $\rho(r)$ first decreases, then reaches a minimum at a point r_B in the bonding region (which is identical with the bond midpoint in the case of homonuclear diatomics) and, finally, increases again to the value it possesses at the second nucleus. At the minimum r_B, the value of $\rho(r)$ is still substantial

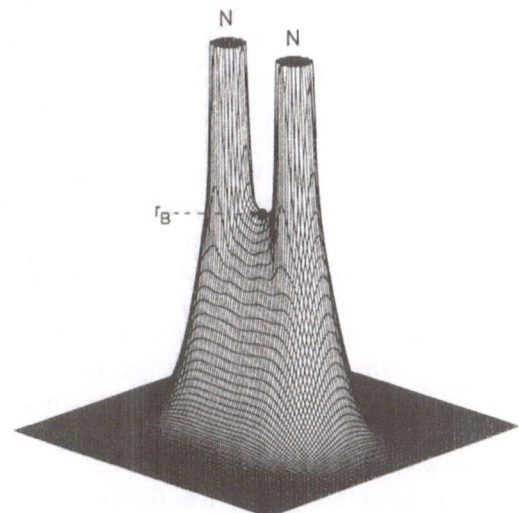

Fig. 1. Perspective drawing of the electron density distribution $\rho(\mathbf{r})$ of N_2 shown with regard to a plane containing the two nuclei (HF/6-31G(d) calculations). In this and the following figures the function value has been cut off above (below) a predetermined value in order to improve the representation

because $\rho(\mathbf{r})$ is at this point (as well as all other points along the internuclear connection line) a maximum in any direction perpendicular to the molecular axis. In other words, *the nuclei are connected by a path of maximum electron density (MED) path*. Any lateral displacement from the MED path leads to a decrease in $\rho(\mathbf{r})$. The position $\mathbf{r_B}$ corresponds to a saddle point of $\rho(\mathbf{r})$ in three dimensions.

The saddle point $\mathbf{r_B}$ is fully characterized by the first and second derivatives of $\rho(\mathbf{r})$ with regard to $\mathbf{r}$. The gradient of $\rho(\mathbf{r})$, $\nabla\rho(\mathbf{r})$, vanishes at $\mathbf{r_B}$. The Hessian matrix of $\rho(\mathbf{r_B})$ is of rank 3, i.e. there are three eigenvalues $\lambda_i = 0$ ($i = 1, 2, 3$). The eigenvalues of the Hessian matrix of $\rho(\mathbf{r})$ give the curvatures of $\rho_B = \rho(\mathbf{r_B})$ along the principal axes. The curvatures λ_1 and λ_2 perpendicular to the MED path are negative while the curvature λ_3 along the MED path is positive due to the minimum of $\rho(\mathbf{r})$ in this direction. This is expressed by the signature $s = \Sigma_i \lambda_i/|\lambda_i|$ which is $+1$ for the saddle point of $\rho(\mathbf{r})$ in the bond region. According to rank and signature, ρ_B classifies as a $(3, +1)$ critical point [26].

Analysis of the electron density distribution $\rho(\mathbf{r})$ of numerous molecules [27–30] has revealed that there exists a one to one relation between MED paths and saddle points $\mathbf{r_B}$ on the one side and chemical bonds on the other side.

This relation is the basis for a definition of a chemical bond [19, 27]: *Two atoms will be bonded if and only if there exists a MED path linking them (necessary condition)*.

This definition is not sufficient since there are MED paths that link the atoms of a van der Waals complex such as He_2. Even at larger distances than the van der Waals distance, the two He atoms are connected by a MED path indicating that there are still small dispersion forces active between the atoms. In order to distinguish closed-shell interactions such as van der Waals interactions, hydrogen bonding or ionic (electrostatic) bonding from covalent bonding, a second criterion is needed. This has been formulated with the aid of the local energy density distribution H(r) [19, 27]:

$$H(\mathbf{r}) = G(\mathbf{r}) + V(\mathbf{r}) \tag{1}$$

Where G(r) is a local kinetic energy density distribution and V(r) is the local potential energy density distribution [31]. If H(r) is negative at r_B, then the local potential energy density distribution V(r) will dominate, and accumulation of electronic charge in the internuclear region will be stabilizing. In this case, one can speak of a covalent bond.

A covalent bond implies that the local energy distribution H(r) is negative at the (3, +1) critical point r_B, (sufficient condition).

The MED path is an image of the covalent bond and, therefore, it is called the *bond path*. For the same reason, the (3, +1) critical point r_B is termed *bond critical point* [32].

If H(r) is zero or positive in the internuclear region, then there will be closed shell interactions between the atoms in question, typical of ionic bonding, hydrogen bonding or van der Waals interactions [19].

On the basis of these definitions one can describe chemical bonding in molecules containing noble gas elements with the aid of the properties of ρ(r). One starts by searching for the bond paths and their associated bond critical points r_B in the molecular electron density distribution. If all bond paths are found, then the properties of ρ(r) along the bond paths will be used to characterize the chemical bonds. For example, the value of ρ_B can be used to determine a bond order, the anisotropy of ρ_B can be related to the π character of a bond, the position of the bond critical point r_B is a measure of the bond polarity and the curvature of the bond path reveals the bent-bond character of a bond [17, 19].

All these features are used to describe bonding in potential Ng compounds. The two conditions for covalent bonding can be applied for each combination of Ng and X. If NgX is calculated to be a potential energy minimum, then the sufficient condition for covalent bonding is utilized to distinguish between van der Waals or electrostatic complexes and chemically bonded molecules NgX.

3.4 The Laplacian of the Electron Density Distribution

The analysis of ρ(r) leads to useful information about chemical bonding. Additional information on the electronic structure of Ng molecules can be gained by analyzing not only ρ(r) but also $\nabla^2\rho(\mathbf{r})$, the Laplacian of the electron density distribution. The Laplacian of a scalar function f indicates where this function concentrates ($\nabla^2 f < 0$) and where it is depleted ($\nabla^2 f > 0$) [33]. For f = ρ(r), the *Laplace concentration* $-\nabla^2\rho(\mathbf{r})$ reveals where the electrons lump together in the molecule [34].

In Fig. 2a, this is illustrated for the N atom in its ⁴S ground state. Contrary to $\rho(\mathbf{r})$, which is characterized by a maximal value at the position of the nucleus and an exponential decay in the off-nucleus direction, the Laplace concentration of N(⁴S) reveals an interesting fine structure. Close to the nucleus there is a peak of high electron concentration, which is surrounded by a sphere of charge depletion. Further away from the nucleus, in the valence region, a second sphere with concentration of negative charge is found (Fig. 2a). Again, this sphere is surrounded by a sphere with (small) depletion of negative charge although this becomes only obvious in a quantitative analysis of $-\nabla^2\rho(\mathbf{r})$. It is appealing to assign the inner sphere of charge concentration to the *1s* electrons and the outer sphere to the valence electrons. Accordingly, one can speak of an inner shell and a valence shell of the Laplacian of $\rho(\mathbf{r})$: *The distribution* $-\nabla^2\rho(\mathbf{r})$ *reflects the shell structure of the electron distribution in an atom.*

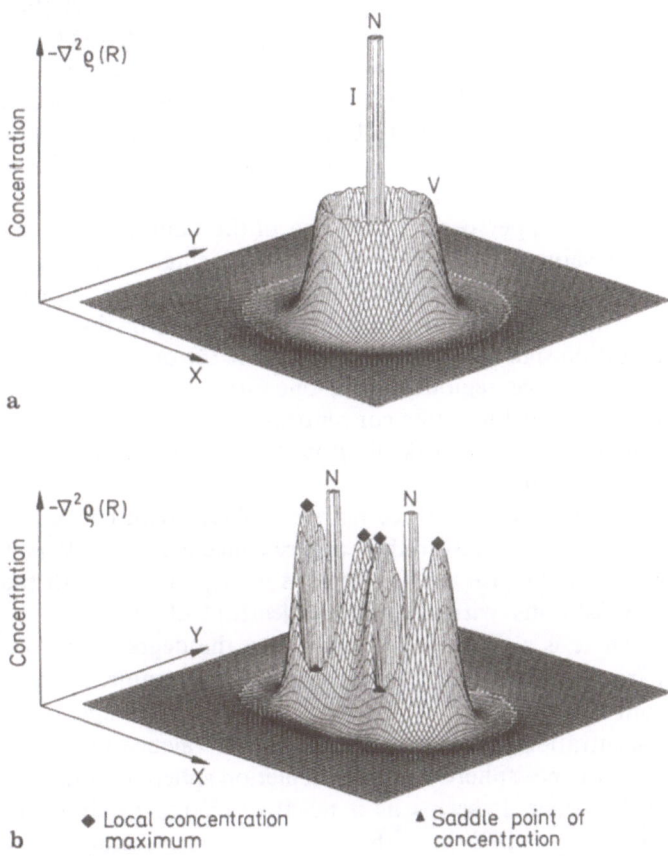

Fig. 2a and b. Perspective drawings of the Laplace concentration, $-\nabla^2\rho(\mathbf{r})$, of (a) the N atom in its ⁴S ground state and (b) the N_2 molecule (HF/6-31G(*d*) calculations). Inner shell and valence shell concentrations are indicated

If two atoms bind, the shell structure of the Laplacian of $\rho(\mathbf{r})$ is deformed in a characteristic way as is shown in Fig. 2b for the N_2 molecule. Electrons lump together in the bonding and in the nonbonding regions. Again, it is very appealing to assign regions with large concentrations of $\rho(\mathbf{r})$ to electron bond and electron lone pairs. The holes in the valence concentrations, which develop when two N atoms form the N_2 molecule, may be considered as positions prone to a nucleophilic attack.

In the case of the N_2 molecule, the valence shell holes are in the direction of the π-orbitals and may be called π-holes to distinguish them from σ-holes in the σ-region. The concentration holes are located in those regions where the LUMO possesses its highest amplitudes. *The size of the hole is a measure of the acceptor ability of the corresponding molecule as is the energy of the LUMO.* Accordingly, both quantities can be used to assess the extent of donor-acceptor interactions between the noble gas element Ng and an acceptor X.

It is important to note that the Laplace concentration reflects the effects of all occupied MOs and, therefore, provides a more reliable description than the frontier MOs that often must be supplemented by the next lower (higher) orbital to the HOMO (LUMO) to reproduce experimental findings. However, it must be born in mind that the analysis of the Laplace concentration as well as the frontier orbital model are both not sufficient to distinguish between bonding and nonbonding situations and to draw a border line between electrostatic and covalent Ng,X bonding.

In Fig. 3, perspective drawings of the Laplace concentration of He, Ne, and Ar are shown. The He atom possesses a single concentration peak surrounded by a sphere of depletion of negative charge (Fig. 3a). For Ne (Fig. 3b), two concentration and two depletion spheres can be distinguished. The inner concentration sphere is at the nucleus while the outer concentration sphere is about in the valence region. Again, one can associate the inner shell with the *1s* electrons and the outer concentration shell with the valence electrons of Ne. Similarly one can speak of an inner shell depletion sphere and a valence shell depletion sphere.

For Ar (Fig. 3c), three pairs of spheres with concentration or depletion of negative charge exist in the Laplace concentration $-\nabla^2\rho(\mathbf{r})$: The most inner can be associated with the *1s* electrons of Ar; the next with the *2s2p* electrons, and the most outer with the valence electrons of Ar.

There is a direct relation between the degree of charge concentration and charge depletion in the valence shell. For example, the more electronegative an atom is, the smaller are the radii of the valence spheres and the higher is the concentration of negative charge in the valence region. With the contraction of the valence sphere the outer depletion sphere becomes more pronounced and much deeper. It seems as if negative charge residing further away from the nucleus is pulled into the valence concentration sphere to increase electron-nucleus attraction and to leave a hole surrounding the valence sphere. The extent of charge concentration and the extent of charge depletion are indeed

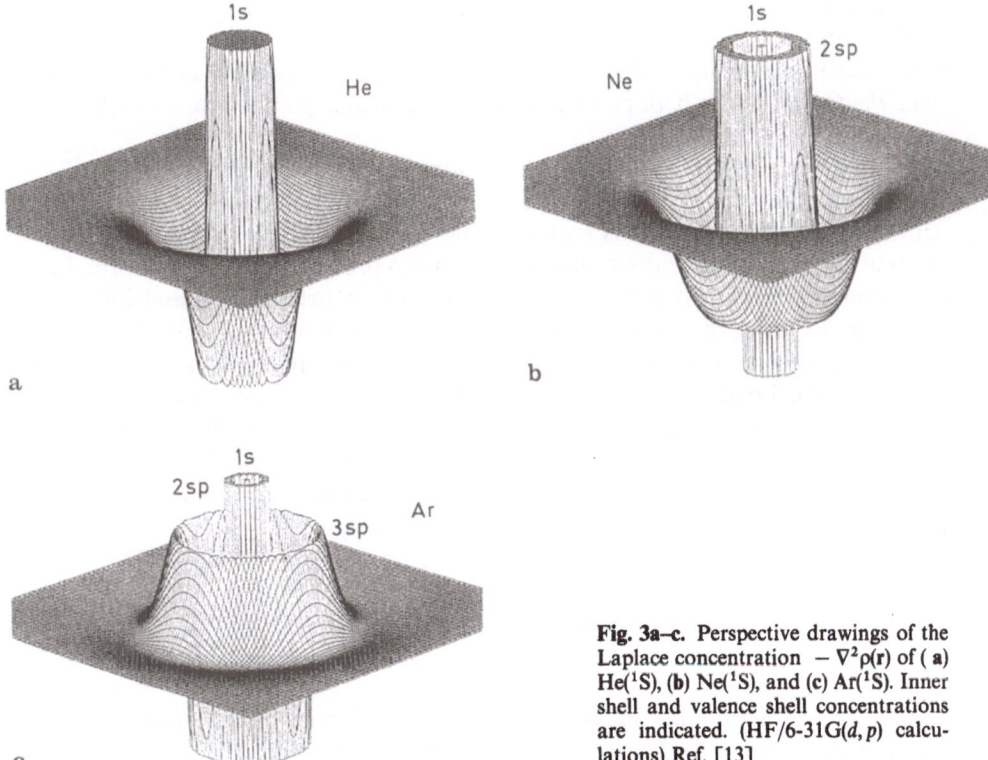

Fig. 3a–c. Perspective drawings of the Laplace concentration $-\nabla^2\rho(\mathbf{r})$ of (**a**) He(^{1}S), (**b**) Ne(^{1}S), and (**c**) Ar(^{1}S). Inner shell and valence shell concentrations are indicated. (HF/6-31G(d, p) calculations). Ref. [13]

related as becomes obvious from the local virial theorem [35]:

$$1/4\ \nabla^2\rho(\mathbf{r}) = 2G(\mathbf{r}) + V(\mathbf{r}) \tag{2}$$

which states that the kinetic energy density distribution $G(\mathbf{r})$ and the potential energy density distribution $V(\mathbf{r})$ add up at each point r in space to the Laplace distribution. Integration over the total space must yield the virial theorem, which says that twice the total kinetic energy T of a molecule is equal to the negative total potential energy V:

$$1/4\ \int\nabla^2\rho(\mathbf{r})d\mathbf{r} = 2\int G(\mathbf{r})d\mathbf{r} + \int V(\mathbf{r})d\mathbf{r} = 2T + V = 0 \tag{3}$$

Equation (3) reveals that fluctuations in the Laplace distribution summed over all space vanish. The more negative charge is concentrated in the valence shell, the deeper is the surrounding depletion shell. This can be seen for He and Ne, where in the latter case the extension of the valence shell from two *2s* electrons to additional six *2p* electrons has a distinct effect on both the value of charge concentration and charge depletion in the valence shell.

For Ar, the *2sp* shell shields to some extent the nucleus. As a consequence, the *3sp* valence shell is less contracted as can be seen from the Laplace concentration shown in Fig. 3c. Also, the outer depletion sphere is rather shallow and less pronounced as that of Ne in accordance with the lower electronegativity of Ar (Table 1). *He and Ne are more related than Ne and Ar since the former atoms both possess deep valence shell depletion spheres while Ar does not.*

Formation of a bond leads to a distortion of both the valence shell concentration and the valence shell depletion sphere. In the direction of the bond, depletion of negative charge is reduced while in other directions depletion may become larger. For He and Ar, these effects are so pronounced that there can be even charge concentration in the interatomic region between Ng and its next neighbor. For Ne, however, charge concentration along a bond does not overcompensate atomic depletion in the outer valence sphere. Charge concentration in the bond region can be even impeded by electron repulsion. Ne possesses $4p\pi$ electrons in linear structures, which cause destabilizing electron interactions if a potential binding partner also possesses $p\pi$ electrons. This has to be born in mind when analyzing the electronic structure of Ng compounds.

4 Diatomic Ions of Helium, Neon and Argon

The chemistry of all noble gas elements Ng is characterized by the fact that, in a molecule, Ng can only donate electrons [192], and a binding partner must be capable of withdrawing electrons from Ng forming partially or completely charged Ng^+. Thus, a suitable binding fragment X in NgX should either be strongly oxidizing, i.e. having a high electron affinity, or it should be a strong Lewis acid, i.e. it should have an energetically low-lying unoccupied orbital (LUMO). Until 1962, no neutral species X was known which had a sufficiently high attracting power to bind the noble gases chemically. However, molecular ions containing noble gas elements have been observed as early as 1925 when Hogness and Lunn [3] attributed a peak found in a mass spectrometer at $m/e = 5$ to the helium hydride ion HeH^+. Thus, we will begin our discussion of light noble gas compounds with cations.

4.1 Noble Gas Ions $NgNg^{n+}$

The first observation of He_2^+, Ne_2^+ and Ar_2^+ was made in 1936 by Tüxen [36], using a mass spectrometer. The formation of Ng_2^+ (Ng = He, Ne, Ar, Kr, Xe) produced by electron impact at 10^{-4} to 10^{-2} mm Hg in a mass spectrometer was later studied by Hornbeck and Molnar in 1951 [37]. It was suggested by Inghram in 1953 [38] that all possible diatomic cross-ions of the noble gas atoms and their isotopes are formed from appropriate mixtures in a mass spectrometer, and most of them have been observed spectroscopically.

Table 2 exhibits the dissociation energies of $NgNg^+$ ions in their ground states. The trends in the bond strengths of heteroatomic ions $NgNg^+$ can be explained by frontier-orbital interaction [39] between the highest occupied orbital (HOMO) of neutral Ng and the lowest unoccupied orbital (LUMO) of Ng^+. For example, in the series $HeNg^+$, which dissociates into $He + Ng^+$, the energy level of the LUMO of Ng^+ increases with atomic size and therefore, orbital interaction decreases because the HOMO–LUMO energy difference becomes larger [39]. Consequently, the dissociation energies for $HeNg^+$ show the order $He > Ne > Ar > Kr$.

For the $XeNg^+$ ions in their electronic ground states, which dissociate into $Xe^+ + Ng$, the order for the binding energy is $Ne < Ar < Kr < Xe$ because the energy level of the HOMO of Ng increases, which causes smaller HOMO–LUMO differences and thus larger orbital interaction. $HeXe^+$ is the only cation in this series for which no experimental D_0 value for dissociation of the ground state is available. In a mass spectrometric study of homonuclear and heteronuclear noble gas molecular ions, Munson et al. [40] were able to obtain appearance potentials for all species except $HeXe^+$, although a signal for the ion was observed. The intensity of $HeXe^+$ was too low to estimate its dissociation energy. From the data in Table 2 it can be concluded that $HeXe^+$ is the weakest bound $NgNg^+$ ion with $D_0 < 0.6$ kcal/mol. The donor-acceptor model also explains the results of scattering experiments by Weise and Mittmann [41] for elastic scatttering of He^+ on Ne, Ar, Kr and Xe. The potential depths correspond to excited states of $NeNg^+$ and show the order $HeNe^+$ (1.2 kcal/mol) $< HeAr^+$ (4.4 kcal/mol) $< HeKr^+$ (5.1 kcal/mol) $< HeXe^+$ (6.5 kcal/mol). This is because the donor ability of Ng increases with atomic size.

The decrease in binding energies for the homoatomic dimers Ng_2^+ with $He > Ne > Ar > Kr > Xe$ is not so simple to explain. At first approximation, the donor strength of Ng increases with atomic size as the acceptor strength of Ng^+ becomes weaker, thus cancelling each other. The difference in the dissociation energy between He_2^+ and Ne_2^+ is clearly larger than for the following diatomics. The relation in the number of electrons in bonding valence orbitals (2) at antibonding electrons (1) is the highest for He_2^+ but is lower (8 to 7) for the

Table 2. Dissociation energies[a] D_0 of diatomic noble gas cations $NgNg^+$ (in kcal/mol)

	He	Ne	Ar	Kr	Xe
He	54.5[a]				
Ne	15.9[b]	31.3[d]			
Ar	1.1[c]	1.8[e]	29.2[f]		
Kr	0.6[j]	1.3[e]	12.2[g]	26.5[h]	
Xe		0.9[e]	4.1[g]	8.9[g]	23.7[i]

[a] Ref. 45; [b] Ref. 172; [c] Ref. 173, slightly lower values have been reported in Refs. 174 and 56; [d] Ref. 175; [e] Ref. 176; [f] Ref. 177; [g] Ref. 178; [h] Ref. 179; [i] Ref. 180; [j] Ref. 181

heavier Ng_2^+ ions. Wadt [42] calculated that the decrease in well depth from Ar_2^+ to Kr_2^+ and Xe_2^+ is mainly because of spin-orbit effects. He_2^+ is clearly the strongest bound diatomic noble gas ion $NgNg^+$.

The dihelium cation He_2^+ is of particular interest not only because of its high binding energy, but possibly also for interstellar chemistry. It has been speculated that He_2^+ may exist in the atmospheres of helium-rich stars and in ionized nebulae in sufficient quantities to affect the chemistry of those objects [43]. An important step toward the identification of this ion in outer space has recently been made by the first report of the vibration-rotation spectrum of the isotope $^3He^4He^+$ by Yu and Wing [44].

It is illuminating to compare the bond strength of $NgNg^+$ with isoelectronic non-noble gas ions. Very similar D_0 values are found when the homonuclear cations Ng_2^+ are compared with isoelectronic X_2^-, especially for halogen anions. There are nearly identical D_0 values for F_2^- (29.5 kcal/mol) and Ne_2^+ (31.3 kcal/mol), Cl_2^- (29.0 kcal/mol) and Ar_2^+ (29.2 kcal/mol), Br_2^- (26.5 kcal/mol) and Kr_2^+ (26.5 kcal/mol), and J_2^- (24.0 kcal/mol) and Xe_2^+ (23.7 kcal/mol) [45]. H_2 does not form a stable molecular anion in the gas phase, but taken the D_0 value for H_2 (103.2 kcal/mol [45]) and the electron affinity of H atom (17.4 kcal/mol [46]), a dissociation energy of 85.8 kcal/mol can be deduced for hypothetical H_2^-. This is much larger than D_0 of F_2^- which agrees with $D_0(He_2^+) \gg D_0(Ne_2^+)$. Isoelectronic reasoning is not valid for the comparison of heteronuclear ions $HeNg^+$ with HX^- because HX^- dissociates into halogen anions $X^- + H$, whereas the fragmentation products of $HeNg^+$ are $He + Ng^+$. For example, HF^- has a purely repulsive ground state curve which dissociates into $H + F^-$ [47].

The difference between unbound Ng_2 and bound Ng_2^+ in the ground state is readily explained by the removal of one electron from the highest lying, strongly antibonding (σ) orbital of the neutral molecule. What happens when two (or more) electrons are taken off? Orbital interactions in Ng_2^{2+} should be even more attractive than in Ng_2^+. However, doubly charged cations encounter Coulomb repulsion which favors dissociation into two singly charged cations. Both forces act in opposite direction which makes the bonding situation for multiply charged cations more complicate than for singly charged species. For the most simple case involving noble gas atoms, i.e. He_2^{2+}, Pauling predicted as early as 1933 [48] on the basis of quantum chemical arguments that He_2^{2+} should be an observable molecule. Only recently has this been proven experimentally by charge-stripping mass spectrometry [49].

Bonding in doubly charged cations XY^{2+} may qualitatively be discussed [4, 50, 51] in the following way which is based on Pauling's investigation of He_2^{2+} [48]. The interactions between cations X^+ and Y^+ are mainly determined by repulsive Coulomb forces which exhibit a $1/r$ curve as shown in Figs. 4a–d. In contrast, the potential energy curve between X^{2+} and Y (and $X + Y^{2+}$) will show a minimum caused by attractive interactions. If mixing of the two states is symmetry allowed, there will be an avoided crossing. Depending on the nature of X and Y, four different cases can qualitatively be distinguished. Figure 4a

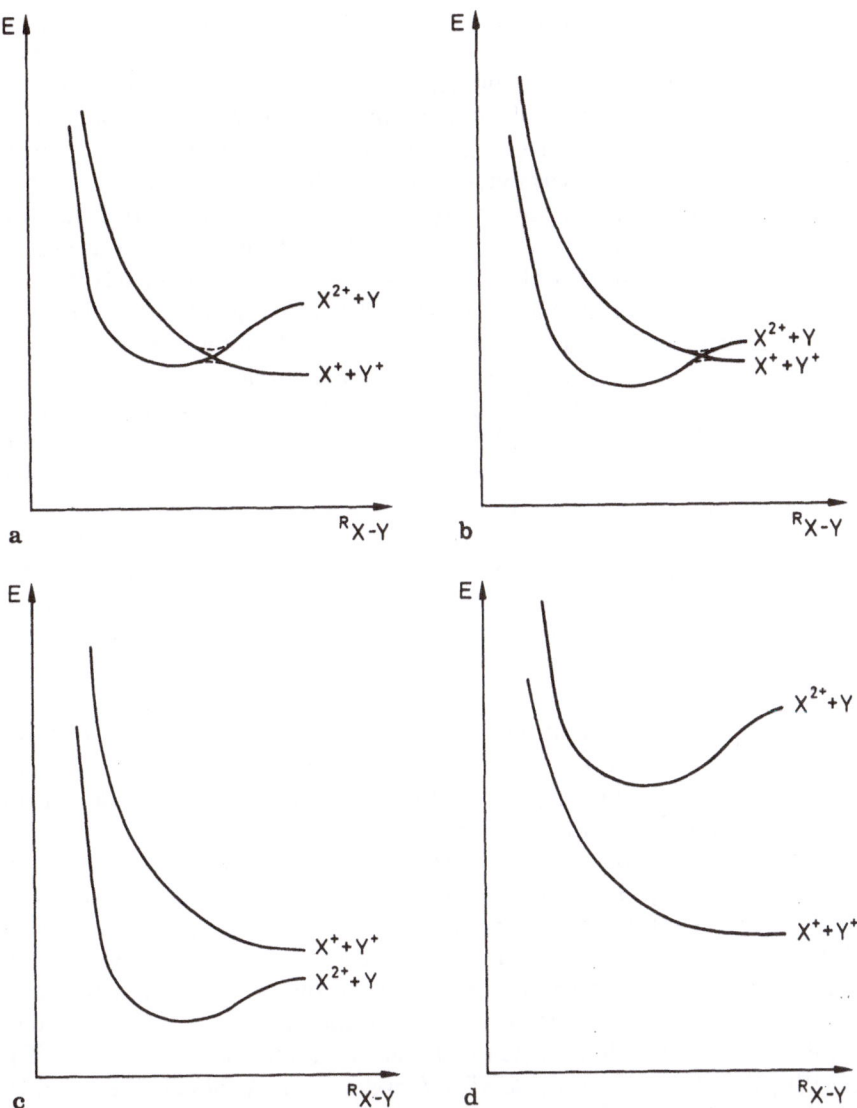

Fig. 4a–d. Schematic representation of the potential energy curves for molecules XY^{2+} dissociating into $X^+ + Y^+$ and $X^{2+} + Y$. (**a**) The dissociation products $X^+ + Y^+$ are lower in energy than $X^{2+} + Y$ and the XY^{2+} minimum energy is higher than $X^+ + Y^+$. XY^{2+} is thermodynamically unstable. (**b**) The same as (a), but the energy difference of the dissociation products is smaller, and the XY^{2+} minimum energy is lower than $X^+ + Y^+$. XY^{2+} is thermodynamically stable. (**c**) The energy of $X^+ + Y^+$ is higher than $X^{2+} + Y$, XY^{2+} is thermodynamically stable. (**d**) The energy of $X^+ + Y^+$ is much lower than $X^{2+} + Y$, XY^{2+} does not form a minimum energy structure. Ref. [4]

shows the situation when $X^+ + Y^+$ are lower in energy than $X^{2+} + Y$ (and $X + Y^{2+}$, only one possibility is shown for simplicity). The energy difference between the two dissociation limits is small enough to allow sufficiently high interactions of the two states which yields a minimum in the ground-state curve, but higher in energy than the dissociation limit. The resulting XY^{2+} species is metastable, i.e. the dissociation is exothermic, but XY^{2+} may be observed if the barrier is high enough. Figure 4b depicts the situation when the energy difference between the dissociation limits is smaller, thus allowing stronger interactions so that the potential energy minimum is lower than the dissociation products. Now XY^{2+} is thermodynamically stable although Coulomb repulsion is released in the charge separation reaction. In Fig. 4c the relative stabilities of the dissociation products are reversed, $X^{2+} + Y$ is lower in energy than $X^+ + Y^+$, and the ground state potential energy curve is completely determined by the interactions of the former product. Finally, Fig. 4d shows the situation where the two curves are too far apart in energy to interact significantly. In this case, XY^{2+} is not observable as a ground state species.

It follows from the above that the well depth of the ground state potential energy curve for XY^{2+} is mainly determined by (i) the strength of the interactions between $X^{2+} + Y$ and (ii) the mixing of the two curves, which depends on the energy difference between the dissociation limits. The curves shown in Figs. 4a–d are, of course, idealized cases. However, it will be seen that they are well-suited to understand experimental and theoretical results of doubly charged cations not only for diatomics. He_2^{2+} is an example for the case shown in Fig. 4a. The energy difference between the two curves at the dissociation limits ($He^+(^2S) + He^+(^2S)$ and $He(^1S) + He^{2+}$) is 29.829 eV ($I_2(He) - I_1(He)$, see Table 1). Bonding in $(X^1\Sigma_g^+)$ He_2^{2+} is quite strong: high-level ab initio calculations [52] predict an interatomic He,He distance of only 0.704 Å, which is even shorter than in He_2^+ (1.018 Å [45]). The barrier for dissociation is calculated as 37.3 kcal/mol, although the dissociation reaction is strongly exoenergetic by 199.6 kcal/mol [52a]. In the case of $(^1\Sigma^+)$ $HeNe^{2+}$, the energy difference between the two dissociation limits ($He^+(^2S) + Ne^+(^2P)$ and $He(^1S) + Ne^{2+}(^1D)$) is 19.579 eV [55]. Theoretical studies predict only a shoulder [53] or an extremely low (< 0.1 kcal/mol) minimum [54] for the $^1\Sigma^+$ ground state curve of $HeNe^{2+}$. Although the two curves are closer in energy than for He_2^{2+}, the mixing does not yield a genuine minimum in the ground state curve because Ne^{2+} is a weaker Lewis acid than He^{2+}. In the case of $(^1\Sigma^+)$ $HeAr^{2+}$, the two curves are very close in energy, the dissociation limits ($He^+(^2S) + Ar^+(^2P)$ and $He(^1S) + Ar^{2+}(^1D)$) are only 4.779 eV apart [55]. Although the interactions between Ar^{2+} and He are even weaker than between Ne^{2+} and He, the much smaller energy difference of the two curves yields strong mixing [56]. Consequently, a rather deep well depth of 45.7 kcal/mol at $r_e = 1.43$ Å is theoretically predicted [57]. Clearly, the $(1)^1\Sigma^+$ state of $HeAr^{2+}$ should be experimentally observable, but not $(1)^1\Sigma^+$ $HeNe^{2+}$.

The ionization energies shown in Table 1 indicate that I_2 of Kr and Xe are lower than I_1 of He. This means that $HeKr^{2+}$ and $HeXe^{2+}$ are examples for the situation shown in Fig. 4c, and they are predicted to be thermodynamically stable. There are no theoretical studies reported for the two dications. Experimental attempts to observe $HeKr^{2+}$ and $HeXe^{2+}$ by double ionization of the respective neutral van der Waals dimers in a mass spectrometer were hampered by the overlap of their mass spectra with those of the much more abundant Xe^{2+} and Kr^{2+} atomic ions [58]. However, in the same experiments done by Helm and coworkers, doubly charged $NeXe^{2+}$ and $ArXe^{2+}$ [58] and also $NeKr^{2+}$ [59] were identified. In the context of the binding model, $NeXe^{2+}$ is an example of the case shown in Fig. 4c, and $ArXe^{2+}$ and $ArKr^{2+}$ are examples for Fig. 4a.

It is tempting to use isoelectronic arguments to discuss the stability of the noble gas dications. The rather strong bond in He_2^{2+} is often compared with H_2. However, the analogy fails for the heteroatomic species, similar as in case of the singly charged species. He_2^{2+}, $HeNe^{2+}$ and $HeAr^{2+}$ are isoelectronic with H_2, HF and HCl, respectively. The bond strengths of the three neutral molecules (H_2: 104.1 kcal/mol; HF: 136.2 kcal/mol; HCl: 103.1 kcal/mol [45]) do not correlate with the stabilities of the isoelectronic noble gas dications, especially not in case of $HeNe^{2+}$/HF. A systematic comparison of isoelectronic hydrogen and helium (+) compounds has shown the limits of this approach [50].

4.2 Noble Gas Hydride Ions NgH^{n+}

The noble gas hydride ions NgH^+ are rather stable species which have been observed and extensively studied by means of mass spectroscopic techniques [60]. The proton affinities shown in Table 3 increase regularly with the sequence He < Ne < Ar < Kr < Xe indicating stronger binding for the heavier noble gases. This can be explained by the increase in donor ability (Lewis basicity) of Ng in the same order. The two-electron system HeH^+ was subject to a quantum mechanical treatment as early as 1936 by Beach [61] when he calculated the potential well depth as 46.6 kcal/mol, which is in excellent agreement with the experimentally derived value for $D_e = 46.1$ kcal/mol [62]. A more sophisticated calculation of the James-Coolidge type by Kolos and Peek using an 83-term function [63] predicts a dissociation energy $D_e = 47.0$ kcal/mol. Very high quality calculations for NeH^+, ArH^+, KrH^+ and XeH^+ have also been made, most notably by Rosmus [64]. The high-resolution vibration-rotation spectra of HeH^+ [65], NeH^+ [66–68], ArH^+ [67, 69] and KrH^+ [70] have been reported. Generally, the NgH^+ cations are well-understood diatomic systems.

In the same theoretical paper by Beach [61], the one-electron system HeH^{2+} was also calculated and it was predicted to have a purely repulsive potential energy curve for the $^2\Sigma^+$ ground state. The same result was found in more recent

Table 3. Proton affinities[a] $(PA)_e$ of noble gas atoms (in kcal/mol)

	He	Ne	Ar	Kr	Xe
$(PA)_e$	46.1	52.1	96.1	106.0	133.4

[a] From Ref. 45

calculations of HeH^{2+} [71]. Also for NeH^{2+} [72], the ($^2\Pi$) electronic ground state is theoretically predicted to be repulsive. The two energetically lowest lying $^2\Pi$ states, which dissociate into $Ne^+(^2P) + H^+$ and $Ne^{2+}(^3P) + H(^2S)$, respectively, apparently do not mix strongly enough to yield a minimum in the ground state curve [72]. The energy difference of the two curves become smaller for the heavier noble gases. For ArH^{2+}, the lowest lying $^2\Pi$ state has a shoulder or a very shallow minimum [73] which, however, is not sufficiently deep to make experimental detection feasible. It seems possible that KrH^{2+} and XeH^{2+} exhibit a minimum in the ground state potential energy curve, which makes them potentially observable as metastable species. There are no theoretical or experimental studies reported on KrH^{2+} and XeH^{2+}.

4.3 Noble Gas Ions NgX^{n+} with First-Row Elements $X = Li–Ne$

The comparative study of the strength of the interactions of helium, neon, and argon atoms with first row atomic ions Li^+ to Ne^+ is an excellent probe to elucidate the principle of bonding between the light noble gases in other elements. For only a few NgX^+ cations have experimentally determined dissociation energies been reported, mainly for Ng = argon, but theoretical results are available for all 24 different species. Also for dications NgX^{2+}, some experimental and theoretical data are reported. Most experimental data for NgX^+ result from atomic scattering experiments which will not be discussed in detail. Very often, the experimental studies were carried out together with theoretical calculations of the potential energy curves, and there is generally good agreement between theory and experiment concerning the dissociation energies of NgX^+ cations.

Atomic scattering results show weak attractive interactions between Li^+ and the light noble gas elements. The reported data for the potential wells are 1.1–1.6 kcal/mol for $HeLi^+$ [74], 2.6 kcal/mol for $NeLi^+$ [74c] and 7.0–7.2 kcal/mol for $ArLi^+$ [103c, 75]. For other NgX^+ systems, experimental data are available only for argon ions (except for $HeNe^+$ and $NeNe^+$ which have already been discussed). $ArBe^+$ has a dissociation energy $D_e = 12.9 \pm 0.15$ kcal/mol [76]. For the $X^1\Sigma^+$ ground state of ArB^+, the D_e value is given as 6.9 kcal/mol and for the $^3\Pi$ excited state, it is 34.6 kcal/mol [77]. The ground states of $ArC^+(X^2\Pi)$, $ArN^+(X^3\Sigma^-)$ and $ArO^+(^4\Sigma^-)$ are bound with D_e values of 21.6 kcal/mol [78], 53.0 kcal/mol [79], and 15.7 kcal/mol [80], respectively. The reported data indicate that there is no systematic increase in binding energies for ArX^+ cations along the first-row elements. For ArF^+, the bond energy has been established experimentally as $D_0 > 38.1$ kcal/mol [81].

There are numerous theoretical studies reported for diatomic cations NgX^+ for He, Ne or Ar and first-row atoms X, but only few of them address the comparison of bond strengths and trends in dissociation energies for various X. Perhaps the first attempt was made by Liebman and Allen [82], who reported ab initio calculations at the Hartree-Fock level for several helium, neon and argon diatomic cations in ground and excited states with boron, nitrogen, oxygen and fluorine. Later Cooper and Wilson [83] performed Hartree-Fock calculations for HeC^+, HeN^+ and HeO^+ as well as higher charged cations. They found that the doubly-charged ions HeX^{2+} are more strongly bound than their singly-charged counterparts. From their data, they concluded that the various trends in stability can be understood in terms of the effective nuclear charges of the two centers which show the order $HeO^{n+} > HeN^{n+} > HeC^{n+}$ [83]. Very recently, Frenking et al. made a complete "first-row sweep" of HeX^{n+} (n = 1, 2) [4], NeX^+ and ArX^+ [5] reporting ab initio calculations including correlation energy for the ground and selected excited states. Their theoretical data agree very well with experimentally observed binding energies and other high-level calculations and will be used to discuss the trends in binding energies.

Table 4 shows the calculated dissociation energies D_e for the ground states of HeX^+, NeX^+ and ArX^+ as reported by Frenking et al. [4, 5]. With two exceptions, the NgX^+ cations dissociate into $Ng + X^+$, the exceptions being ArF^+ and $ArNe^+$ (the latter is not really an exception, because it should be considered as an NeX^+ species). In the two cases, the energetically lowest lying dissociation products are Ar and X^+ (X = F, Ne). The data in Table 4 show that for a given X there is an increase in binding energy $HeX^+ < NeX^+ < ArX^+$. $ArNe^+$ is only seemingly an exception because the dissociation limit is different than for the respective He and Ne analog. The trend is observed if the dissociation energy D_e for $ArNe^+$ into $Ar + Ne^+$ is considered (134.6 kcal/mol [5]). The increase in bond strength for He < Ne < Ar can be expected because the polarizability increases in the same order, which means stronger charge-induced dipole interactions, and also the ionization energy decreases for He < Ne < Ar, which means an increase in Lewis basis strength (donor-acceptor interactions).

More information can be gained from the trends in Ng, X^+ binding energies along the first-row atoms X = Li–Ne, which are not so trivial. The trends are

Table 4. Theoretically predicted[a] dissociation energies D_e (in kcal/mol) and equilibrium distances r_e (in Å) for the ground states of HeX^+, NeX^+ and ArX^+

X	D_e			r_e		
	He	Ne	Ar	He	Ne	Ar
Li	1.5	3.0	5.7	2.062	1.986	2.441
Be	0.3	0.9	10.9	3.132	1.856	2.045
B	0.5	1.2	5.6	2.912	2.474	2.738
C	1.1	3.0	20.6	2.406	2.077	2.059
N	4.1	9.2	48.7	1.749	1.767	1.836
O	0.8	1.2	10.0	2.473	2.273	2.292
F	1.4	3.6	49.6	2.123	1.960	1.637
Ne	10.4	29.8	1.8	1.406	1.717	2.410

[a] Refs. 4 and 5

nearly the same for the three noble gases He, Ne and Ar. For HeX$^+$ and NeX$^+$, there is a decrease in D$_e$ from X = Li to Be, but an increase for ArX$^+$. The three systems show a regular increase in binding energy from X = boron to carbon and nitrogen. The calculated binding energies for HeN$^+$ and NeN$^+$ are the highest in the two series, and the dissociation energy for ArN$^+$ is only slightly smaller than the highest value in the ArX$^+$ series, i.e. ArF$^+$. For the three cases, the D$_e$ value is significantly smaller for NgO$^+$ than for NgN$^+$. There is a regular increase in binding energy from NgO$^+$ to NgF$^+$ and NgNe$^+$ for the helium, neon and argon contains (if for ArNe$^+$ the dissociation into Ar + Ne$^+$ is considered).

Clearly, these results do not exhibit the trends for the dissociation energies of NgX$^+$ cations which are expected when the effective charges at X are considered. An explanation for the irregular behavior of the dissociation energies has been given by Frenking et al. [4, 5, 50] based on the donor-acceptor model described in Sect. 3.2. Figure 5 shows schematically the valence orbitals of the first row ions X^{n+} (n = 1, 2) in the ground and selected excited state for the situation of an approaching Ng atom. In this situation, the triply degenerate $2p$ AOs split into doubly degenerate $p(\pi)$ AOs and a $p(\sigma)$ orbital. Then, the dominant orbital interactions between Ng(^{1}S) and X$^+$ in its ground state involve the following orbitals of X$^+$. For Li$^+$(^{1}S) and Be$^+$(^{2}S), it is the $2s$ AO, which is empty in Li$^+$ and singly occupied in Be$^+$. This orbital is lower in energy in Be$^+$ than in Li$^+$ invoking stronger interactions. However, HOMO–LUMO interactions with singly occupied orbitals are weaker than with empty orbitals, and they may even be repulsive when the HOMO–LUMO overlap is very small [84].

In the case of He and Ne, the binding energy is weaker with Be$^+$ than with Li$^+$. For Ar, the interactions with Be$^+$ are stronger than with Li$^+$ because argon is a better donor and the HOMO–LUMO overlap becomes larger. In case of B$^+$(^{1}S), C$^+$(^{2}P) and N$^+$(^{3}P), the dominant orbital interaction with Ng involves the empty $p(\sigma)$ AO. Because the energy of the orbital decreases with increasing nuclear charge, there is a regular increase in dissociation energy D$_e$ for the three series of NgX$^+$ cations with B < C < N. As noted above, there is a substantial decrease in ground state binding energy in all three cases from NgN$^+$ to NgO$^+$ (Table 4). Figure 5 shows that the $p(\sigma)$ AO, which is empty in N$^+$(^{3}P), is singly occupied in the ground state of O$^+$(^{4}S), causing a significant decrease in stabilizing interaction with the HOMO of Ng. The ground states of O$^+$, F$^+$ and Ne$^+$ all have a singly occupied $p(\sigma)$ AO as lowest lying orbital, which can interact favorably with the HOMO of Ng. Consequently, there is a regular trend for the three NgX$^+$ series in dissociation energies O < F < Ne (see above for the apparent anomaly of ArNe$^+$). However, the result for ArF$^+$ differs from that of NeF$^+$ and HeF$^+$ in that the ground state of ArF$^+$ has $^1\Sigma^+$ symmetry. (X$^1\Sigma^+$) ArF$^+$ is not the product of Ar(^{1}S) + F$^+$(^{3}P), but rather dissociates into Ar$^+$(^{2}P) + F(^{2}P).

Argon is the lightest noble gas for which there is still an element besides the noble gases, i.e. fluorine, that has a higher ionization energy (17.422 ev [85]).

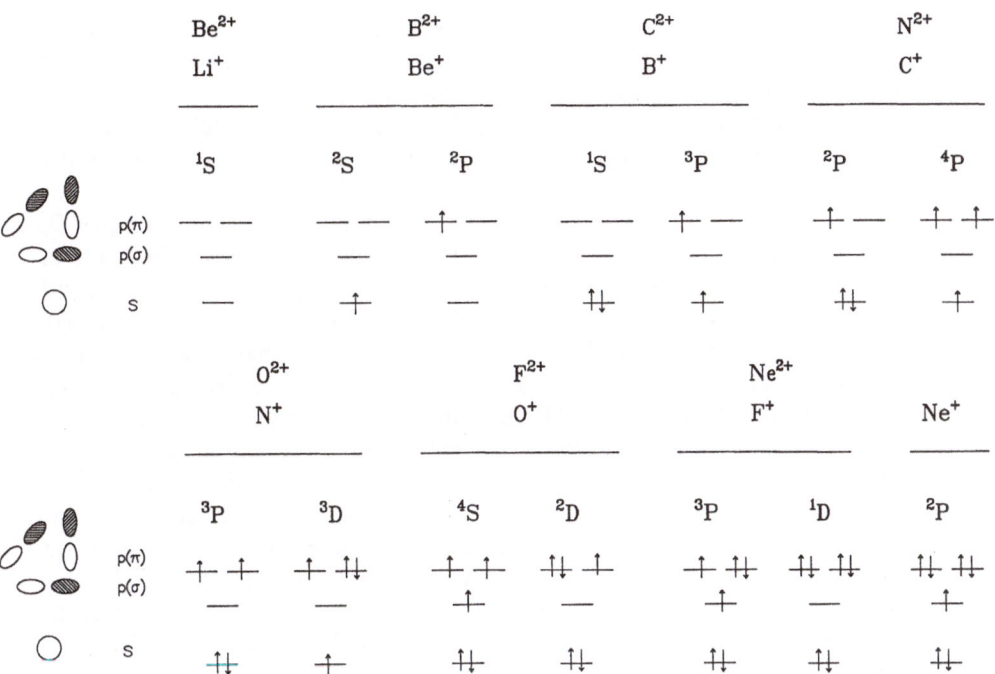

Fig. 5. Schematic representation of the valence atomic orbitals of first-row atomic ions in the ground and selected excited states shown for the situation of an approaching Ng atom. Ref. [4]

This peculiarity is the reason that ArF^+ salts were suggested early [86, 87] as the best candidates for isolating neutral argon compounds. This has been supported by more recent, accurate calculations of the Ar, F^+ bond energy and by estimating the stabilization energies of suitable salt compounds, for which SbF_6^- is suggested as the best candidate counter anion [6]. However, the systematic comparison of NgX^+ cations shows that ArN^+ as the strongest bound cation of the "normal" dissociating species is nearly as strongly bound in its ground state as ArF^+.

Associated HeN^+ was experimentally observed by Jonkman and Michl [88] in cluster ions $(HeN)^+(N_2)_n$ (which may have been $(HeN_3)^+(N_2)$) via secondary ion mass spectrometry. In the same paper, the observation of ArN^+, ArN_2^+ and ArN_4^+ was reported [88].

The orbital model discussed above not only provides a rationale for the trends in dissociation energies found for the NgX^+ ground states, it also explains the often dramatic increase in bond strengths and shortening in the bond lengths for some excited states. Figure 5 exhibits several excited states of X^+ where one electron is excited from the HOMO of the X^+ ground state. For example, the *2s* AO is doubly occupied in the ^1S ground state of B^+, but it is half empty in the ^3P excited state which allows stronger HOMO–LUMO interactions. In agreement with this, the corresponding $^3\Pi$ state of NgB^+ is stronger bound for Ng = He [4], Ne and Ar [5]. For all excited states of X^+ shown in Fig. 5, the corresponding state of NgX^+ has a shorter interatomic distance and higher dissociation energy than the ground state [4, 5].

The same reasoning has also successfully been applied to HeX^{2+} dications [4]. Here, the binding interactions are between He and X^{2+}. Depending on the second ionization energy of X, dissociation of HeX^{2+} will yield either X^{2+} or X^+, besides He and He^+, respectively. The calculated dissociation energies D_e and the interaction energies IA between He and X^{2+} are shown in Table 5. There are two first-row elements, beryllium and carbon, which have a second ionization energy lower than I_1 of He (18.211 eV for Be, 24.383 eV for C; [85]). Consequently, $HeBe^{2+}$ and HeC^{2+} are energetically stable in their ground states with dissociation energies D_e of 21.1 and 17.7 kcal/mol, respectively [4]. $HeBe^{2+}(X^1\Sigma^+)$ and $HeC^{2+}(X^1\Sigma^+)$ are examples of the case shown in Fig. 4c. Boron has a slightly higher I_2 (25.154 eV, [85]) than I_1 of helium. The interaction energy of B^{2+} with He is sufficiently strong to yield a minimum in the $^2\Sigma^+$ ground state potential energy curve that is lower than the dissociation limit $He^+(^2S) + B^+(^1S)$. $HeB^{2+}(X^2\Sigma^+)$ is energetically stable by 16.6 kcal/mol [4] and represents a case as shown in Fig. 4b. HeN^{2+} $(X^1\Pi)$ and HeO^{2+} $(X^3\Sigma^-)$ are predicted to be unstable, but have a minimum in the ground state curve as schematically shown in Fig. 4a. $HeLi^{2+}(X^2\Sigma^+)$, $HeF^{2+}(X^4\Sigma^-)$, and $HeNe^{2+}$ $(X^3\Pi)$ are calculated with purely repulsive ground state curves; these systems are examples for Fig. 4d. Like the singly charged HeX^+ cations, bound HeX^{2+} dications exhibit significantly shorter interatomic distances and higher interaction energies IA (which are in case of HeX^+ identical with dissociation energies D_e) in the calculated excited states than in the ground states (Table 5). The calculated trends in HeX^{2+} from X = Li to Ne and the differences between

Table 5. Calculated dissociation energies D_e (in kcal/mol), interaction energies IA between He and X^{2+} (in kcal/mol) and equilibrium distances r_e (in Å) of HeX^{2+} dications[a]

Struct.	Symm.	D_e	IA	r_e
$HeLi^{2+}$	$X^2\Sigma^+$	—	—	Diss.
$HeBe^{2+}$	$X^1\Sigma^+$	+ 20.1	+ 20.1	1.453
HeB^{2+}	$X^2\Sigma^+$	+ 16.6	+ 26.8	1.339
	$^2\Pi$	+ 18.6	+ 64.5	1.191
HeC^{2+}	$X^1\Sigma^+$	+ 17.7	+ 17.7	1.575
	$^3\Pi$	− 74.2	+ 63.0	1.167
HeN^{2+}	$X^2\Pi$	− 71.1	+ 45.4	1.321
	$^4\Sigma^-$	− 151.4	+ 122.7	1.060
HeO^{2+}	$X^3\Sigma^-$	− 151.4	+ 93.6	1.164
	$^3\Pi$	− 288.2	+ 223.3	1.003
HeF^{2+}	$X^4\Sigma^-$	—	—	Diss.
	$^2\Pi$	− 197.3	+ 135.4	1.044
$HeNe^{2+}$	$X^3\Pi$	—	—	Diss.
	$^1\Sigma^+$	− 214.1	+ 235.3	1.025

[a] Taken from Ref. 4

ground and excited states have been explained using frontier orbital interactions between He and X^{2+} in the same manner as the HeX^+ species [4].

Experimental information concerning the stability of noble gas first row dications NgX^{2+} is very scarce. No data are available for HeX^{2+} species. Very recently, Jonathan et al. [89] reported results of charge-stripping mass spectrometry for several diatomic dications containing one inert gas atom. Using the singly charged cations NgX^+ with $Ng = Ne$, Ar, Kr, Xe, and $X = C$, N, O, the doubly charged species NeN^{2+}, ArN^{2+}, ArO^{2+}, KrO^{2+}, XeC^{2+}, XeN^{2+} and XeO^{2+} were observed in the charge-stripping experiments. The NgN^{2+} cations seemed to be the most stable and NgC^{2+} the least stable species [89]. A possible explanation for the failure to detect NeC^{2+} and NeO^{2+}, while NeN^{2+} had been observed, was later given by Frenking and Koch [21]. The explanation is based on the stabilities of the precursor ions NeX^+ which show the trend $N > C > O$, as shown above. However, the observed species were not always in their electronic ground states. From the translational energy release measured in the unimolecular dissociation reactions it was concluded that in some cases, for example ArO^{2+} and KrO^{2+}, the observed cations correspond to excited states and not to the electronic ground states [89].

Detailed information on the stabilities of some NgX^{2+} dications with first-row elements X are available from theoretical studies, but many systems have not been studied yet. Besides the comparative study of Frenking et al. [4] which covers a complete first-row sweep for HeX^{2+} dications, ab initio calculations are reported for only a few NgX^{2+} systems. Results for $HeBe^{2+}$ are available from several studies [90] which give similar values for the dissociation energy as noted above (18.9 kcal/mol). $NeBe^{2+}$ has been calculated together with $HeBe^{2+}$ by Hayes and Gole [90a] and it was predicted that the ground state of $NeBe^{2+}$ is 5.4 kcal/mol less bound than $HeBe^{2+}$. Like for helium, I_1 of Ne is higher than I_2 of beryllium. Thus, $NeBe^{2+}(X^1\Sigma^+)$ dissociates into $Ne(^1S) + Be^{2+}(^1S)$. The greater stability of $HeBe^{2+}$ over $NeBe^{2+}$ was explained by the assumption that the binding energy in the dications be mainly due to charge induced dipole interactions between Ng and Be^{2+} which follows r_e^{-4}. This result does not agree with the donor-acceptor model used by Frenking et al. for noble gas compounds [4, 5, 7, 21, 50]. Neon is a better donor than helium and HOMO–LUMO interactions with Be^{2+} should therefore be stronger in $NeBe^{2+}$. A recent high-level ab initio calculation shows that the dissociation energy D_o for $NeBe^{2+}$ is 35.3 kcal/mol, considerably higher than $HeBe^{2+}$ which is calculated at the same level of theory with a D_o value of 17.2 kcal/mol [90d].

For HeC^{2+}, Hartree-Fock SCF calculations were reported for the $X^1\Sigma^+$ ground state by Harrison et al. [91] and by Cooper and Wilson [83] who also calculated HeN^{2+} and HeO^{2+}. The potential energy curves of the low-lying electronic states of HeC^{2+} [8] and NeC^{2+} [9] were calculated by Koch et al. using a complete active space SCF (CASSCF) approach. From these calculations it is predicted that the $(X^1\Sigma^+)$ ground state of HeC^{2+} is thermodynamically stable with $D_o = 16.6$ kcal/mol [8] while $NeC^{2+}(X^1\Sigma^+)$ has a barrier for the exothermic dissociation into $Ne^+(^2P) + C^+(^3P)$ of 37.3 kcal/mol.

NeC^{2+} dissociates as shown in Fig. 4a, while HeC^{2+} is an example for Fig. 4c. The low-lying electronic states of NeN^{2+} have been studied at the CASSCF+CI level by Koch et al. [10]. The $X^2\Pi$ ground state is predicted to be metastable with a barrier for dissociation into $Ne^+(^2P)+N^+(^3P)$ of 21.7 kcal/mol.

4.4 Noble Gas Ions NgX^{n+} with Metal Atoms X Having Z Above 10

Two of the eight first-row atoms, beryllium and carbon, have a second ionization energy I_2 lower than I_1 of helium. As discussed above, $HeBe^{2+}$ and HeC^{2+} are quite strongly bound. Also HeB^{2+} is thermodynamically stable, because I_2 of B is only slightly higher than I_1 of He. Looking at the ionization energies of the elements it can be recognized that there are several elements, especially metals, which have I_n values, even for n = 3, lower than I_1 of He and also I_1 of Ne. That means that triply charged ions NgX^{3+} are thermodynamically stable for many metals X if Ng is helium or neon. Table 6 shows a list of atoms which have low-lying third and fourth ionization energies. It seems conceivable that even quadruply charged ions HeX^{4+} ions might be detected.

Several HeX^{2+} and HeX^{3+} ions with X being metal atoms such as V, MO, Rh, Pt, Ir, Ta, W, Re were observed experimentally by Müller, Tsong and co-workers [92] by means of time-of-flight mass spectrometry. While no direct information about the bond strength of the highly charged molecules could be extracted from the experiments, the rather long life time, which must be longer than the typical flight time of up to 100 microseconds, indicates a substantial well depth in the potential energy curve of the observed species. However, a deep potential well does not automatically ensure a long lifetime for helium cations.

Table 6. Ionization energies I_n^a (in eV)

Element	n	I_n	Element	n	I_n
Al	2	18.828	La	3	19.175
	3	28.447	Ce	3	20.20
Sc	3	24.76	Pr	3	21.62
Ti	2	13.58	Tm	3	23.71
	3	27.491	Yb	3	25.2
V	2	14.65	Hf	3	23.3
	3	29.310		4	33.3
Y	3	20.52	Bi	2	16.69
Zr	3	22.99		3	25.56
	4	34.34	Th	3	20.0
Nb	2	14.32		4	28.8
	3	25.04			

[a] Ref. 85

By an ingenious analysis of the observed travel time of $HeRh^{2+}$ cations and the dissociation rate into $He + Rh^{2+}$, Tsong and Liou [92i] conclude that the dissociation is mainly due to tunneling of the He atom through the potential barrier. Tunneling phenomena are critically dependent on the mass of the tunneling atom and the width of the potential barrier. The small size of the helium atom makes the dissociation via tunneling an important mechanism.

Concerning the binding interactions in "metal halides" HeX^{n+}, CASSCF calculations by Hotokka et al. [93] predict $D_e = 4.4$ kcal/mol for dissociation of $HeTi^{2+}$ into $He + Ti^{2+}$. For the triply charged HeV^{3+}, which dissociates into $He^+ + V^{2+}$, a barrier for the dissociation is calculated at 17.2 kcal/mol [93]. Calculations were also performed for the $HeAl^{3+}$ cation. The barrier for dissociation into $He^+ + Al^{2+}$ is predicted as 27.4 kcal/mol [93]. The very high ionization energy of helium, which prohibits any "normal" chemistry for the lightest noble gas, is a favorable condition for the remarkable stability of these highly charged cations.

The possibility of binding helium in highly charged cations has also been recognized by Radom and coworkers [94, 95]. They calculated $HeSi^{n+}$ with $n = 1 - 4$ and found computationally [94] that the $HeSi^{4+}$ tetracation has the shortest bond (1.550 Å) of the four charged species, although the exothermicity of the dissociation into $He^+(^2S) + Si^{3+}(^2S)$ is enormous: 385.7 kcal/mol. I_4 of Si is 45.141 eV [85]. The attractive interactions are explained using molecular orbital arguments [94]. There are only two valence electrons in $HeSi^{4+}$ which occupy the 4σ bonding orbital. In $HeSi^{3+}$ and $HeSi^{2+}$, the 5σ-antibonding orbital is singly and doubly occupied, respectively. This yields longer interatomic distances especially for $HeSi^{2+}$ (2.568 Å), although $HeSi^{2+}$ is thermodynamically stable towards dissociation into $He + Si^{2+}(^1S)$, albeit only by 2 kcal/mol [94]. $HeSi^{4+}$ is an example for the situation shown in Fig. 4a, with the lower dissociation limit being He^+ and Si^{3+} and the higher limit as $He + Si^{4+}$.

4.5 The Nature of Bonding in Diatomic Noble Gas Ions

Figure 6 shows a contour line diagram of the electron density distribution of the van der Waals complex He_2 (He,He distance 2.74 Å). The electron density at the midpoint between the two He atoms is just 0.008 e/Å^3, quite different from the values found for a typical covalent bond between first-row elements (1–5 e/Å^3). Despite the smallness of $\rho(r)$ in the internuclear region, the He nuclei are linked by a MED path and the midpoint is the position of a $(3, -1)$ critical point (Fig. 6). As pointed out above this does not imply the existence of a chemical bond. The energy density $H(r)$ is positive at the $(3, -1)$ critical point, which means that the kinetic energy rather than the potential energy dominates in the internuclear region. There is no chemical bond between the He atoms.

If one electron is removed to yield He_2^+, the equilibrium distance between the two He nuclei decreases to 1.08 Å. This is caused by the contraction of the

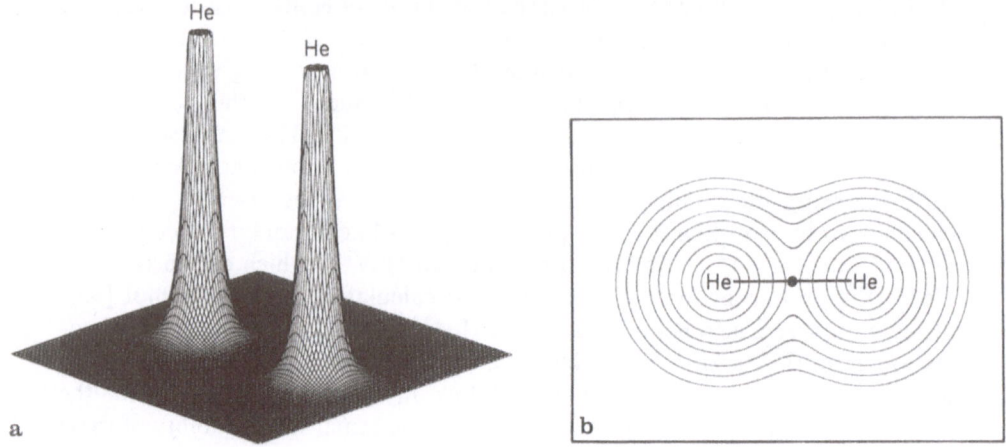

Fig. 6a and b. Perspective drawing (**a**) and contour line diagram (**b**) of the electron density distribution $\rho(r)$ of He_2 at its van der Waals distance (2.74 Å) shown with regard to a plane that contains the nuclei. MED path and (3, − 1) critical point are indicated in the contour line diagram. (HF/6-31G(p) calculations)

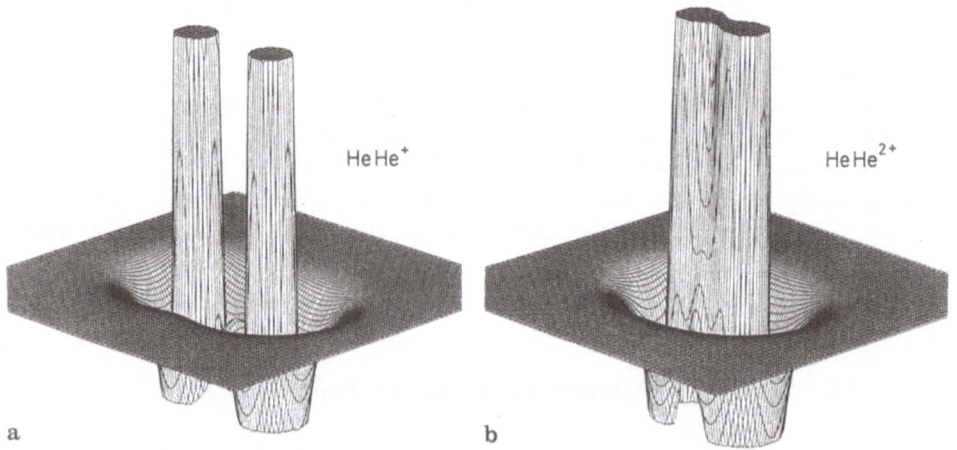

Fig. 7a and b. Perspective drawings of the Laplace concentration, $-\nabla^2 \rho(r)$, of (**a**) He_2^+ and (**b**) He_2^{2+} ions at their equilibrium distance shown with regard to a plane containing the nuclei

orbitals due to the positive charge. Electronic charge is drawn closer to the nuclei which is reflected by the calculated Laplace concentration. The nuclei are more shielded, while at the same time repulsion between the He electrons is decreased. This leads to a reduction of the internuclear distance relative to that of the van der Waals complex He_2.

There is a MED path connecting the two nuclei of He_2^+. The electron density ρ_B at the (3, − 1) critical point, the midpoint between the two He atoms, is 1.2 e/Å^3, which is typical of a covalent bond. More convincing is the result that the energy density H_B is negative (-0.73 hartree/Å^3), indicating that the kinetic energy has become smaller than the potential energy in the internuclear region. A covalent bond has been formed. This description is in line with the common understanding that bonding is accompanied by a decrease in kinetic energy [18].

If a second electron is removed as in He_2^{2+}, the internuclear distance decreased to 0.70 Å for the same reasons described for He_2^+. The electron density at the $(3, -1)$ critical point is 4.4 e/Å^3, while the energy density H_B is -4.4 hartree/Å^3, both indicating covalent bonding.

The Laplace concentrations of the two molecules He_2^+ and He_2^{2+} reflect these trends nicely as can be seen from Fig. 7. For the monocation, the Laplace concentration is increased in the bond region relative to that of the separated atoms, which possess in the outer valence region a sphere with depletion of negative charge. However, depletion is not totally compensated in the cation: the Laplace concentration is still negative, i.e. $-\nabla^2\rho(\mathbf{r}) < 0$. For the dication, the Laplace concentration becomes positive, i.e. $-\nabla^2\rho(\mathbf{r}) > 0$. Obviously, covalent bonding implies an increase of the Laplace concentration in the bonding region relative to the Laplace concentration of the noninteracting atoms.

Mono- and dications Ng_2^{n+} possess covalent bonds comparable to those of other homonuclear diatomics such as H_2, N_2, or F_2. This, however, is not necessarily true for cations and dications NgX^{n+} with X being different from Ng. Since the electronegativity of Ng is higher than that of X, positive charge is preferentially located at X and Ng will be polarized by the positive charge at X. Hence, Ng and X are attracted by charge-induced dipole interactions proportional to the amount of positive charge q at X and the value of the static electric dipole polarizability α of Ng. Predictions on the stability of NgX^{n+} can be made by using Eq. (4), which estimates the electrostatic interaction energy ΔE_{elec} from the known values of q, α, and the Ng, X distance r [4].

$$\Delta E_{elec} = -0.5\alpha q^2/r^4 \tag{4}$$

Whether or not the stability of NgX^{n+} is enhanced by additional covalent interactions depends on the electronic structure of X^{n+}. As mentioned above, a covalent Ng,X bond can only be achieved by donation of Ng electrons to X which implies that X must be a sufficiently strong electron acceptor. Inspection of Table 7 reveals that for the electronic ground states of HeX^+ none of the first-row elements X is capable of withdrawing electrons from He thereby establishing a covalent bond. The electron density ρ_B at the $(3, -1)$ critical point is extremely small, ranging from 0.01 (X = Be) to 0.09 e/Å^3 (X = F). Only $HeNe^+(X^3\Sigma^-)$ has a somewhat larger value of ρ_B (0.26 e/Å^3, Table 7). All H_B values are larger than zero, indicating electrostatic interactions rather than covalent bonds.

Nevertheless, covalent bonding between He and X^+ is possible for the monocations HeX^+. It requires a rearrangement of electrons in the valence shell of X^+ thus yielding an excited electronic state of X^+ and, subsequently, HeX^+. The electron rearrangement must be such that the electron acceptor ability of

X^+ increases. This is best achieved by creating a s- or $p\sigma$-hole in the valence shell that is prone to accept s-electrons from He. An s- or σ-hole in connection with a sufficiently large nuclear charge is the prerequisite for covalent He,X bonding (Table 7).

All excited states of HeX^+ listed in Table 7 possess a He,X bond with covalent character according to the properties of the calculated electron and energy distribution [4]. The ρ_B and H_B data in Table 7 suggest that the He,B bond is the weakest Ng,X bond, actually more electrostatic than covalent. The strongest covalent bond is found for $HeN^+(^3\Pi)$, which is in line with the frontier orbital model (Sect. 4.3) and the Laplace description of He,X interactions.

Investigation of the Laplace concentration $-\nabla^2\rho(\mathbf{r})$ of HeX^+ provides additional information. In Fig. 8, perspective drawings and contour line diagrams of $-\nabla^2\rho(\mathbf{r})$ are given for both the $X^3\Sigma^-$ ground state and the $^3\Pi$ excited state of HeN^+. The valence shell concentration of $N^+(^3P)$ is highly anisotropic. There are concentration lumps in the direction of the singly occupied $2p(\pi)$ orbitals and deep concentration holes in the direction of the unoccupied $2p(\sigma)$

Table 7. Nature of bonding in HeX^{n+} ions according to ab initio calculations[a]

HeX^{n+} (state)	Electronic structure of X^{n+}	Hole	ρ_b [$e\text{Å}^{-3}$]	H_b [hartree Å^{-3}]	D_e [kcal/mol]	Nature of bonding
		Ground States of HeX^+ and HeX^{2+}				
$HeLi^+(X^1\Sigma^+)$	$1s^2$	$2s$	0.04	0.02	1.5	electrostatic
$HeBe^+(X^2\Sigma^+)$	$K\,2s$	$2s$	0.01	0.00_2	0.3	electrostatic
$HeB^+(X^1\Sigma^+)$	$K\,2s^2$	$2p$	0.02	0.01	0.5	electrostatic
$HeB^{2+}(X^2\Sigma^+)$	$K\,2s$	$2s$	0.57	-0.19	26.8	covalent
$HeC^+(X^2\Pi)$	$K\,2s^2 2p$	$2p$	0.05	0.02	1.1	electrostatic
$HeC^{2+}(X^1\Sigma^+)$	$K\,2s^2$	$2p$	0.55	-0.11	17.7	covalent
$HeN^+(X^3\Sigma^-)$	$K\,2s^2 2p^2$	$2p$	0.26	0.07	4.1	electrostatic
$HeN^{2+}(X^2\Pi)$	$K\,2s^2 2p$	$2p$	1.19	-0.80	48.2	covalent
$HeO^+(X^4\Sigma^-)$	$K\,2s^2 2p^3$	$2p$	0.04	0.01	0.8	electrostatic
$HeO^{2+}(X^3\Sigma^-)$	$K\,2s^2 2p^2$	$2p$	1.96	-1.86	98.4	covalent
$HeF^+(^3\Pi)$	$K\,2s^2 2p^4$	$2p$	0.09	0.03	1.4	electrostatic
$HeF^{2+}(^2\Pi)$	$K\,2s^2 2p^3$	$2p$	2.82	-2.93	143.0	covalent
$HeNe^+(X^2\Pi)$	$K\,2s^2 2p^5$	$2p$	0.80	0.01	10.4	(electrostatic)
$HeNe^{2+}(X^1\Sigma^+)$	$K\,2s^2 2p^4$	$2p$	2.92	-2.58	240.1	covalent
		Excited States of HeX^+ and HeX^{2+}				
$HeB^+(^3\Pi)$[b]	$K\,2s2p$	$2s$	0.34	-0.08	5.6	(covalent)
$HeB^{2+}(^2\Pi)$	$K\,2p$	$2s$	0.82	-0.16	64.5	covalent
$HeC^+(^4\Sigma^-)$[b]	$K\,2s2p^2$	$2s$	0.98	-0.76	29.3	covalent
$HeC^{2+}(^3\Pi)$[b]	$K\,2s2p$	$2s$	1.24	-1.39	67.0	covalent
$HeN^+(^3\Pi)$	$K\,2s2p^3$	$2s$	2.06	-3.51	68.9	covalent
$HeN^{2+}(^4\Sigma^-)$	$K\,2s2p^2$	$2s$	2.16	-4.06	127.9	covalent
$HeO^+(^2\Pi)$	$K\,2s2p^4$	$2s$	1.51	-1.21	4.4	covalent
$HeO^{2+}(^3\Pi)$	$K\,2s2p^3$	$2s$	2.88	-3.59	223.3	covalent
$HeF^+(X^1\Sigma^+)$	$K\,2s2p^5$	$2s$	2.63	-2.72	47.5	covalent

[a] Values of ρ_b and H_b from HF/6-31G(d, p) calculations [4]. For HeX^{2+}, IA values are given. [b] A minimum rather than a maximum was found in a direction perpendicular to the internuclear axis indicating a poor description of $\rho(\mathbf{r})$.

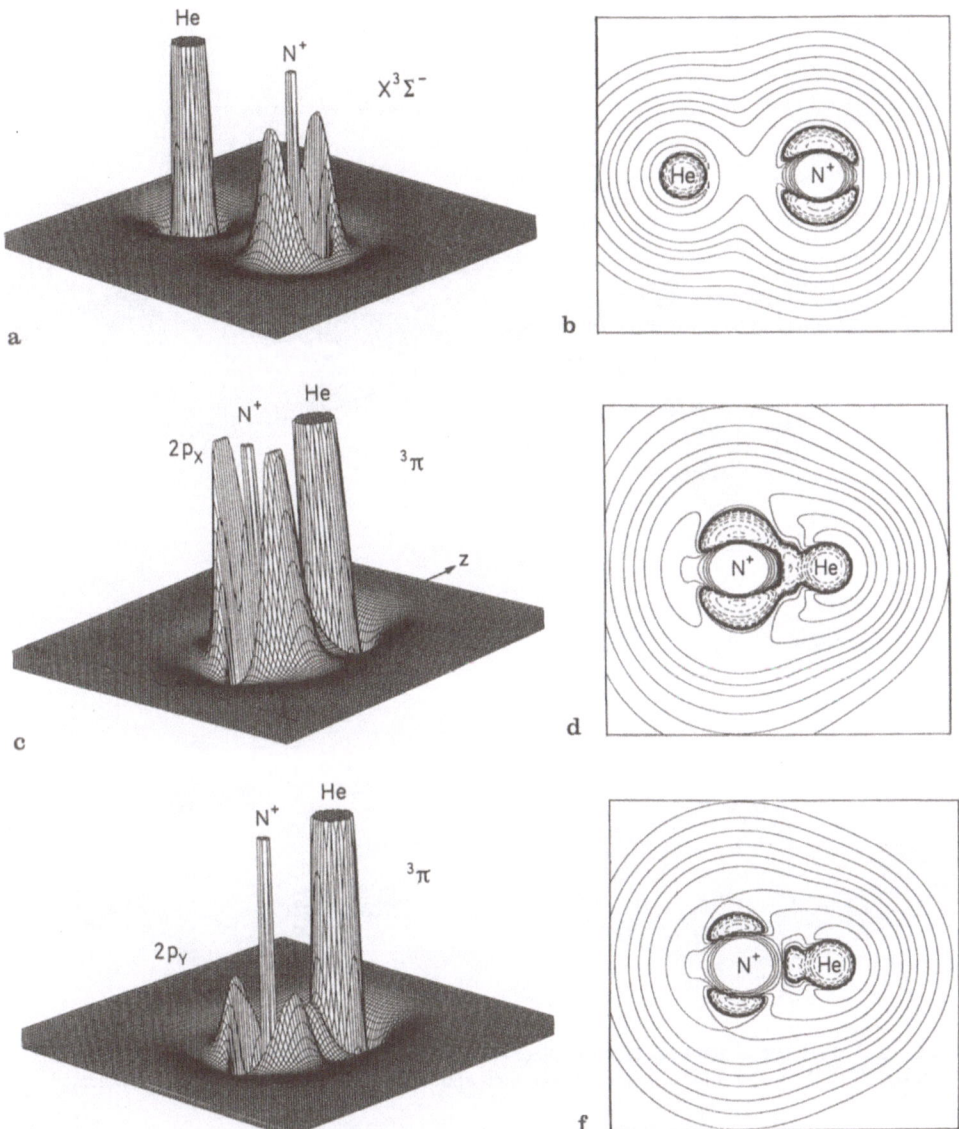

Fig. 8a–f. Perspective drawings and contour line diagrams of the Laplace concentration $-\nabla^2\rho(\mathbf{r})$ of HeN$^+$. Dashed contour lines are in regions of concentration of negative charge, solid contour lines in regions of charge depletion. Inner-shell concentration of N$^+$ is not shown in the contour line diagrams. (**a**) and (**b**) HeN$^+$(X$^3\Sigma^-$) in xz-plane. (**c**) and (**d**) HeN$^+$($^3\Pi$) in xz-plane. (**e**) and (**f**) HeN$^+$($^3\Pi$) in yz-plane. (HF/6-31G(d, p) calculations). Ref. [4]

orbital. If a van der Waals complex is formed with He (Fig. 8a, b), the noble gas atom approaches N$^+$(^{3}P) in the direction of the $p\sigma$-holes. The nucleus of N is less shielded in this direction and, accordingly, the electrons of He "feel" the attracting positive charge of the nucleus most strongly in this direction.

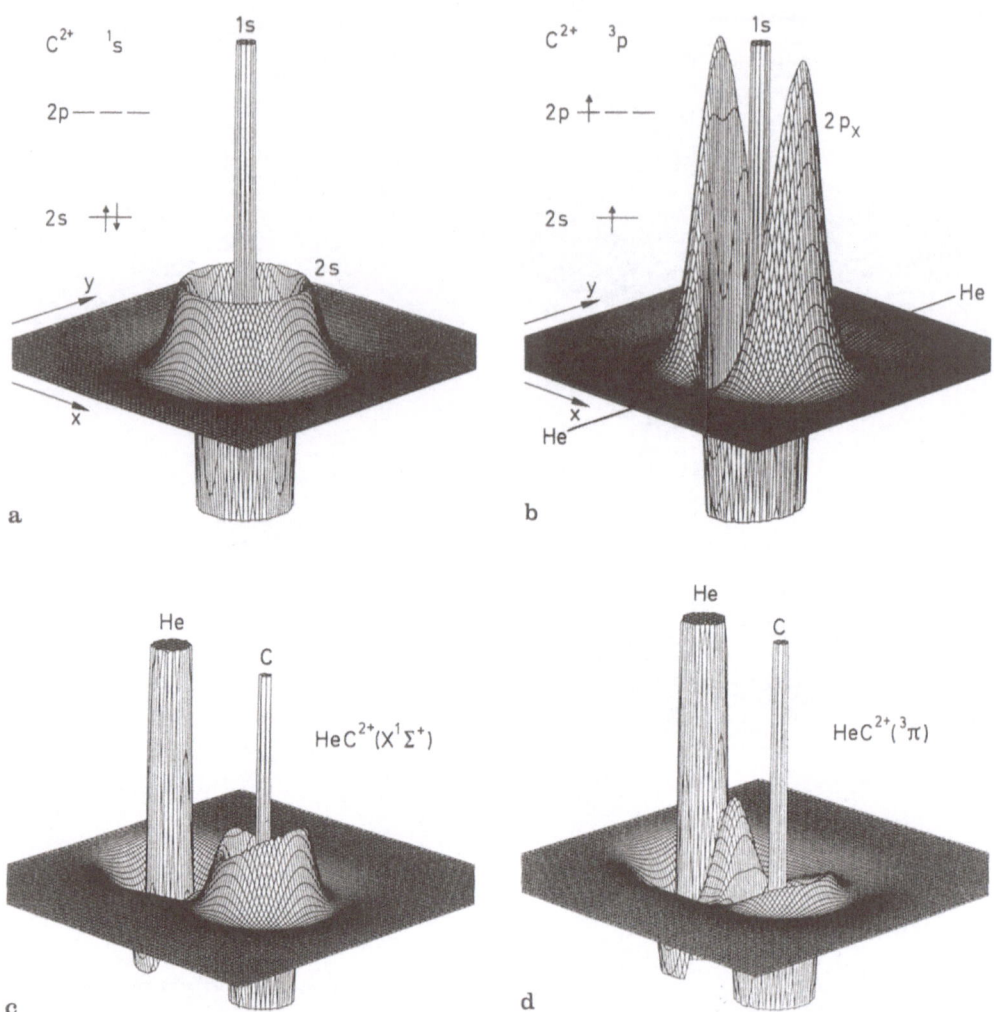

Fig. 9a–d. Perspective drawings of the Laplace concentration, $-\nabla^2\rho(\mathbf{r})$, of (a) $C^{2+}(^1S)$, (b) $C^{2+}(^3P)$, (c) $HeC^{2+}(X^1\Sigma^+)$, and (d) $HeC^{2+}(^3\Pi)$. Inner shell and valence shell concentrations are indicated

The contour line diagram in Fig. 8b reveals that the electronic structure of both He and $N^+(^3P)$ are hardly distorted in the $X^3\Sigma^-$ ground state of HeN^+. The $p\sigma$-hole slightly widens on the side of the He atom (the elliptic concentration areas are shifted slightly to the back) such as to increase stabilizing electrostatic interactions. This is in line with the description given above.

When the ground state ^3P of N^+ is excited to $N^+(^3D)$ the concentration holes in the valence shell of N^+ deepen significantly due to the transfer of a $2s$ electron to a $2p$ orbital. Inspection of Fig. 8c shows that the acceptor ability of N^+ is significantly enhanced. The electronic charge is concentrated in the internuclear region of $HeN^+(^3\Pi)$ because negative charge is shifted from He to the σ-hole at $N^+(^3D)$. A semipolar bond between He and $N^+(^3D)$ is formed. In the contour line diagram of the Laplace concentration of $HeN^+(^3\Pi)$ the semipolar bond appears as a droplet like appendix of the spherical He concentration (Fig. 8d). This pattern of $-\nabla^2\rho(r)$ has been found to be typical of all semipolar covalent He,X bonds investigated so far [4, 7, 13].

Increase of the positive charge at atom X enhances the acceptor ability for X^{2+}. All HeX^{2+} dications are covalently bonded according to the calculated ρ_B and H_B values listed in Table 7. This holds also for X^{2+} ions with isotropical charge distribution, i.e. species without holes in the valence sphere. The Laplace concentration of $C^{2+}(^1S)$ is shown in Fig. 9a. Although isotropical, the charge concentration in the valence shell is so small that the C nucleus is insufficiently shielded and an oncoming He atom is attracted by $C^{2+}(^1S)$. A weakly covalent bond is formed ($D_e = 17.9$ kcal/mol) which shows up in the Laplace concentration only by distortion in the valence shell concentration of $C^{2+}(^1S)$ (Fig. 9c). The charge concentration is more pronounced along the bond axis rather than perpendicular to it.

Again, covalent bonding is much stronger when an s-hole is available, for example in the ^3P excited state of C^{2+} (Fig. 9b). Due to the excitation of a $2s$ electron to a $2p$ orbital, the carbon nucleus is deshielded in the direction of the unoccupied $2p$ orbitals. There are deep concentration holes in these directions, while in the direction of the occupied $2p$ orbital there are large concentration lumps. Electronic charge of He is shifted toward the concentration holes upon collision of He and $C^{2+}(^3P)$. A strong covalent semipolar bond can be formed between He and $C^{2+}(^3P)$. This is clearly reflected by the Laplace concentration of $HeC^{2+}(^3\Pi)$, the calculated ρ_B and H_B values, and the dissociation energy (Table 7). Similar descriptions can be given for the ground and excited states of other HeC^{2+} ions. Donor-acceptor interactions increase with

a) the nuclear charge,
b) the total charge of X and
c) the enlargement of the σ-holes.

Covalent semipolar bonds result if an s-hole is available in the valence shell of X. This is in line with the donor-acceptor description given above.

What are the differences between the helium ions HeX^{n+} and the neon and argon analogues NeX^{n+} and ArX^{n+}? This question has been answered by basically applying the same analysis as used for the HeX^{n+} ions [5]. However, a simplified energy analysis may already provide important informations. The stabilization arising from charge-induced dipole interactions can be estimated from Eq. (4) using calculated equilibrium distances and dipole polarizabilities α taken from the literature. The resulting electrostatic interaction energies ΔE_{elec} are larger than the actual electrostatic attraction, because Eq. (4) is not strictly valid at small internuclear distances [96]. Nevertheless, covalent bonding can be assumed if the calculated D_e values are significantly larger than the electrostatic attraction energy ΔE_{elec}.

In Table 8 the results of this analysis are summarized for NgX^+ ions in their electronic ground states. There are no covalent bounded HeX^+ ions, since $D_e - \Delta E_{elec}$ is in all cases negative. NeN^+ ($D_e - \Delta E_{elec} = 2.5$ kcal/mol) and ArC^+ ($D_e - \Delta E_{elec} = 5.5$ kcal/mol) are predicted as slightly covalent, ArN^+ ($D_e - \Delta E_{elec} = 25$ kcal/mol) and ArF^+ ($D_e - \Delta E_{elec} = 37$ kcal/mol) seem to be clearly covalent. All other NgX^+ ions are electrostatically bound without significant covalent contribution according to the energy criterion. These predictions are confirmed by the analysis of $\rho(\mathbf{r})$ and $H(\mathbf{r})$ (Table 9). Electrostatic interactions increase with the polarizability of Ng. Since the value of α differs not very much for He and Ne, the D_e values are comparable. But for Ar with α being five times as large as the values for Ne and He, electrostatic interactions gradually change to covalent bonding.

If the positive charge of X is increased to X^{2+}, or if X is excited to a state with pronounced s-holes, then the corresponding NeX^+ and Ar^+ species are covalently bound for the same reasons discussed for HeX^+. The semipolar bond between Ng and X^{n+} is formed by sharing of the two $2p\sigma$ electrons of Ne or Ar. Alternatively, two $2p\pi$ electrons of Ne or Ar could be donated to X^+. However, the acceptor ability of X with regard to π electrons is much lower. There is no π-donation by Ne or Ar towards X^+.

It is interesting to note that the $p\pi$ electrons of Ne and Ar may lead to some destabilization of NgX^{2+}. In Fig. 10, contour line diagrams and perspective drawings of $-\nabla^2\rho(\mathbf{r})$ are shown for the $^1\Sigma^+$ ground state and the $^3\Pi$ excited state of ArC^{2+} [5]. The ground state corresponds to $C^{2+}(^1S)$ which does not possess any p electrons; the excited state corresponds to $C^{2+}(^3P)$, which possesses a σ-hole but also a $2p$ electron (compare with Fig. 9). In the ArC^{2+} states, the valence shell concentration of the Ar atom is distorted in a characteristic way. There are holes in the direction of the bond while there are concentration lumps in the nonbonding region of Ar. Distortions in the valence shell concentration of the C atom are complementary to those at the Ar atom, i.e. the

Table 8. Nature of bonding in NgX^+ monocations according to Eq. (4)[a]

X	He	Ne	Ar
Li	electrostatic	electrostatic	electrostatic
Be	electrostatic	electrostatic	electrostatic
B	electrostatic	electrostatic	electrostatic
C	electrostatic	electrostatic	covalent
N	electrostatic	covalent	covalent
O	electrostatic	electrostatic	electrostatic
F	electrostatic	electrostatic	covalent
Ne	electrostatic	covalent	electrostatic

[a] Electrostatic bonding is assumed for $\Delta E^{elec} > D_e$, covalent bonding for $\Delta E^{elec} < D_e$.

Table 9. Nature of bonding in NeX$^+$ and ArX$^+$ ions according to ab initio calculations[a]

NgX$^+$ (state)	Electronic structure of X$^+$	Hole	ρ_b [eÅ^{-3}]	H_b [hartree Å^{-3}]	D_e [kcal/mol]	Nature of bonding
A. NeX$^+$ ions						
NeLi$^+$(X$^1\Sigma^+$)	$1s^2$	$2s$	0.10	0.06	2.8	electrostatic
NeBe$^+$(X$^2\Sigma^+$)	K $2s$	$2s$	0.16	0.04	− 0.1	electrostatic
NeB$^+$(X$^1\Sigma^+$)	K $2s^2$	$2p$	0.10	0.01	1.1	electrostatic
($^3\Pi$)[b]	K $2s2p$	$2s$	0.30	− 0.10	7.3	covalent
NeC$^+$(X$^2\Pi$)	K $2s^22p$	$2p$	0.23	− 0.01	3.0	electrostatic
($^4\Sigma^-$)[b]	K $2s2p^2$	$2s$	0.59	− 0.51	21.7	covalent
NeN$^+$(X$^3\Sigma^-$)	K $2s^22p^2$	$2p$	0.50	− 0.07	9.2	covalent
NeO$^+$(X$^4\Sigma^-$)	K $2s^22p^3$	$2p$	0.14	0.04	1.2	electrostatic
($^2\Pi$)	K $2s2p^4$	$2s$	0.92	− 0.20	6.3	covalent
NeF$^+$($^3\Pi$)	K $2s^22p^4$	$2p$	0.30	0.07	3.6	electrostatic
(X$^1\Sigma^+$)	K $2s^22p^5$	$2s$	1.29	− 0.34	39.8	covalent
NeNe$^+$(X$^2\Pi$)	K $2s^22p^5$	$2p$	0.65	0.14	29.8	(electrostatic)
B. ArX$^+$ ions						
ArLi$^+$(X$^1\Sigma^+$)	$1s^2$	$2s$	0.07	0.03	5.8	electrostatic
ArBe$^+$(X$^2\Sigma^+$)	K $2s$	$2s$	0.27	− 0.01	10.7	electrostatic
ArB$^+$(X$^1\Sigma^+$)	K $2s^2$	$2p$	0.12	0.01	5.0	electrostatic
($^3\Pi$)	K $2s2p$	$2s$	0.61	− 0.41	35.2	covalent
ArC$^+$(X$^2\Pi$)	K $2s^22p$	$2p$	0.53	− 0.14	20.6	covalent
($^4\Sigma^-$)	K $2s2p^2$	$2s$	1.15	− 1.26	56.5	covalent
ArN$^+$(X$^3\Sigma^-$)	K $2s^22p^2$	$2p$	0.98	− 0.49	48.7	covalent
ArO$^+$(X$^4\Sigma^-$)	K $2s^22p^3$	$2p$	0.30	0.05	10.0	electrostatic
($^2\Pi$)	K $2s2p^4$	$2s$	1.53	− 1.08	50.5	covalent
ArF$^+$(X$^1\Sigma^+$)	K $2s^22p^5$	$2s$	1.59	− 1.02	43.2	covalent
($^3\Pi$)	K $2s^22p^4$	$2p$	0.43	0.01	16.6	electrostatic
ArNe$^+$(X$^2\Pi$)	K $2s^22p^5$	$2p$	0.15	0.02	1.8	electrostatic

[a] Values of ρ_b and H_b from HF/6-31G(d, p) calculations [5]; [b] A maximum rather than a saddle point was found. [5]

concentration lumps are in the direction of the molecular axis while the holes are in the nonbonding region. The distortions are the result of a transfer of $3p\sigma$ electrons from Ar to C^{2+}, either into a $2p$ (C^{2+}, ^{1}S) or $2s$ (C^{2+}, ^{3}P) orbital. In the latter case, the charge transfer is more pronounced and the concentration holes at Ar are much deeper (Fig. 10c) than in the former case (Fig. 10a). Figure 10 also shows that ArC^{2+}($^3\Pi$) suffers from $2p\pi$, $3p\pi$ repulsion (see concentration lumps in the nonbonding regions of C and Ar in Fig. 10c) which is absent in ArC^{2+}($^1\Sigma^+$) (Fig. 10a). Because of the difference in $p\pi$-$p\pi$ electron repulsion, the Ng,C bond strength in the triplet states is almost four times as large than in the singlet state if Ng = He, but only twice as large for Ng = Ne, and less than twice as large for Ng = Ar. In the latter case it is also important that donation from a $3p\sigma$ orbital to a $2s$ ($2p\sigma$) orbital is less efficient than from a $2p\sigma$ to a $2s$ ($2p\sigma$) orbital.

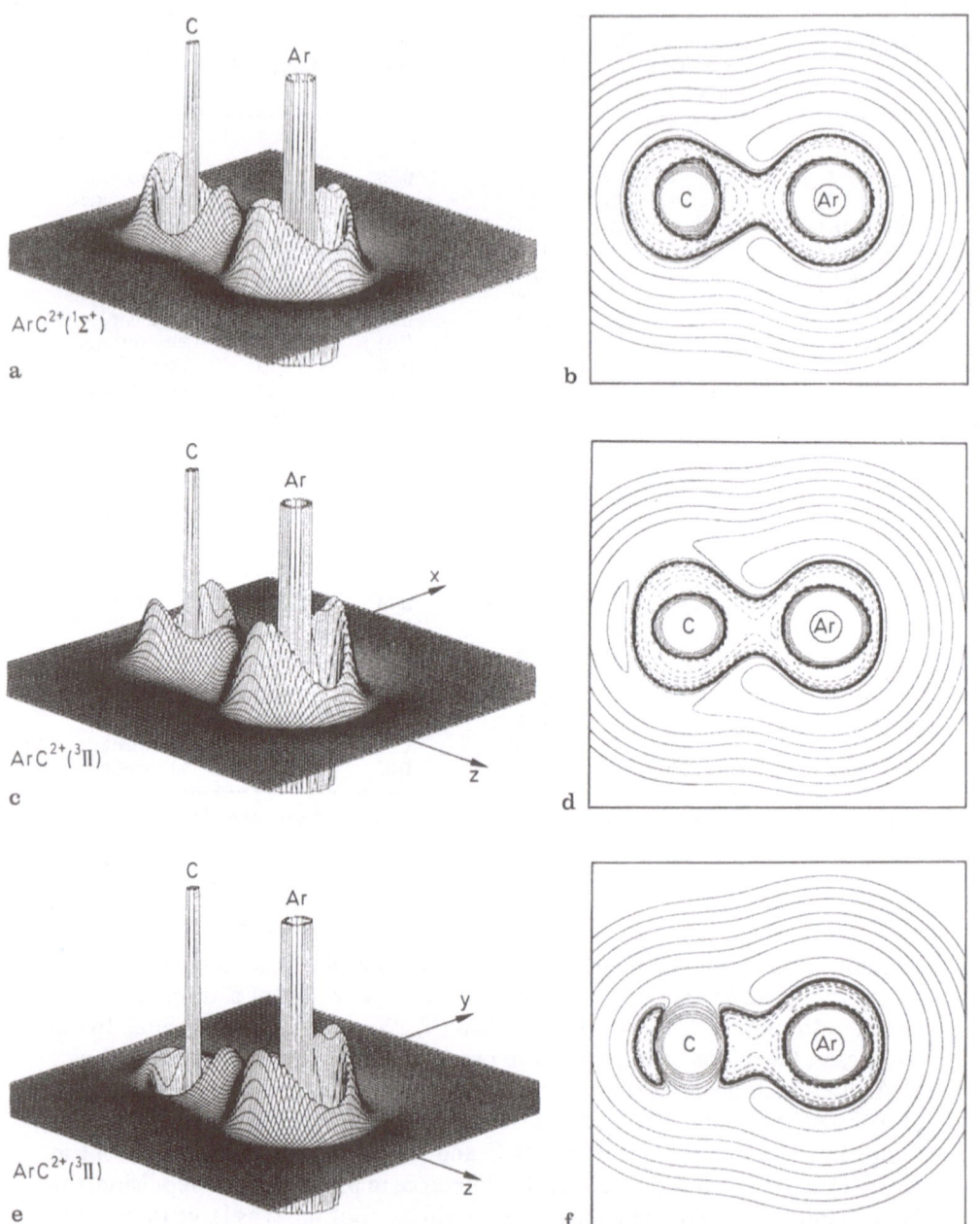

Fig. 10a–f. Contour line diagrams and perspective drawings of the Laplace concentration, $-\nabla^2\rho(\mathbf{r})$, of ArC^{2+}: **(a, b)** $^1\Sigma^+$ state, **(c, d)** $^3\Pi$ state, direction of filled π-orbital and **(e, f)** $^3\Pi$ state, direction of unfilled π-orbital. In the contour line diagrams inner shell concentrations are no longer shown. Ref. [13]

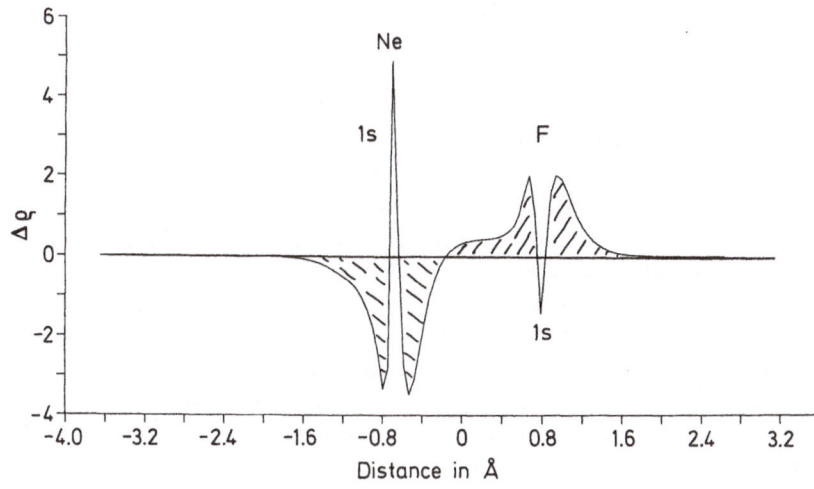

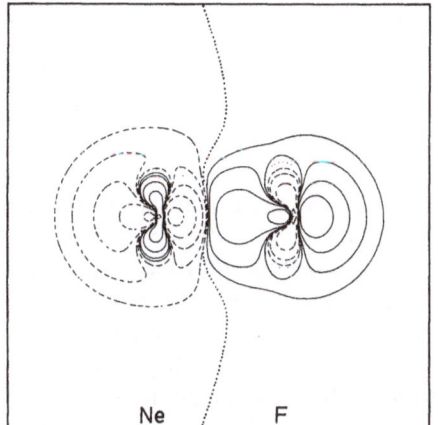

Fig. 11. Profile plot along the internuclear axis (*above*) and contour line diagram (*below*) of the difference electron density $\Delta\rho(\mathbf{r}) = \rho(\mathbf{r})^{molecule} - \rho(\mathbf{r})^{promolecule}$ calculated for $NeF^+(X^1\Sigma^+)$. (HF/6-31G(d) calculations). Ref. [5]

The bonding mechanism in NgX^{n+} and ArX^{n+} becomes even more clear when calculating the difference electron density distribution $\Delta\rho(\mathbf{r})$. In Fig. 11, a profile plot and a contour line diagram of $\Delta\rho(\mathbf{r})$ of $NeF^+(X^1\Sigma^+)$ are given [5]. Both diagrams indicate that electron density is transferred from Ne to F^+. It accumulates in the bonding and nonbonding region of F^+. Ne becomes partially positive while F^+ looses some of its positive charge. As a consequence, negative charge residing in the $2p\pi$ or the $1s$ orbitals, which are not involved in the charge transfer from Ne to F^+, is contracted toward the Ne nucleus but becomes more diffuse at the F nucleus. Also, some of the $2p\pi$ charge may be transferred from F to Ne atom via back donation. As a consequence, $\Delta\rho(\mathbf{r})$ is positive (negative) in the region of the $2p\pi$ and $1s$ orbitals of Ne ($2p\pi$ and $1s$ orbitals of F). This confirms that the formation of a covalent Ng,X bond is different for Ng = Ne or Ar than for Ng = He.

5 Polyatomic Ions of Helium, Neon and Argon

Most research of light noble gas chemistry has focussed on diatomic ions, because only very few polyatomic cations have been observed experimentally. Under the conditions of a mass spectrometer, where most gas-phase experiments are performed, it is difficult to detect more complex molecules containing He, Ne and Ar. The experimental technique of flowing-afterglow mass spectrometry [193] helped to investigate the formation of more complex noble gas ions. For example, it was found by Bohme et al. [98] that Ar_2^+ reacts with CO and forms, among other products, the triatomic cation $ArCO^+$. Under the same condition, the formation of $HeCO^+$ and $NeCO^+$ from CO and He_2^+ of Ne_2^+, respectively, was not observed [99]. In a similar way, Munson et al. [99] report the formation of $ArCO^+$ and ArN_2^+ in a mass spectrometer, while the helium and neon analogs could not be detected in analog reactions. In recent years, theoretical studies of small noble gas ions have been reported [7, 13, 83, 90d, 95, 97] and these may help experimentalists to tailor their experiments to more promising candidates. The results of these studies let us expect that more focussed experiments guided by theory will find new cations of He, Ne and Ar.

The presentation of the reported research on polyatomic cations will be made analogous to the diatomic species. As shown above, atomic ions X^{n+} may bind a light noble gas element with a binding energy up to 96.1 kcal/mol (ArH^+). We will first ask the question how many noble gas atoms may the atomic ions X^{n+} bind in polyatomic cations Ng_mX^{n+}. Then, we will proceed to cations which are formally formed from diatomic ions XY^{n+} and He, Ne or Ar. The reader will see that the available information on polyatomic cations is well understood using this scheme and that it is possible to make an "educated guess" about the stability of unknown species without even performing quantum mechanical calculations.

5.1 Ng_mH^+ Cations

Of all the atomic cations X^+ in the ground state, hydrogen($+$) forms the strongest bonds with Ng. Therefore, it can be expected that the proton should be capable of attracting more than one noble gas atom in hydrogen-bonded cluster ions Ng_mH^+. In fact, the formation of Ng_2H^+ (H = He, Ne, Ar) using the flowing afterglow technique has been reported by Adams et al. [100]. The triatomic molecules were formed in reaction (1):

$$Ng_2^+ + H_2 \longrightarrow Ng_2H^+ + H \tag{1}$$

The reaction rates of reaction (1) have been measured for the three light noble gases but only for Ar_2H^+ was the bond energy deduced as $> 3.82 - D(Ar_2)$ eV [100]. Theoretical studies at the Hartree-Fock level have been reported by Matcha, Milleur and coworkers for He_nH^+ [101], Ne_nH^+ [102] and Ar_nH^+

[103]. The results for Ng_2H^+ are shown in Chart 1. All three ions have a linear geometry and a $^2\Sigma_g^+$ ground state. This dissociation energy of the triatomic species into $NgH^+ + Ng$ is nearly the same (10–12 kcal/mol) for the three species. A more recent calculation of He_2H^+ by Dykstra [104] using self consistent electron pair (SCEP) and double-substitution coupled-cluster (CCD) methods gave slightly higher values of $D_e = 13.2$ kcal/mol (SCEP) and 13.3 kcal/mol (CCD). The lower D_e value for Ar_2H^+ may be caused by the relatively low theoretical level. Thus, the interactions between NgH^+ and Ng are much weaker than between Ng and H^+. This is because NgH^+ is a weaker Lewis acid than a proton. The bonds in Ng_2H^+ are significantly longer and weaker than in NgH^+. Milleur et al. also calculated [101] that He_3H^+ and He_4H^+ are energetically unstable toward dissociation into He_2H^+ and helium. The same has been assumed for Ar_nH^+ without explicit calculations being carried out [102]. For the argon analogs it was speculated [103] that stable Ar_nH^+ clusters with n > 2 may be possible because of the greater bond strength in ArH^+ than for NeH^+ and HeH^+. However, the nearly identical dissociation energies for the reaction $Ng_2H^+ \rightarrow NgH^+ + Ng$ for Ng = He, Ne and Ar argues against this. While the donor ability increases for He < Ne < Ar, there is a parallel decrease in Lewis acidity for $HeH^+ > NeH^+ > ArH^+$ which seems to cancel each other, thereby explaining the similar D_e values. It seems unlikely that thermodynamically stable Ng_nH^+ cluster ions exist for the light noble gases with n > 2.

Matcha and coworkers [105] also reported theoretical results based on Hartree-Fock SCF calculations for the mixed triatomic cation $HeNeH^+$. They

Chart 1. Optimized interatomic distances at MP2/6-31G(d, p) for NgH^+ and $NgHNg^+$ ions (in Å) and calculated dissociation energies D_e at MP4(SDTQ)/6-31G(2df, 2pd)//MP2/6-31G(d, p) for the loss of Ng atom.

He———H^+ 0.770	$D_e = 44.5$ kcal/mol[a]
	$\Delta = 11.0$ kcal/mol
He———H———He^+ 0.926	$D_e = 55.5$ kcal/mol[a]
Ne———H^+ 0.973	$D_e = 49.6$ kcal/mol[b]
	$\Delta = 11.0$ kcal/mol
Ne———H———Ne^+ 1.143	$D_e = 61.1$ kcal/mol[b]
Ar———H^+ 1.292	$D_e = 82.6$ kcal/mol[c]
	$\Delta = 9.3$ kcal/mol
Ar———H———Ar^+ ·1.586	$D_e = 91.9$ kcal/mol[c]

[a] Ref. 101; [b] Ref. 102; [c] Ref. 103

predicted dissociation energies of 9.4 kcal/mol for the loss of He and 14.5 kcal/mol for the loss of Ne. These numbers, which are smaller (for He loss) and larger (for Ne loss) compared with the D_e values for He_2H^+ and Ne_2H^+ agree nicely with the donor-acceptor model. Ne is a better donor than He and HeH^+ is a better acceptor than NeH^+ which explains the calculated reaction energies. No experimental reports are known to us about mixed cations $NgNg'H^+$.

5.2 Ng_mX^{n+} Cations with First-Row Elements X

In 1970, Clampiti and Jefferies [106] reported their molecular beam experiments for various ion clusters. For Ne_mLi^+, they found the largest peak for m = 1 followed by nearly identical signals for m = 2, 3, 4, 5, 6. The yield of ion clusters decreases abruptly beyond He_6Li^+. The same result was found for hydrogen clusters $(H_2)_mLi^+$. A remarkable feature of the mass spectrum was the relative insignificance of clusters with m > 6. It was concluded [106] that "six molecules around an ion probably constitute a complete shell". This result is important in comparison with theoretical studies of Ng_mBe^{2+} ions.

As noted above, I_2 of beryllium is lower than I_1 of helium and neon, which causes $HeBe^{2+}$ and $NeBe^{2+}$ to the thermodynamically stable by ca. 18 kcal/mol [4, 90] and 37 kcal/mol [90d], respectively. Harrison et al. [90b] found computationally that the binding energy for the stepwise attachment of the first (18.6 kcal/mol) helium atom is only slightly higher than the second (17.7 kcal/mol) He forming He_2Be^{2+}. Also the bond lengths in $HeBe^{2+}$ (1.434 Å) and linear He_2Be^{2+} (1.439 Å) are nearly the same [90b]. This is strikingly different to what is reported for the hydrogen analogs He_nH^+ discussed above [101]. Schleyer [90d] has extended these studies to helium, neon and argon clusters Ng_mBe^{2+} with m = 8 for He, m = 6 for Ne and m = 3 for Ar. The calculated association energies are shown in Table 10. As can be seen from the predicted binding energies, cluster ions He_mBe^{2+} and Ne_mBe^{2+} are stable toward the loss of a noble gas atom with m ≤ 6. These clusters all have the highest symmetry possible, i.e. linear for m = 2, trigonal for m = 3, tetrahedral for m = 4, trigonal bipyramidal for m = 5 and octahedral for m = 6. Schleyer [90d] speculated that the break in calculated binding energies for He_mBe^{2+} from m = 4 to m = 5 may be a hint toward octet rule consideration. However, such a break is found for Ne_mBe^{2+} with only the addition of a third and, to a lesser degree, for a fourth atom. In the case of Ar_mBe^{2+}, there is a significant decrease in binding energy even for the second and third argon. This result rather points toward the size of the noble gas atoms as the most important factor. Also, the change from negative to positive association energies for He_mBe^{2+} clusters for m > 6 shows some similarity to the experimental results for He_mLi^+ species which hints that the optimum shell size may be 6. For the Ar_mBe^{2+} clusters it should be kept in mind that the calculated association energies do not refer to the most favorable dissociation products. Since I_1 of Ar

Table 10. Calculated association energies for He_nBe^{2+}, Ne_nBe^{2+}, and Ar_nBe^{2+} clusters (in kcal/mol)[a]

	Ng		
Reactions	He	Ne	Ar
$Be^{2+} + Ng \longrightarrow NgBe^{2+}$	-18.2	-37.1	-69.7
$NgBe^{2+} + Ng \longrightarrow Ng_2Be^{2+}$	-17.2	-35.3	-49.4
$Ng_2Be^{2+} + Ng \longrightarrow Ng_3Be^{2+}$	-15.2	-25.1	-24.1
$Ng_3Be^{2+} + Ng \longrightarrow Ng_4Be^{2+}$	-13.0	-20.7	
$Ng_4Be^{2+} + Ng \longrightarrow Ng_5Be^{2+}$	-6.1	-17.0	
$Ng_5Be^{2+} + Ng \longrightarrow Ng_6Be^{2+}$	-6.5	-18.9	
$Ng_6Be^{2+} + Ng \longrightarrow Ng_7Be^{2+}$	$+2.6$		
$Ng_7Be^{2+} + Ng \longrightarrow Ng_8Be^{2+}$	$+5.6$		

[a] Taken from Ref. 90d

is lower than I_2 of Be, $ArBe^{2+}$ will dissociate into $Ar^+ + Be^+$ for which no dissociation energy is given.

The principle of donor-acceptor interactions in noble gas cations is further exemplified by the theoretical study of Radom and coworkers [95a,c] on the series He_nC^{n+}. Figure 12 shows the optimized geometries for the four cations. The intriguing result of the theoretical studies is the rather short He,C bond length for the triply and quadruply charged species He_3C^{3+} and He_4C^{4+}. The latter molecule, which is isoelectronic with methane CH_4, has also been calculated by Schleyer [90d]. He_3C^{3+} and He_4C^{4+} are "explosive" molecules with highly exothermic dissociation reactions of 246 and 383 kcal/mol, respectively [95c]. However, the barriers for these processes are calculated as 36.3 kcal/mol for He_3C^{3+} and 17.2 kcal/mol for He_4C^{4+}, which are sufficiently large to make experimental observation feasible [95c], although the experimental obstacles to proving the theoretical predictions are enormous.

The calculated interatomic distances shown in Fig. 12 for He_nC^{n+} can be rationalized when the interactions between n helium atoms and the atomic ion C^{n+} are considered. The acceptor strength (Lewis acidity) of C^{n+} increases with n, C^{4+} has no valence electrons and is a very strong Lewis acid. However, the higher charge means increasing Coulomb repulsion. While the increase in attractive orbital interactions dominates for He_3C^{3+} compared with He_2C^{2+}, the two opposing effects seem to nearly cancel each other when going from He_3C^{3+} to He_4C^{4+} as indicated by the nearly constant bond length.

The dihelium carbene dication He_2C^{2+} has also been studied theoretically by the groups of Frenking and Cremer [7, 11] together with the nitrogen and oxygen analogs He_2N^{2+} and He_2O^{2+}. The interatomic distances for the ground states of He_2X^{2+} decrease with $C > N > O$ just like the diatomic HeX^{2+} (see above). The calculated geometries are shown in Fig. 13.

He_2C^{2+} is isoelectronic with methylene CH_2 which is known to have a $^3B_1$ ground state 9.0 kcal/mol lower than the $^1A_1$ excited state [107]. In the case of He_2C^{2+}, the $^1A_1$ state is predicted to be 63.8 kcal/mol lower than the $^3B_1$ state which is also shown in Fig. 13 [7]. The latter state has a much shorter He, C distance than the $^1A_1$ ground state. The same result has been found for the $^1\Sigma^+$ ground state and $^3\Pi$ excited state of diatomic HeC^{2+} (see Sect. 4.3) and can be explained in the same fashion. HOMO–LUMO interactions of $He(^1S)$ with the $2p$ LUMO of $C^{2+}(^1S)$ are much weaker than with the half empty AO of $C^{2+}(^3P)$ (Fig. 2). But the stronger binding interactions do not compensate for the excitation energy which is the reason that the weakly bound $^1A_1$ state of He_2C^{2+} is the ground state and the stronger bound $^3B_1$ state is the excited state.

He_2C^{2+} has not been observed experimentally (nor have He_2N^{2+} and He_2O^{2+}) but since it is predicted to be stable towards dissociation into the atomic products, it might be found in molecular beam experiments. The stabilization energy of the $^1A_1$ ground state relative to $C^{2+}(^1S) + 2He$ is 30.6 kcal/mol (Table 11) [7, 11, 13]. Thus, the association energies for helium atoms with C^{2+} in the ground state are 17.0 kcal/mol for the first He and 13.6 kcal/mol for the second He (Table 11), which indicates similar sequence as found for He_mBe^{2+} (Table 10) [90b,d]. No studies have been reported for He_mC^{2+} with $m > 2$.

The neon and argon analogs of Ng_2X^{2+} have also been studied theoretically by Frenking et al. [13]. The calculated geometries and dissociations energies are

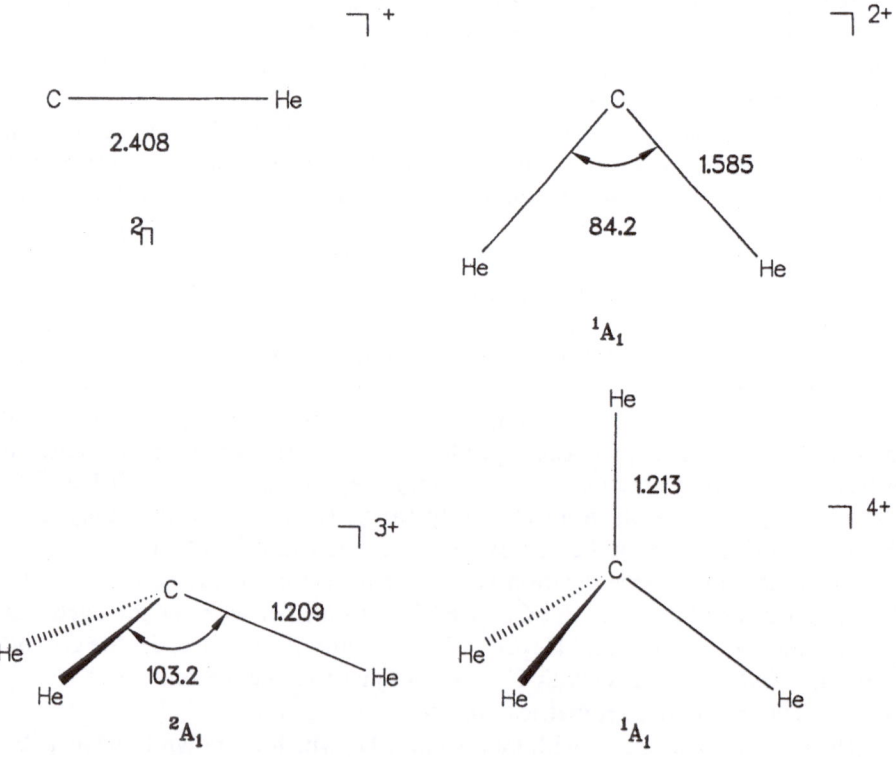

Fig. 12. Optimized geometries (MP4/6-311G(d, p)) for CHe_n^{n+} cations [95a]. Distances in Å, angles in °

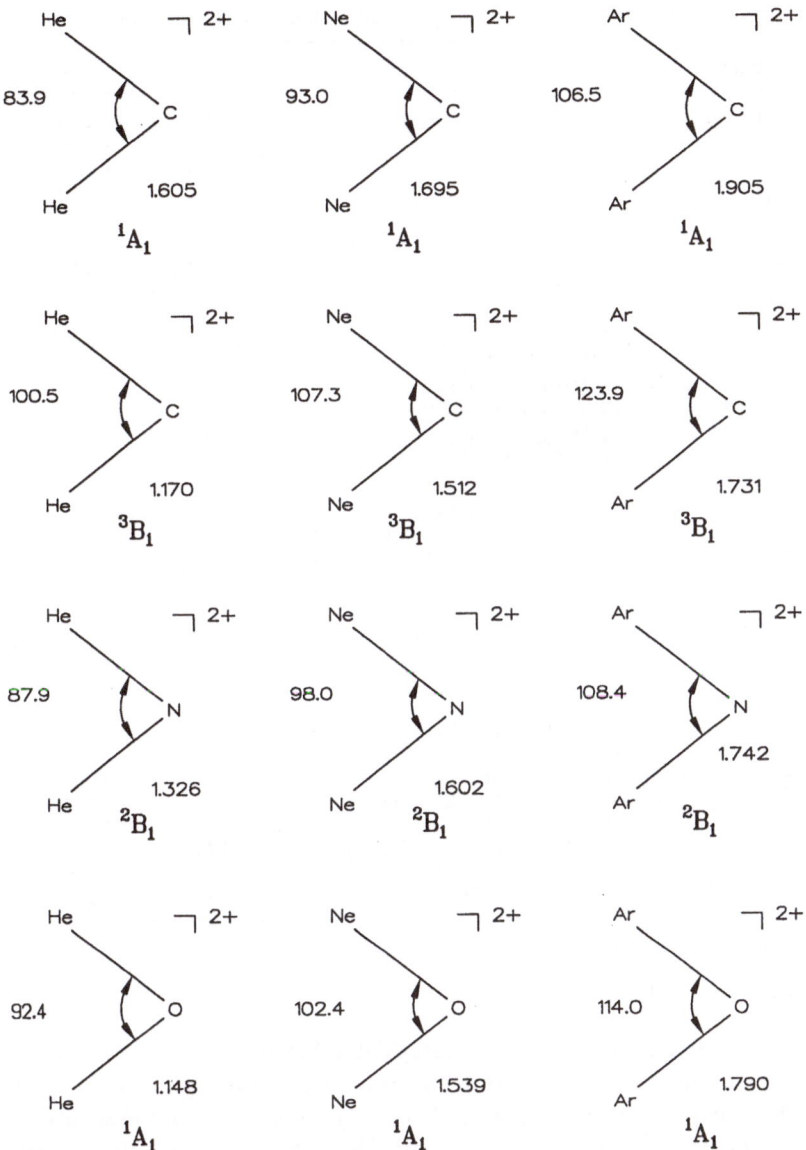

Fig. 13. Optimized geometries (MP2/6-31G(d, p)) for Ng_2X^{2+} cations [13]. Distances in Å, angles in °

shown in Fig. 13 and Table 11, respectively. The results are in total agreement to what could have been predicted from the donor-acceptor model. For Ng_2C^{2+}, the Ng,C interatomic distance is always longer in the $^1A_1$ ground state than the $^3B_1$ excited state. Accordingly, the dissociation energies for the fragmentation into Ng + C^{2+} in the corresponding electronic states are significantly higher for

Table 11. Calculated dissociation energies (kcal/mol) for Ng_2X^{2+} structures[a]

Reaction	D_e	D_o
$He_2C^{2+}(^1A_1) \longrightarrow HeC^{2+}(^1\Sigma^+) + He(^1S)$	+ 14.7	+ 13.6
$\longrightarrow C^{2+}(^1S) + 2He(^1S)$	+ 32.6	+ 30.6
$\longrightarrow C^+(^2P) + He^+(^2S) + He(^1S)$	+ 37.4	+ 35.4
$He_2C^{2+}(^3B_1) \longrightarrow HeC^{2+}(^3\Pi) + He(^1S)$	+ 50.1	+ 47.2
$\longrightarrow C^{2+}(^3P) + 2He(^1S)$	+ 117.4	+ 112.5
$\longrightarrow C^+(^2P) + He^+(^2S) + He(^1S)$	− 23.4	− 28.3
$Ne_2C^{2+}(^1A_1) \longrightarrow NeC^{2+}(^1\Sigma^+) + Ne(^1S)$	+ 27.4	+ 27.0
$\longrightarrow C^{2+}(^1S) + 2Ne(^1S)$	+ 71.9	+ 70.6
$\longrightarrow C^+(^2P) + Ne^+(^2P) + Ne(^1S)$	+ 3.9	+ 2.6
$Ne_2C^{2+}(^3B_1) \longrightarrow NeC^{2+}(^3\Pi) + Ne(^1S)$	+ 53.6	+ 52.6
$\longrightarrow C^{2+}(^3P) + 2Ne(^1S)$	+ 143.0	+ 140.8
$\longrightarrow C^+(^2P) + Ne^+(^2P) + Ne(^1S)$	− 70.7	− 72.9
$Ar_2C^{2+}(^1A_1) \longrightarrow ArC^{2+}(^1\Sigma^+) + Ar(^1S)$	+ 49.2	+ 48.8
$\longrightarrow C^{2+}(^1S) + 2Ar(^1S)$	+ 178.6	+ 177.2
$\longrightarrow C^+(^2P) + Ar^+(^2P) + Ar(^1S)$	− 26.7	− 28.1
$Ar_2C^{2+}(^3B_1) \longrightarrow ArC^{2+}(^3\Pi) + Ar(^1S)$	+ 89.2	+ 88.1
$\longrightarrow C^{2+}(^3P) + 2Ar(^1S)$	+ 294.5	+ 292.3
$\longrightarrow C^+(^2P) + Ar^+(^2P) + Ar(^1S)$	− 56.9	− 59.1
$He_2N^{2+}(^2B_1) \longrightarrow HeN^{2+}(^2\Pi) + He(^1S)$	+ 37.6	+ 35.2
$\longrightarrow N^{2+}(^2P) + 2He(^1S)$	+ 86.2	+ 82.4
$\longrightarrow N^+(^3P) + He^+(^2S) + He(^1S)$	− 30.4	− 34.0
$Ne_2N^{2+}(^2B_1) \longrightarrow NeN^{2+}(^2\Pi) + Ne(^1S)$	+ 48.6	+ 47.7
$\longrightarrow N^{2+}(^2P) + 2Ne(^1S)$	+ 133.4	+ 131.1
$\longrightarrow N^+(^3P) + Ne^+(^2P) + Ne(^1S)$	− 55.9	− 58.2
$Ar_2N^{2+}(^2B_1) \longrightarrow ArN^{2+}(^2\Pi) + Ar(^1S)$	+ 70.4	+ 67.9
$\longrightarrow N^{2+}(^2P) + 2Ar(^1S)$	+ 276.4	+ 272.8
$\longrightarrow N^+(^3P) + Ar^+(^2P) + Ar(^1S)$	− 49.5	− 53.1
$He_2O^{2+}(^1A_1) \longrightarrow O^{2+}(^1D) + 2He(^1S)$	+ 160.0	+ 154.9
$\longrightarrow O^+(^2D) + He^+(^2S) + He(^1S)$	− 66.3	− 71.4
$Ne_2O^{2+}(^1A_1) \longrightarrow O^{2+}(^1D) + 2Ne(^1S)$	+ 211.5	+ 208.9
$\longrightarrow O^+(^2D) + Ne^+(^2P) + Ne(^1S)$	− 87.6	− 90.2
$Ar_2O^{2+}(^1A_1) \longrightarrow O^{2+}(^1D) + 2Ar(^1S)$	+ 408.8	+ 407.0
$\longrightarrow O^+(^3P) + 2Ar^+(^2P)$	− 25.8	− 27.6

[a] Taken from Ref. 13

the $^3B_1$ state than for the $^1A_1$ state (Table 11). Interaction energies between Ng and C^{2+} in the corresponding electronic states are significantly higher for the $^3B_1$ state than for the $^1A_1$ state (Table 11). Interaction energies between Ng and C^{2+} in Ng_2C^{2+} become higher with He < Ne < Ar for both electronic states. The numbers for the $^1A_1$ ground state are: 32.6 kcal/mol (He), 71.9 kcal/mol (Ne), 178.6 kcal/mol (Ar); for the $^3B_1$ state the results are: 117.4 kcal/mol (He), 143.0 kcal/mol (Ne), 294.5 kcal/mol (Ar). The stability differences between the $^1A_1$ ground state and $^3B_1$ excited state of Ng_2C^{2+} are 63.8 kcal/mol (He), 56.7 kcal/mol (Ne) and 31.1 kcal/mol (Ar). As before, the difference between He and Ne is less than between Ne and Ar. However, the $^1A_1$ ground state of He_2C^{2+} is distinguished from the Ne and Ar analogs by the energetically lowest lying atomization reaction. Unlike He_2C^{2+}, which dissociates endothermic into $C^{2+} + 2He$, the ground states of Ne_2C^{2+} and Ar_2C^{2+} dissociate exothermic

(Ar) or nearly thermoneutral (Ne) into $Ng^+(^2P) + C^+(^2P)$ (Table 11). Somewhat paradoxic, the weaker bound He_2C^{2+} is stable toward dissociation into the atoms, whereas the stronger bound Ne_2C^{2+} and Ar_2C^{2+} are much less stable (Ne) or unstable (Ar). This is because I_1 of He is higher than I_2 of C, but I_1 of Ne and Ar are lower than I_2 of C.

The results for Ng_2N^{2+} and Ng_2O^{2+} are straightforward. The dissociation reactions into the respective $Ng^+ + N^+$ and $Ng^+ + O^+$ ions are exothermic in all cases. The interaction energies between Ng and X^{2+} increase in Ng_2X^{2+} for X with the sequence $C < N < O$ and for Ng in the order $He < Ne < Ar$ (Table 11). This is also reflected in the calculated geometries (Fig. 13). Again, larger differences in the interaction energies are found between Ne and Ar than between He and Ne (Table 11).

5.3 Ng_mXY^{n+} Ions

Our knowledge about binding interactions in light noble gas compounds has been extended significantly in the last couple of years by the application of modern ab initio methods to small cations which are formed from diatomic species XY^{n+} (X, Y = first-row atoms) and one or two Ng atoms. A summary of the optimized geometries is shown in Fig. 14. Table 12 shows some relevant dissociation energies.

A qualitative understanding for the structures and stabilities of Ng_mXY^{n+} ions is achieved in terms of donor-acceptor interactions between Ng and XY^{n+} in the respective electronic state. Figure 15 shows schematically the molecular orbitals of several XY^{n+} species. We will begin with molecules XY^{n+} with 8 valence electrons.

Wilson and Green [97] calculated the structures and stabilities of $HeCN^+$ at the Hartree-Fock SCF level using various basis sets. They also performed preliminary calculations for $NeCN^+$. For $HeCN^+$, they predict the bond lengths for He,C as approx. 1.10–1.17 Å and for C–N as 1.13–1.16 Å. The dissociation energy D_e for He loss was calculated between 35 and 46 kcal/mol. In case of $HeCN^+$, the Ne,C interatomic distance is predicted as being 1.593 Å. The dissociation energy D_e for $NeCN^+$ is significantly smaller (13.8 kcal/mol) compared with $HeCN^+$, but it was assumed that with a better basis set, the D_e value for the Ne,CN^+ bond would be similar to the He,CN^+ bond. Thus, $HeCN^+$ and $NeCN^+$ are predicted at the Hartree-Fock level to be bound rather strongly. However, a totally different result is found when correlation energy is included in the calculations! Using Møller-Plesset perturbation theory, Frenking et al. [13] report that $HeCN^+$ in the $^1\Sigma^+$ ground state has a very long He,C-distance of 2.515 Å, and the He,C binding energy is only $D_e = 0.9$ kcal/mol (Fig. 14, Table 12). The isomer $HeNC^+$ is predicted with the same reaction energy for dissociation of He, i.e. $D_e = 0.9$ kcal/mol. Also for the neon analogs weak interaction energies and rather long distances for the neon bonds are predicted. As shown in Table 12, the dissociation energies D_e are

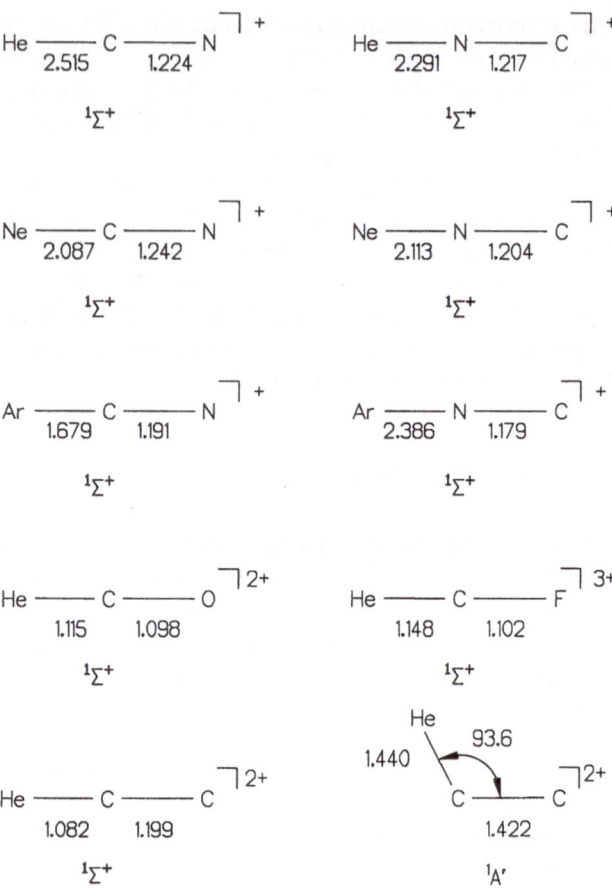

Fig. 14. Optimized geometries (MP2/6-31G(*d, p*)) for He, Ne, and Ar cations [7, 13]. Distances in Å, angles in °

3.6 kcal/mol for NeCN$^+$ and NeNC$^+$. Frenking et al. [13] show that the much larger stabilization energies reported by Wilson and Green [97] are an artefact of the Hartree-Fock level. Much higher stabilization energies are reported for ArCN$^+$ and ArNC$^+$ than for the helium and neon analogs. As shown in Table 12, the computed dissociation energy for ArCN$^+$ is $D_e = 35.7$ kcal/mol and for ArNC$^+$ it is $D_e = 17.5$ kcal/mol. This points towards covalent interactions in the argon compounds, but not in the helium and neon structures. As discussed in Sect. 5.5, the analysis of the electron density distribution reveals that ArCN$^+$ is indeed covalently bound, whereas the attractive interactions in ArNC$^+$ are caused by charge-induced dipole interactions.

The valence molecular orbitals of CN$^+$ in the X$^1\Sigma^+$ ground state are schematically shown in Fig. 15 (XY(8e)). The 3σ, 4σ and the degenerate 1π MO are occupied, and the lowest unoccupied MO is the 5σ orbital. Thus, HOMO–LUMO interactions with Ng involve the 5σ LUMO of CN$^+$. Because

$$\left[\text{He} - \text{C} - \text{C} - \text{He}\right]^{2+} \quad \left[\text{Ne} - \text{C} - \text{C} - \text{Ne}\right]^{2+} \quad \left[\text{Ar} - \text{C} - \text{C} - \text{Ar}\right]^{2+}$$

He—C—C—He   1.085 1.197 $^1\Sigma_g^+$

Ne—C—C—Ne   1.416 1.193 $^1\Sigma_g^+$

Ar—C—C—Ar   1.648 1.203 $^1\Sigma_g^+$

He 1.187 / 128.2 — C — 1.282 — C — He $[\ ]^{2+}$ ^1A_g

Ne 1.523 / 119.7 — C — 1.340 — C — Ne $[\ ]^{2+}$ ^1A_g

Ar 1.649 / 172.0 — C — 1.205 — C — Ar $[\ ]^{2+}$ ^1A_g

He 1.359 / 90.0 — N — 1.228 — N — He $[\ ]^{2+}$ ^1A_g

He 2.231 / 114.4 — O — 1.007 — O — He $[\ ]^{2+}$ $^1A'$

He 1.099 / 143.5 — C — 1.226 — C — 1.075 , 163.9 — H $[\ ]^{+}$ $^1A'$

Ne 1.507 / 141.5 — C — 1.238 — C — 1.077 , 152.3 — H $[\ ]^{+}$ $^1A'$

Ar 1.659 / 169.5 — C — 1.212 — C — 1.069 , 174.2 — H $[\ ]^{+}$ $^1A'$

this orbital is dominantly located at the carbon atom, the $ArCN^+$ isomer is clearly more stable than $ArNC^+$. The identical dissociation energies for the He and Ne isomers indicate that the weak binding is mainly caused by charge-induced dipole interactions and that orbital interactions are negligibly small.

The same orbital scheme as for CN^+ is found for isoelectronic CO^{2+}, CF^{3+} and CNe^{4+}. The interactions between He and the multiply charged cations have been reported by Radom and coworkers [95b,c]. $HeCNe^{4+}$ is predicted not to be a minimum on the potential energy hypersurface. For $HeCO^{2+}$ and $HeCF^{3+}$ rather short helium bond lengths are calculated (see Fig.14). This is reasonable, because the higher nuclear charge in the multiply charged XY^{n+} cations yields lower-lying orbitals, which in turn caused stronger orbital interactions with He. The interaction energies between He and $(^1\Sigma^+)CO^{2+}$ are calculated including correlation energy as $D_o = 43.2$ kcal/mol [95b]. $HeCO^{2+}$ has also been calculated by Cooper and Wilson [83] who reported a well depth of approx.

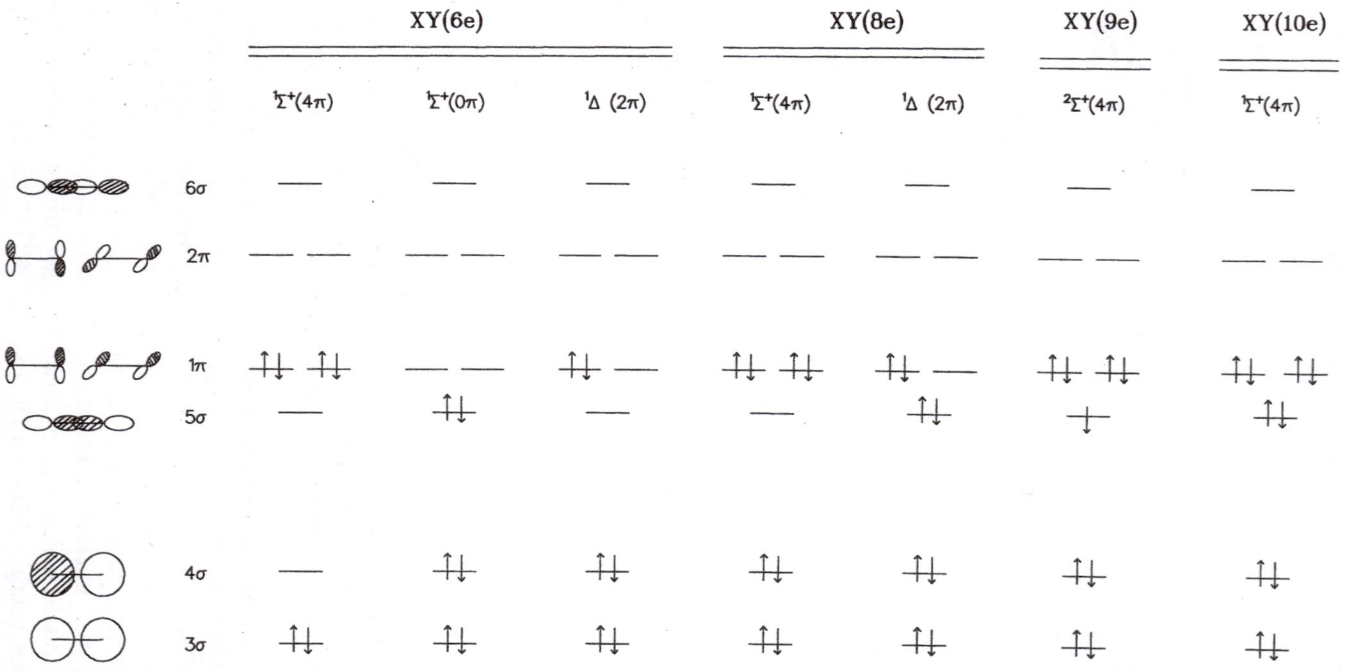

Fig. 15. Schematic representation of some electronic states of XY diatomics with 6, 8, 9, and 10 electrons. In the case of the Δ-states only one possible electron configuration is shown

Table 12. Theoretically predicted dissociation energies D_e and D_o (in kcal/mol)[a]

Reaction	D_e	D_o
$HeCN^+(^1\Sigma^+) \longrightarrow CN^+(^1\Sigma^+) + He(^1S)$	+ 0.9	+ 0.6
$HeNC^+(^1\Sigma^+) \longrightarrow CN^+(^1\Sigma^+) + He(^1S)$	+ 0.9	+ 0.4
$NeCN^+(^1\Sigma^+) \longrightarrow CN^+(^1\Sigma^+) + Ne(^1S)$	+ 3.6	+ 2.0
$NeNC^+(^1\Sigma^+) \longrightarrow CN^+(^1\Sigma^+) + Ne(^1S)$	+ 3.6	+ 2.2
$ArCN^+(^1\Sigma^+) \longrightarrow CN^+(^1\Sigma^+) + Ar(^1S)$	+ 35.7	+ 34.6
$ArNC^+(^1\Sigma^+) \longrightarrow CN^+(^1\Sigma^+) + Ar(^1S)$	+ 17.5	
$HeCO^{2+}(^1\Sigma^+) \longrightarrow CO^{2+}(^1\Sigma^+) + He(^1S)$	+ 45.9[b]	+ 43.2[b]
$\longrightarrow CO^+(^2\Sigma^+) + He^+(^2S)$	+ 19.8[b]	+ 17.7[b]
$HeCF^{3+}(^1\Sigma^+) \longrightarrow CF^{2+}(^1\Sigma^+) + He^+(^2S)$	− 228.6[b]	− 232.4[b]
$HeCC^+(^2\Sigma^+) \longrightarrow CC^+(^2\Sigma_u^+) + He(^1S)$	+ 34.0[d]	− 28.8[d]
$\longrightarrow CC^+(^2\Pi_u) + He(^1S)$	− 32.9[d]	− 39.1[d]
$HeCC^{2+}(^1\Sigma^+) \longrightarrow CC^{2+}(^1\Sigma_g^+, 4\pi) + He(^1S)$	+ 34.0[d]	− 28.8[d]
$\longrightarrow CC^{2+}(^1\Sigma_g^+, 0\pi) + He(^1S)$	− 32.9[d]	− 39.1[d]
$HeCC^{2+}(^1A') \longrightarrow CC^{2+}(^2\Sigma_g^+, 0\pi) + He(^1S)$	− 34.0[d]	− 28.8[d]
$HeCCHe^{2+}(^1\Sigma_g^+) \longrightarrow 2He(^1S) + CC^{2+}(^1\Sigma_g^+, 4\pi)$	+ 174.6	
$\longrightarrow 2He(^1S) + CC^{2+}(^1\Sigma_g^+, 0\pi)$	+ 18.9	+ 10.0
$\longrightarrow He(^1S) + He^+(^2S) + CC^+(^2\Pi_u)$	+ 86.0	+ 78.6
$\longrightarrow 2HeC^+(^2\Pi)$	− 103.6	− 112.6
$HeCCHe^{2+}(^1A_g) \longrightarrow 2He(^1S) + CC^{2+}(^1\Delta_g)$	+ 58.2	
$\longrightarrow 2He(^1S) + CC^{2+}(^1\Sigma_g^+, 0\pi)$	+ 22.8	+ 14.2
$\longrightarrow He(^1S) + He^+(^2S) + CC^+(^2\Pi_u)$	+ 89.9	+ 82.8
$\longrightarrow 2HeC^+(^2\Pi)$	− 99.7	− 108.6
$NeCCNe^{2+}(^1\Sigma_g^+) \longrightarrow 2Ne(^1S) + CC^{2+}(^1\Sigma_g^+, 4\pi)$	+ 169.6	
$\longrightarrow 2Ne(^1S) + CC^{2+}(^1\Sigma_g^+, 0\pi)$	+ 13.9	+ 8.3
$\longrightarrow Ne(^1S) + Ne^+(^2P) + CC^+(^2\Pi_u)$	+ 8.2	+ 4.1
$\longrightarrow 2NeC^+(^2\Pi)$	− 116.0	− 121.5
$NeCCNe^{2+}(^1A_g) \longrightarrow 2Ne(^1S) + CC^{2+}(^1\Delta_g)$	+ 75.6	
$\longrightarrow 2Ne(^1S) + CC^{2+}(^1\Sigma_g^+, 0\pi)$	+ 34.2	+ 28.4
$\longrightarrow Ne(^1S) + Ne^+(^2P) + CC^+(^2\Pi_u)$	+ 28.5	+ 24.2
$\longrightarrow 2NeC^+(^2\Pi)$	− 95.6	− 101.6
$ArCCAr^{2+}(^1\Sigma_g^+) \longrightarrow 2Ar(^1S) + CC^{2+}(^1\Sigma_g^+, 4\pi)$	+ 336.7	
$\longrightarrow 2Ar(^1S) + CC^{2+}(^1\Sigma_g^+, 0\pi)$	+ 173.4	+ 168.2
$\longrightarrow Ar(^1S) + Ar^+(^2P) + CC^+(^2\Pi_u)$	+ 34.9	+ 31.2
$\longrightarrow 2ArC^+(^2\Pi)$	+ 14.3	+ 9.4
$ArCCAr^{2+}(^1A_g) \longrightarrow 2Ar(^1S) + CC^{2+}(^1\Delta_g)$	+ 222.8	
$\longrightarrow 2Ar(^1S) + CC^{2+}(^1\Sigma_g^+, 0\pi)$	+ 173.3	+ 168.1
$\longrightarrow Ar(^1S) + Ar^+(^2P) + CC^+(^2\Pi_u)$	+ 34.8	+ 31.1
$\longrightarrow 2ArC^+(^2\Pi)$	+ 14.2	+ 9.3

[a] Unless otherwise noted, data are taken from Ref. 13; [b] Ref. 95 b; [c] Ref. 83; [d] Ref. 7

63 kcal/mol at the Hartree-Fock level. However, the energetically lowest lying dissociation products of $HeCO^{2+}$ are $He^+ + CO^+(^2\Sigma^+)$. But even for the energetically favored charge separation reaction, the $HeCO^{2+}$ dication is predicted not only to be kinetically stable but also thermodynamically stable by 17.7 kcal/mol [95b,c]. The dissociation of $HeCO^{2+}$ into $He^+ + CO^+$ is another example of Fig. 4b. Fragmentation of $HeCF^{3+}$ ($^1\Sigma^+$) into $He^+ + CF^{2+}(^2\Sigma^+)$ is

highly exothermic by 232.4 kcal/mol [95b,c]. Nevertheless, a sufficiently high (29.5 kcal/mol) barrier makes the observation even of the trication feasible. The fragmentation of $HeCF^{3+}$ into $He^+ + CF^{2+}$ is an example for Fig. 4a, whereas $HeCNe^{4+}$ is an example for Fig. 4d.

The 8-valence electron systems CN^+, CO^{2+}, CF^{3+} and CNe^{4+} may form stable ions with noble gas atoms not only in the $^1\Sigma^+$ state but also in other electronic states. Figure 15 shows schematically the $^1\Delta$ state (XY(8e)) of the diatomics when approached by another atom [108]. In the $^1\Delta$ state, the LUMO is the $1\pi(Y)$ orbital (the $1\pi(X)$ is occupied). Because this orbital is usually higher in energy than the 5σ MO, HOMO–LUMO interactions with Ng will be weaker. Also, the resulting $NgXY^{n+}$ species is expected to be non-linear, because the LUMO has π symmetry. This has been found for the interactions between He and another 8-valence electron system XY^{n+}, i.e. NN^{2+}. Cooper and Wilson [83] calculated a bond angle of 92.3° for the non-linear equilibrium structure of $HeNN^{2+}$ with a He,N bond length of only 1.50 Å. The N,N distance is 1.44 Å, which indicates that the corresponding state of NN^{2+} has not $^1\Sigma_g^+$ symmetry. The interaction energy between He and NN^{2+} is computed at the Hartree-Fock level as approx. 86 kcal/mol [83]. No data were presented for the dissociation of $HeNN^{2+}$ into $He^+ + NN^+$, which is the energetically lowest-lying fragmentation reaction. Also the electronic state of the calculated species was not given. Since there is a balance between the excitation energy of the acceptor cation and the donor-acceptor interaction with Ng, it is not always easy to predict the energetically lowest-lying state of $NgXY^{n+}$ molecules. A striking example is presented below.

After discussion the interactions between Ng and 8-valence electron systems XY^{n+}, we now turn to XY^{n+} species with 6-valence electrons. In particular, we will discuss interactions between Ng and CC^{2+}. Koch et al. [7] investigated theoretically the structures of $HeCC^{2+}$ in several electronic states. They found a strongly $^1\Sigma^+$ state with a He,C distance of only 1.082 Å and an interaction energy of 89.9 kcal/mol. The corresponding $^1\Sigma_g^+(4\pi)$ state of C_2^{2+} is schematically shown in Fig. 15 (XY(6e) $^1\Sigma^+(4\pi)$). The short He,C bond and linear geometry can be rationalized with strong HOMO–LUMO interactions involving the low-lying 4σ LUMO [109]. A second energy minimum with a non-linear geometry was found for $HeCC^{2+}$ ($^1A'(2\pi)$) with a longer He,C bond (1.440 Å) and a bond angle of 93.6° (Fig. 14), which is similar to what is found for $HeNN^{2+}$. This structure corresponds to the $^1\Delta_g$ state of CC^{2+} which is also shown in Fig. 15 (XY(6e), $^1\Delta(2\pi)$). Here, the LUMO is the higher-lying 5σ orbital, which explains the weaker interactions in the bent form of $HeCC^{2+}$. The energetically lowest-lying state of CC^{2+} is the $^1\Sigma_g^+(0\pi)$ state, which has an even higher-lying (1π) LUMO (Fig. 15, XY(6e) $^1\Sigma^+(0\pi)$). Not surprisingly, no bound $HeCC^{2+}$ structure was found with the $X^1\Sigma_g^+(0\pi)$ ground state of CC^{2+}. The $^1A'(2\pi)$ state of $HeCC^{2+}$ is computed to be 51.0 kcal/mol lower in energy than the stronger bound $^1\Sigma^+$ state [7]. Also triplet states were investigated and the lowest lying state of $HeCC^{2+}$ is predicted to have $^3A''(1\pi)$ symmetry [7]. The model of donor-acceptor interaction has also been used to explain the structures and stabilities of the triplet states of $HeCC^{2+}$ and $HeCCHe^{2+}$ [7, 50].

Koch et al. [7] studied also the interactions of CC^{2+} with two He atoms. When both He are attached to the same carbon, vinylidene analogs He_2CC^{2+} are formed. When He is bound to the terminal carbon atoms, acetylenic structures $HeCCHe^{2+}$ are found. The $^1\Sigma_g^+$ state of $HeCCHe^{2+}$ has a very short He,C bond length of 1.085 Å (Fig. 14) and the dissociation energy D_e into CC^{2+} ($^1\Sigma_g^+(4\pi)$) + 2 He is very high, 174.6 kcal/mol (Table 12) [7, 50]. The dissociation energy corresponds to a He,C binding energy of 87.3 kcal/mol. The strong binding interaction compensates even for the excitation energy of CC^{2+} from the $X^1\Sigma_g^+(0\pi)$ state to the $^1\Sigma_g^+(4\pi)$ state. The dissociation of $HeCCHe^{2+}$ ($^1\Sigma_g^+$) into the ground state of $CC^{2+}(^1\Sigma_g^+,0\pi)$ + 2He is still endothermic by 10.0 kcal/mol (Table 12) [7, 50]. The $^1\Sigma_g^+$ state is also lower in energy than the triplet states of $HeCCHe^{2+}$ which were calculated by Koch et al. [7]. However, the linear singlet form of $HeCCHe^{2+}$ is not the lowest lying bound state!! A later investigation of $NgCCNg^{2+}$ (Ng = He, Ne, Ar) by Frenking et al. [13] showed that the $^1\Delta_g$ state of CC^{2+} may bind two helium atoms forming the trans-planar ^1A_g state of $HeCCHe^{2+}$ shown in Fig. 14. The ^1A_g state of $HeCCHe^{2+}$ is predicted to be 4.2 kcal/mol lower in energy than the linear $^1\Sigma_g^+$ state. The He,C bonds are slightly longer and weaker in the trans-form, but the weaker He,C binding interactions are compensated by the deexcitation energy of CC^{2+} from the $^1\Sigma_g^+(4\pi)$ state to $^1\Delta_g$ (Fig. 15, XY(6e)).

Also for $NeCCNe^{2+}$, the trans-planar ^1A_g state is calculated to be lower in energy than the linear $^1\Sigma_g^+$ form by 20.2 kcal/mol [13]. In the case of $ArCCAr^{2+}$, a linear and slightly bent structure were optimized (Fig. 14) which are nearly degenerate in energy. The calculation of the potential energy hypersurfaces of $HeCCHe^{2+}$ and $NeCCNe^{2+}$ at MP2/6-31G(d, p) as a function of the NgCC angle gave a maximum energy structure that was only slightly higher than the corresponding linear structures. At MP4(SDTQ)/6-311G(2df, 2pd), the energy of the "transition states" were even lower than for the linear forms [13]. This result indicates that the linear structures of $HeCCHe^{2+}$ and $NeCCNe^{2+}$ may not even be true minima on the potential energy surface.

Considering the linear and trans-bent forms of $NgCCNg^{2+}$ as products of Ng and CC^{2+} in the $^1\Sigma_g^+(4\pi)$ and $^1\Delta_g$ state, respectively, the following binding energies per Ng,C bond are computed: for the linear forms 87.3 kcal/mol (He,C), 84.8 kcal/mol (Ne,C), and 168.4 kcal/mol (Ar,C); for the trans-bent forms 29.1 kcal/mol (He,C), 37.8 kcal/mol (Ne,C) and 111.4 kcal/mol (Ar,C) (Table 12). The slightly lower bond strength in the linear form of $NeCCNe^{2+}$ than calculated for linear $HeCCHe^{2+}$ can be explained by the p-π repulsion in the former structure. In the linear and trans-bent forms of $NgCCNg^{2+}$, Ar is much stronger bound than He an Ne. $ArCCAr^{2+}$ is theoretically predicted to be stable not only kinetically, but also thermodynamically. Unlike $HeCCHe^{2+}$ and $NeCCNe^{2+}$, the charge separation reaction into $2NgC^+$ is endothermic for Ng = Ar by 9.3 kcal/mol. The corresponding dissociation reactions of $HeCCHe^{2+}$ and $NeCCNe^{2+}$ are exothermic by 108.6 kcal/mol and 101.6 kcal/mol, respectively (Table 12).

Koch et al. [7] also investigated the changes in the noble gas structures for the series $HeCCHe^{2+}$, $HeNNHe^{2+}$, and $HeOOHe^{2+}$. Figure 14 shows that the helium bonds are much longer for HeNNHe^{2+} and especially HeOOHe^{2+}. This can be explained by donor-acceptor interactions between He and the respective diatomic dication. Figure 15 shows the molecular orbitals of the diatomics with 4 π-electrons (XY with 6, 8, 10 electrons, $^1\Sigma^+(4\pi)$). The lowest-lying orbital of CC^{2+}($^1\Sigma_g^+(4\pi)$) is the 4σ MO [109]. NN$_2^{2+}$ and O$_2^{2+}$ have two and four more electrons, respectively. HOMO–LUMO interactions involve the 5σ MO of N$_2^{2+}$ and the 2π MO of O$_2^{2+}$ [109], which are higher in energy (Fig. 15). The predicted geometry for HeNNHe^{2+} by Koch et al. [7] is in agreement with the calculated structure of HeNN^{2+} by Cooper and Wilson [83].

There are very few experimental reports about light noble gas ions Ng$_m$XY^{n+}. For the 9-valence electron systems CO$^+$ and NN$^+$, interactions with noble gases have been studied in a mass spectrometer by Munson et al. [99]. Besides the corresponding krypton and xenon compounds, ArNN$^+$ and ArCO$^+$ were observed, but not the helium and neon analogs. ArNN$^+$ had already earlier been detected in a mass spectrometer by Kaul and Fuchs [110]. Assuming that NN$^+$ and CO$^+$ were in the $^2\Sigma^+$ ground state ($^2\Sigma_g^+$ for NN$^+$), interactions with Ng should be weaker than between Ng and the 8-valence electron system CN$^+$ in the $^1\Sigma^+$ ground state. This is, because the 5σ orbital is empty in CN$^+$, but singly occupied in NN$^+$ and CO$^+$ (Fig. 15, XY(9e) $^2\Sigma^+(4\pi)$). Teng and Conway [111] determined the dissociation energy for the Ar,NN$^+$ bond experimentally using a mass spectrometer as $D_0 = 24.5$ kcal/mol, significantly less than what is calculated for ArCN$^+$ ($D_0 = 34.6$ kcal/mol [13]). Theoretical studies for NgNN$^+$ are needed to determine the electronic state of the experimentally observed ArNN$^+$ ion and the binding energies for HeNN$^+$ and NeNN$^+$.

5.4 Other Light Noble Gas Ions

There are not many studies reported for molecular ions of the light noble gases that do not fall into one of the prior categories. Related to the acetylenic dications NgCCNg^{2+} are the singly charged acetylene analogs NgCCH$^+$, which have also been studied theoretically. Cooper and Wilson [83] calculated HeCCH$^+$ at the Hartree-Fock level with a linear geometry and a He,C interatomic distance of 1.24 Å. Koch et al. [7] performed higher-level calculations and found that the linear structure is not a minimum on the potential energy surface when correlation energy is included. Rather, a *trans*-bent ($^1A'$) structure was found as minimum with a much shorter (1.099 Å) He,C bond. Also for NeCCH$^+$ and ArCCH$^+$, the energy minimum structure is predicted [13] to have a *trans*-bent geometry with bond lengths of 1.507 Å (Ne, C) and 1.659 Å (Ar, C) (Fig. 14). The linear $^1\Sigma^+$ form of NgCCH$^+$ may be thought as the product of Ng and CCH$^+$ in the $^1\Sigma^+(4\pi)$ state. However, this is not the ground state of CCH$^+$. There are three singlet states [112, 113] lower in energy than the $^1\Sigma^+(4\pi)$ state, $^1\Pi$, $^1\Delta$ and $^1\Sigma^-(2\pi)$. The $^1\Pi$ state appears to be the energetically lowest-lying linear singlet state of CCH$^+$ [113]. The interaction energies be-

Table 13. Calculated reaction energies D_e and D_0 (in kcal/mol) for NgCCH$^+$ structures[a]

Reaction	D_e	D_o
HeCCH$^+$(^{1}A$'$) $\longrightarrow$ He(^{1}S) + CCH$^+$($^1\Pi$)	15.4	12.5
NeCCH$^+$(^{1}A$'$) $\longrightarrow$ Ne(^{1}S) + CCH$^+$($^1\Pi$)	7.1	5.9
ArCCH$^+$(^{1}A$'$) $\longrightarrow$ Ar(^{1}S) + CCH$^+$($^1\Pi$)	55.7	54.3

[a] Taken from Ref. 13

tween Ng and ($^1\Pi$) CCH$^+$ are predicted as 15.4 kcal/mol for He [114], 7.1 kcal/mol for Ne and 55.7 kcal/mol for Ar (Table 13) [13]. Again, much stronger interactions are found for argon than for helium and neon. The decrease in binding energy from HeCCH$^+$ to NeCCH$^+$ may be caused by the π-repulsion between Ne and CCH$^+$, which does not exist between He and CCH$^+$ in HeCCH$^+$.

A very interesting ab initio study of β-decay in OHT, NH$_2$T, and CH$_3$T has been reported by Ikuta et al. [115]. β-decay of tritium yields (besides a neutrino particle υ) the helium isotope ^{3}He$^+$. Because of the relatively short half-life of tritium (12.26 years), tritiated compounds may be thought of as a convenient source of helium-containing molecular ions:

$$RT \longrightarrow RHe^+ + \beta^- + \upsilon \qquad (2)$$

Ikuta et al. [115] calculated the transition probabilities in reaction (2) for the three tritiated compounds listed above, and they calculated the potential curves for the product ions containing He at the Hartree-Fock level. The transition probabilities to the ground state daughter ions were predicted as ca. 0.61 for all the parent molecules studied. OHHe$^+$ and NH$_2$He$^+$ were calculated with small barriers for helium dissociation, but the energy release upon relaxation of the daughter ion from the parent geometry to the equilibrium structure was higher than the dissociation energy in both cases. For CH$_3$He$^+$, a repulsive potential curve for the helium dissociation was computed. The results of Ikuta et al. [115] are in essential agreement with earlier Hartree-Fock calculations by Cooper and Wilson [83] who predict approximate well depths for OHHe$^+$, NH$_2$He$^+$, and CH$_3$He$^+$ as ca. 4.3 kcal/mol, 0.7 kcal/mol and 0.1 kcal/mol, respectively. The theoretical calculations show that it is very unlikely that helium cations might be observed via β-decay of OHT, NH$_2$T and CH$_3$T.

The relative abundances of the ionic fragment from the decay of CH$_3$T in a mass spectrometer have been measured by Snell and Pleasonton [116]. They detected small amounts of CH$_3$He$^+$ with 0.06 $\pm$ 0.01% abundancy, the main product being CH$_3^+$. A recent higher-level calculation including correlation energy predicts a small barrier for He dissociation of < 0.25 kcal/mol for CH$_3$He$^+$ and a r_{CHe} equilibrium distance of 2.053 Å [95a]. It is surprising that the very weakly bound CH$_3$He$^+$ has been detected, albeit with very low abundancy, as the product of β-decay of CH$_3$T [116]. There are other helium-hydrocarbon cations which have been experimentally observed in low abundancies in a mass spectrometer via β-decay of the tritium parent compounds [117]. Perhaps the highest yield was achieved from tritiated toluenes, giving up to $0.4 \pm 0.2\%$ $C_7H_7He^+$ [117, 118]. The experiments have mostly been performed to yield gaseous hydrocarbon cations, and the investigated molecules are not expected to bind helium strongly. Higher yields of helium molecular cations may be expected for structures such as $HeCCH^+$, which are predicted by theory to have much stronger helium bonds. Such experiments have been suggested in several theoretical studies [7, 95].

β-decay of molecules containing radioactive isotopes has successfully been employed to produce molecular ions containing another noble gas element, i.e. xenon. Using ^{131}I-labeled compounds RI, xenon-organic cations have been studied in a mass spectrometer by Carson and White [119]. For example, ^{131}I-labeled CH_3I undergoes β-decay of CH_3Xe^+ with the molecular ions remaining intact in 69% of the decays [119a]. The much higher abundancy of the xenon molecular ion compared with the helium analog can be explained by the significantly higher Xe,C bond strength, which has been determined experimentally as 43 ± 8 kcal/mol [120]. Also, the kinetic energy release resulting from the β-decay is deposited to a higher degree in the He,C bond than the Xe,C bond in CH_3Ng^+. Also CH_3Kr^+ has been produced in a mass spectrometer via β-decay of ^{82}Br-labeled CH_3Br. Here, the abundancy of CH_3Kr^+ was only 0.4% [119b]. The Kr–C bond strength in CH_3Kr^+ has experimentally been estimated as 21 ± 15 kcal/mol [120].

5.5 The Nature of Bonding in Polyatomic Noble Gas Ions

A description of the electronic structure and the nature of bonding in polyatomic noble gas ions follows exactly the same lines pursued in the case of diatomic noble gas molecules (Sect. 4.5). Once the electronic wave function has been determined by ab initio calculations, the electron density distribution can be calculated, the MED path and $(3, -1)$ critical points can be found, and the properties of $\rho(r)$ and $H(r)$ can be evaluated at the critical points of $\rho(r)$. This has been done for Ng_2X^{2+} (X = C, N, O), $NgCCNg^{2+}$, $NgCCH^+$, and $NgCN^+$ ions. We will review in this section the results of these investigations [7, 13].

Ng_2X^{2+} **ions:** All NgX bonds, with one exception, are semipolar covalent bonds. According to the ρ_B and H_B data summarized in Table 14, they fulfil necessary and sufficient condition for covalent bonding. The only exception is $He_2C^{2+}(^1A_1)$, which possesses electrostatic bonds. The exceptional electronic structure of $He_2C^{2+}(^1A_1)$ is due to the electronic configuration of $C^{2+}(^1S)$, which does not possess any σ-holes in its valence shell concentration. However, its positive charge is high enough to establish a covalent bond with one He atom (see Sect. 4.5). This is possible by a distortion of the valence shell of $C^{2+}(^1S)$ as can be seen from the perspective drawing of the Laplace concentration of $HeC^{2+}(X^1\Sigma^+)$ given in Fig. 9c.

Table 14. Nature of bonding in Ng_2X^{2+} and $NgCCNg^{2+}$ ions according to ab initio calculations[a]

Dication	State	Bond	ρ_b [eÅ^{-3}]	H_b [hartree Å^{-3}]	B_e[b] [kcal/mol]	Nature of bonding
He_2C^{2+}	$^1A_1$	He, C	0.49	− 0.1	16.3	electrostatic
He_2C^{2+}	$^3B_1$	He, C	1.16	− 1.2	58.7	covalent
Ne_2C^{2+}	$^1A_1$	Ne, C	0.71	− 0.3	35.7	covalent
Ne_2C^{2+}	$^3B_1$	Ne, C	0.89	− 1.0	71.5	covalent
Ar_2C^{2+}	$^1A_1$	Ar, C	0.90	− 0.4	89.3	covalent
Ar_2C^{2+}	$^3B_1$	Ar, C	1.28	− 1.1	147.3	covalent
He_2N^{2+}	$^2B_1$	He, N	1.13	− 0.7	43.1	covalent
Ne_2N^{2+}	$^2B_1$	Ne, N	0.96	− 0.4	66.7	covalent
Ar_2N^{2+}	$^2B_1$	Ar, N	1.36	− 0.9	138.2	covalent
He_2O^{2+}	$^1A_1$	He, O	1.97	− 1.9	80.0	covalent
Ne_2O^{2+}	$^1A_1$	Ne, O	1.17	− 0.4	105.7	covalent
Ar_2O^{2+}	$^1A_1$	Ar, O	1.17	− 0.5	204.4	covalent
$HeCCHe^{2+}$	$^1\Sigma_g^+$	He, C	1.38	− 1.3	87.3	covalent
$HeCCHe^{2+}$	^1A_g	He, C	1.25	− 1.2	89.2	covalent
$NeCCNe^{2+}$	$^1\Sigma_g^+$	Ne, C	0.97	− 1.0	84.8	covalent
$NeCCNe^{2+}$	^1A_g	Ne, C	0.80	− 0.9	94.9	covalent
$ArCCAr^{2+}$	$^1\Sigma_g^+$	Ar, C	1.53	− 2.0	168.3	covalent
$ArCCAr^{2+}$	^1A_g	Ar, C	1.53	− 2.0	168.4	covalent

[a] From Ref. 13; [b] B_e = Bond energy, using one half of the D_e values for Ng dissociation listed in Tables 11 and 12

A second He atom can not be bound in the same way. If two He atoms approach $C^{2+}(^1S)$ from different sides, then the net distortion of this valence shell will be much smaller than in the case of $HeC^{2+}(X^1\Sigma^+)$ as can be seen from Fig. 16. As a consequence, the stability of $He_2C^{2+}(^1A_1)$ is due to electrostatic interactions such as charge-induced dipole attraction rather than covalent bonding [7].

Since the donor ability of Ng increases from He to Ar, both $Ne_2C^{2+}(^1A_1)$ and $Ar_2C^{2+}(^1A_1)$ possess covalent bonds with increasing bond energies. This increase can also be observed for the other Ng_2C^{2+} ions (Table 14).

Another feature of bonding, which is nicely reflected in the electron density distribution $\rho(r)$ at the (3, − 1) critical point of the Ng,C bond, is the dependence of the strength of the bond in the electronic structure of X^{2+} (see also Sect. 4.5). For example, the value of $\rho_b(Ng, C)$ is higher of the $^3B_1$ states than for the $^1A_1$ states of Ng_2C^{2+}. This increase reflects the fact that $C^{2+}(^3P)$ contrary to C^{2+} (^3P) possesses distinct σ-holes in its valence shell configuration, which are prone to accept electronic charge from Ng thereby establishing semipolar bonds with Ng. This interpretation is in line with calculated dissociation energies and geometries.

Bonding in Ng_2N^{2+} and Ng_2O^{2+} can be explained in the same way. In Fig. 17a and b, perspective drawings of the Laplace concentration of $N^{2+}(^2P)$ and $O^{2+}(^1D)$ are shown. Both dications possess concentration holes in their valence shell, which can be filled with electrons of a suitable donor. The holes correspond to low lying unoccupied 2p orbitals, which split into 2pσ and 2pπ

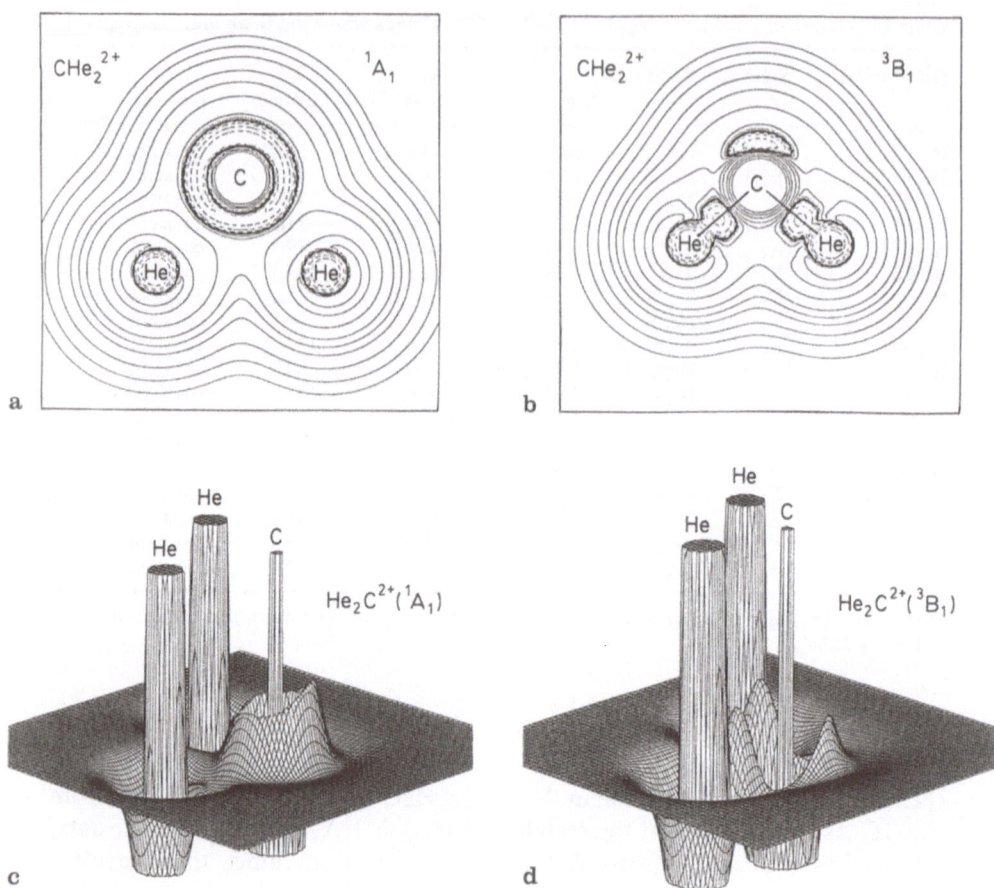

Fig. 16a–d. Contour line diagrams and perspective drawings of the Laplace concentration, $-\nabla^2\rho(r)$, of He_2C^{2+}: (**a, b**) $^1A_1$ state, (**c, d**) $^3B_1$ state, given with regard to the molecular plane. In the contour line diagrams, inner shell concentrations are no longer shown

components upon approach of an Ng atom. The seize of the $2p\sigma$ holes can be estimated by the difference $\nabla^2\rho(r)_{max} - \nabla^2\rho(r)_{min}$ where the subscripts max and min denote maximum and minimal concentration values in the valence shell. According to this difference the seize increases with the atomic number, i.e. going from $C^{2+}(^1S)$ to $O^{2+}(^1D)$ leads to a stronger electron acceptor, which binds Ng atoms better. In MO language, one can say that the $2p\sigma$ MO becomes lower in energy and, hence, frontier orbital interactions between donor and acceptor increase.

This is exactly what is reflected by the calculated bond properties of ions Ng_2X^{2+}. All ions with X = N and X = O possess covalent bonds according to the data collected in Table 14 [13].

NgCCNg^{2+} ions: The analysis of bonding in NgX^{n+} and Ng_2X^{2+} ions reveals that the electronic structure of the acceptor determines whether or not

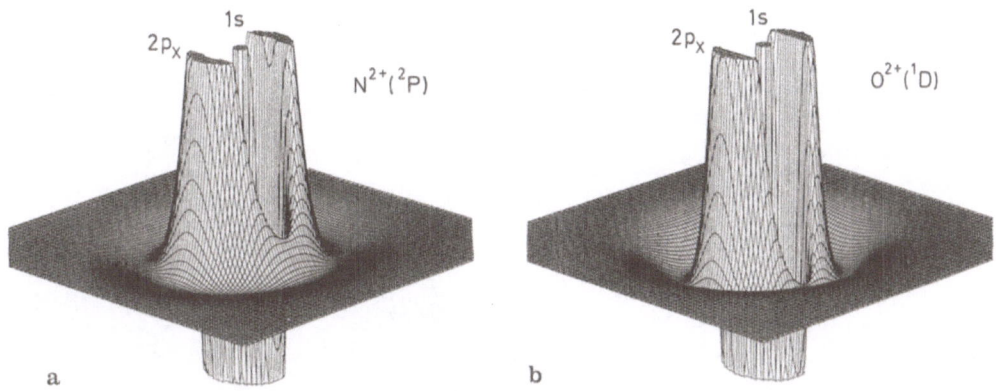

Fig. 17a and b. Perspective drawings of the Laplace concentration, $-\nabla^2\rho(r)$, of (a) $N^{2+}(^2P)$, and (b) $O^{2+}(^1D)$ influenced by an approaching Ng atom. Note that ion states change in this situation to $^2\Pi$ and $^1\Delta$, respectively. Inner shell and valence shell concentrations are indicated. (HF/6-31G(d) calculations)

a covalent bond is formed with Ng. σ-Holes in the valence shell, i.e. low-lying unoccupied σ-MOs, are essential for the nature of the Ng,X interactions. The same observations have also been made in the case of the Ng acceptor CC^{2+}. As explained already above (see also Fig. 18) the CC^{2+} ion is an excellent electron acceptor in its $^1\Sigma_g^+(4\pi)$ excited state, but it does not interact at all with Ng atoms in its $^1\Sigma_g^+(0\pi)$ ground state [7, 13].

The reason for the different reactivities of the two states can nicely be illustrated by the Laplace concentrations given in Fig. 18. In the case of the ground state, there are four concentration lumps along the molecular axis, two in the internuclear region and two in the lone pair regions. They correspond to σ-electrons that shield the nuclei with regard to each other and with regard to an attacking nucleophile. In this way, nuclear repulsion is reduced and the dication is stabilized. In addition, its ability of reacting with a σ-electron donor is diminished. CC^{2+} in its ground state is only a (weak) π-acceptor as is reflected by the π-holes at the nuclei. However, this π-acceptor ability is irrelevant for the bonding of Ng atoms.

In the excited $^1\Sigma_g^+(4\pi)$ state of CC^{2+}, the concentration lumps are in regions where one should expect the four π-electrons, while concentration holes appear in the lone pair regions along the internuclear axis. That means that the concentration holes qualify as σ-holes that impart to the CC^{2+} ion distinct σ-electron acceptor ability. CC^{2+} in its $^1\Sigma_g^+(4\pi)$ state possesses the right electron configuration to undergo reaction with Ng atoms. Noble gas compounds with one or two Ng atoms can be formed as already discussed above.

There are no longer concentration lumps in the internuclear region of $CC^{2+}(^1\Sigma_g^+(4\pi))$. Accordingly, nuclear repulsion is high and the stability of the dication is low. On the other side, the energy gain by bonding Ng atoms can be so large that it compensates the excitation energy of CC^{2+} ions as was discussed in Sect. 5.3.

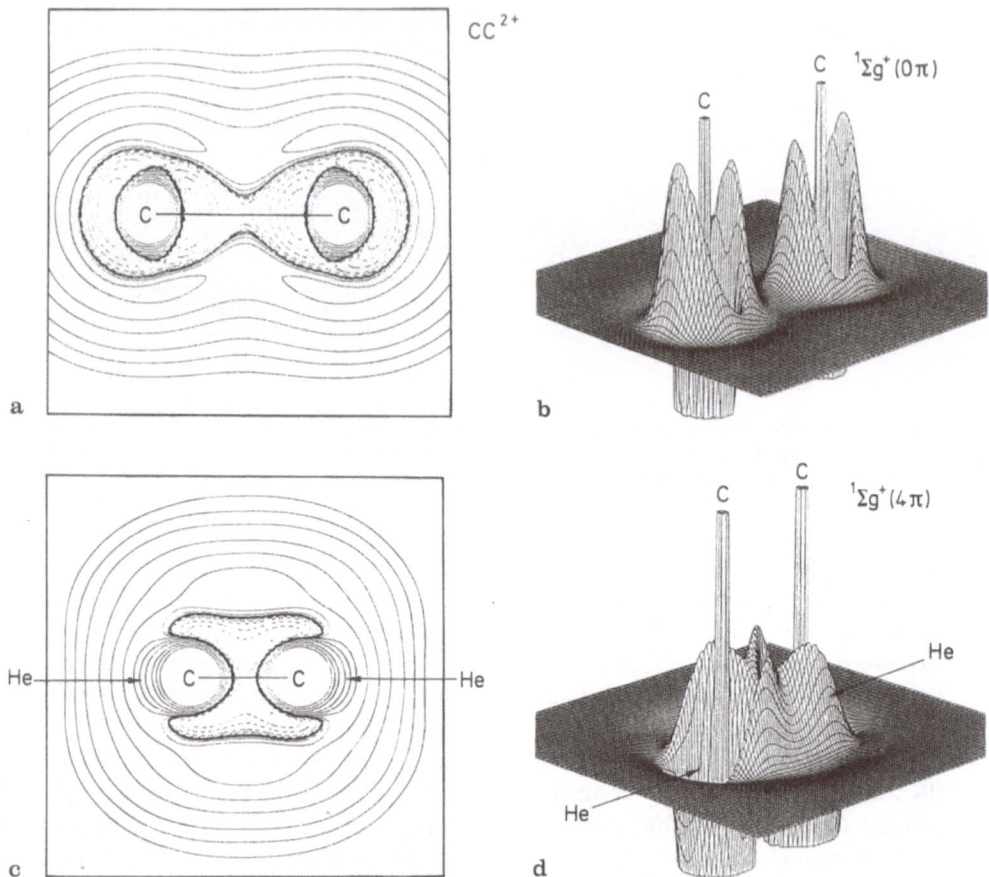

Fig. 18a–d. Contour line diagrams and perspective drawings of the Laplace concentration, $-\nabla^2\rho(r)$, of CC^{2+}: (**a, b**) $^1\Sigma^+(0\pi)$ ground state, (**c, d**) $^1\Sigma^+(4\pi)$ excited state with regard to the plane containing the nuclei. In the contour line diagrams inner shell concentrations are no longer shown. (HF/6-31G(d) calculations). Ref. [7]

All Ng,C bonds of cations $NgCCNg^{2+}(^1\Sigma_g^+)$ investigated so far are covalent. This is suggested by the properties of the electron and the energy density evaluated at the bond critical points (Table 14). The density data also reflect the fact that the Ne,C bond is actually weaker than the He,C bond due to $2p\pi$-π_u electron repulsion absent in the He compound.

Analysis of the electron density distribution confirms that Ng,C bonding in $HeCCHe^{2+}$ is due to s-electron donation from He while in $NeCCNe^{2+}$ and $ArCCAr^{2+}$ it is due to $p\sigma$-electron donation from Ng. Again, this is nicely reflected by the calculated Laplace concentrations (Fig. 19).

Figure 19a depicts a perspective drawing of the calculated Laplace concentration of $ArCCAr^{2+}$. There are concentration holes in the valence shell of the Ar atoms where one would expect the $3p\sigma$-electrons, however concentration

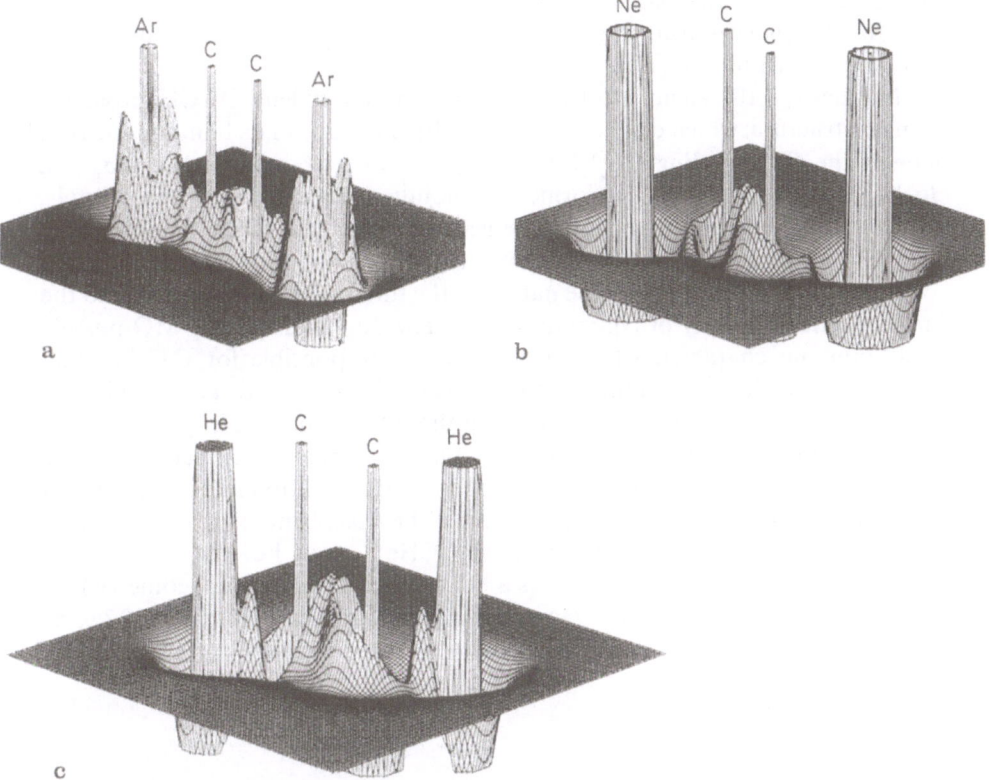

Fig. 19a–c. Perspective drawings of the Laplace concentration, $-\nabla^2\rho(\mathbf{r})$, of **(a)** ArCCAr^{2+}(^{1}Ag), **(b)** NeCCNe^{2+}(^{1}Ag), and **(c)** HeCCHe^{2+}(^{1}A$_g$). Inner shell and valence shell concentrations are indicated. (HF/6-31G(d, p) calculations). Ref. [13]

lumps in the regions of the $3p\pi$-electrons. Furthermore, there is a distinct increase of electron concentration in the Ar,C bond region indicative of a semipolar covalent bond between noble gas atom and carbon. Similar features can be observed in the Laplace concentration of NeCCNe^{2+} although the Ne atoms are still surrounded in the molecule by the outer depletion sphere, which is now less pronounced than in the free atom (Fig. 19b).

For HeCCHe^{2+} the semipolar He,C bond is reflected by an appendix like deformation of the s-electron distribution of He in the bonding region which is exactly opposite of the σ-hole at CC^{2+}. As noted above this is typical of all semipolar He,X bonds and becomes easy detectable in the contour line diagrams of the Laplace concentration.

While the electronic structure of the acceptor allows reasonable predictions on its possibilities to bind Ng atoms, it does not provide a basis to predict the geometry of the most stable NgCCNg^{2+} form calculated by ab initio theory. For example, it is difficult to predict the *trans*-bent form of HeCCHe^{2+} and

NeCCNe^{2+} on the basis of the electronic structure of the acceptor CC$^{2+}(^1\Sigma_g^+(4\pi))$. However, this can be rationalized by simple MO arguments as has been shown in Ref. [13].

Normally, a 10-valence electron system such as acetylene, HCCH, possesses a linear structure, which can be rationalized by the shape of its bonding σ-MOs, namely the C,C bonding ($2\sigma_g$) MO and the two C,H bonding MOs ($2\sigma_u$ and $3\sigma_g$) [121]. The later MO is formed by bonding overlap of $2p\sigma$(C) orbitals, which means that bending of acetylene leads to a decrease of C,H bonding and, hence, to an increase of the molecular energy.

In the case of HeCCHe^{2+}, the nature of the three σ-MOs changes due to the larger electronegativity of Ng relative to C. The $2\sigma_g$- and the $2\sigma_u$-MO possess He,C bonding character, while the $3\sigma_g$-MO is responsible for C,C bonding (Fig. 20). Accordingly, bending of the molecule leads only to an insignificant increase of the energy of the $3\sigma_g$-MO. On the other hand, the in-plane π_u-MO can overlap in a bonding fashion with σ-type He orbitals upon bending of the molecule, thus stabilizing the nonlinear form. Since antibonding overlap between non-nearest neighbors is minimized in the *trans*-bent form, this structure represents the energy minimum of the HeCCHe^{2+} ions. For the same reason HeCC^{2+} in its ground state possesses a bent rather than a linear geometry [7].

In the case of the NeCCNe^{2+} ion bending leads also to a reduction of $2p\pi$-π_u repulsion in the linear form. This is clearly reflected in the relative energies of

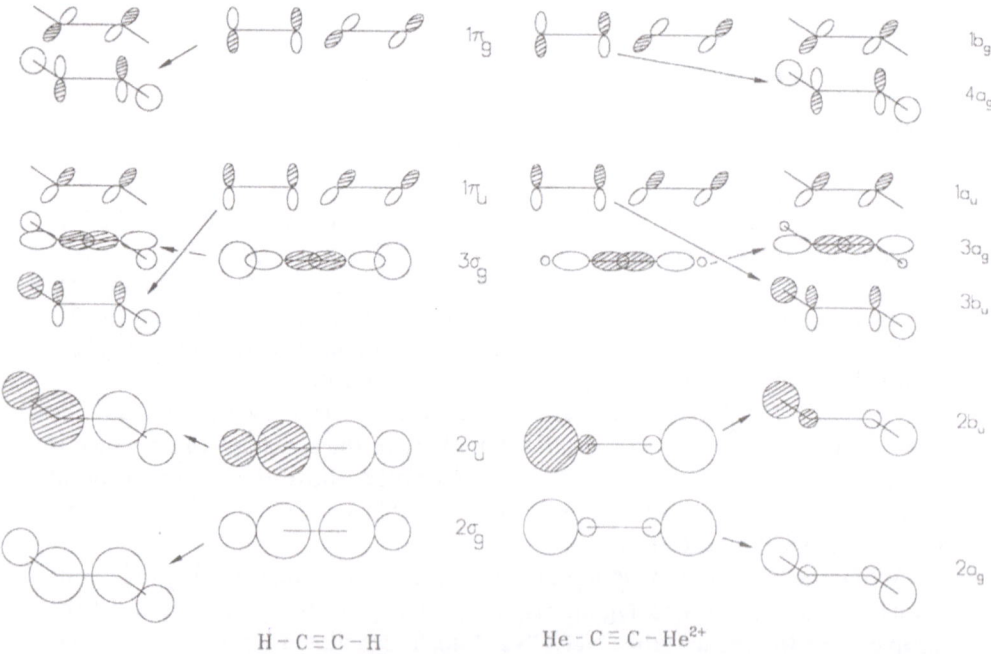

Fig. 20. Orbital correlation diagrams for bending of (a) acetylene and (b) HeCCHe^{2+}. Ref. [13]

linear and bent forms. In the case of $ArCCAr^{2+}$, this energy difference is negligible thus indicating that $2p\pi(C)\text{-}3p\pi(Ar)$ electron repulsion is less important than $2p\pi(C)\text{-}2p\pi(Ne)$ electron repulsion [13].

$NgCCH^+$ ions: The investigation of the NgX^{n+}, Ng_2C^{2+}, and $NgCCNg^{2+}$ ions clearly shows that σ-holes in the valence shell of the acceptor are more important than positive charges. This led Frenking and coworkers to investigate the acceptor ability of suitable monocations such as the HCC^+ ion [7]. In Fig. 21, the Laplace concentration of the ground state and two excited states of the HCC^+ ion are depicted. They show that only the excited $^1\Sigma^+(4\pi)$ state with its deep σ-concentration hole at the terminal C atom should be suited to bind a Ng atom. This has been confirmed by ab initio calculations [7]. The contour line diagram of the Laplace concentration of $HeCCH^+$ is given in Fig. 22. It shows the droplet like appendix in the charge concentration of the He atom that is typical of a semipolar covalent He,C bond. The covalent nature of the He, C bond is confirmed by the properties of electron and energy density distribution in the bonding region (see Table 15). Necessary and sufficient condition for covalent bonding are fulfilled [7, 13].

As noted in Sect. 5.3, $HeCCH^+$ possesses a *trans*-bent geometry, which can be rationalized in the same way as the nonlinear structures of $NgCCNg^{2+}$ discussed above. The MO correlation diagram in Fig. 20 suggests that due to the larger electronegativity of He compared to that of H the HeCC bending angle should be much smaller than the HCC angle. This is confirmed by the calculated bond angles shown in Fig. 4.

In the case of $NeCCH^+$ bending of the molecule leads also to a reduction of electron repulsion between the lone pair electrons of Ne and the π-electrons of the CC unit. Again, $p\pi\text{-}p\pi$ electron repulsion is stronger for the Ne compound than the Ar compound (see above). As a consequence, the ArCC angle is close to $180°$ while the NeCC angle is smaller than the HeCC angle (see calculated geometries in Fig. 14).

Two of the three $NgCCH^+$ ions, namely $HeCCH^+$ and $ArCCH^+$, possess covalent Ng,C bonds (see Table 15). For $NeCCH^+$, however, the analysis of the

Table 15. Nature of bonding in NgX^+ monocations according to ab initio calculations[a]

Dication	State	Bond	ρ_b [eÅ^{-3}]	H_b [hartree Å^{-3}]	D_e [kcal/mol]	Nature of bonding
$HeCCH^+$	$^1A'$	He, C	1.17	-0.9	15.4	covalent
$NeCCH^+$	$^1A'$	Ne, C	$-$[b]	$-$[b]	7.1	electrostatic
$ArCCH^+$	$^1A'$	Ar, C	1.36	-1.9	55.7	covalent
$HeCN^+$	$^1\Sigma^+$	He, C	0.04	0.0	0.9	electrostatic
$HeNC^+$	$^1\Sigma^+$	He, N	0.08	0.0	0.9	electrostatic
$NeCN^+$	$^1\Sigma^+$	Ne, C	0.21	-0.0	3.6	electrostatic
$NeNC^+$	$^1\Sigma^+$	Ne, N	0.24	0.0	3.6	electrostatic
$ArCN^+$	$^1\Sigma^+$	Ar, C	1.38	-1.7	35.7	covalent
$ArNC^+$	$^1\Sigma^+$	Ar, N	0.27	0.0	17.5	electrostatic

[a] From Ref. 13; [b] No (3, -1) critical point found. Characterization of the bond has been done on the basis of the Laplacian of $\rho(\mathbf{r})$ and the D_e value.

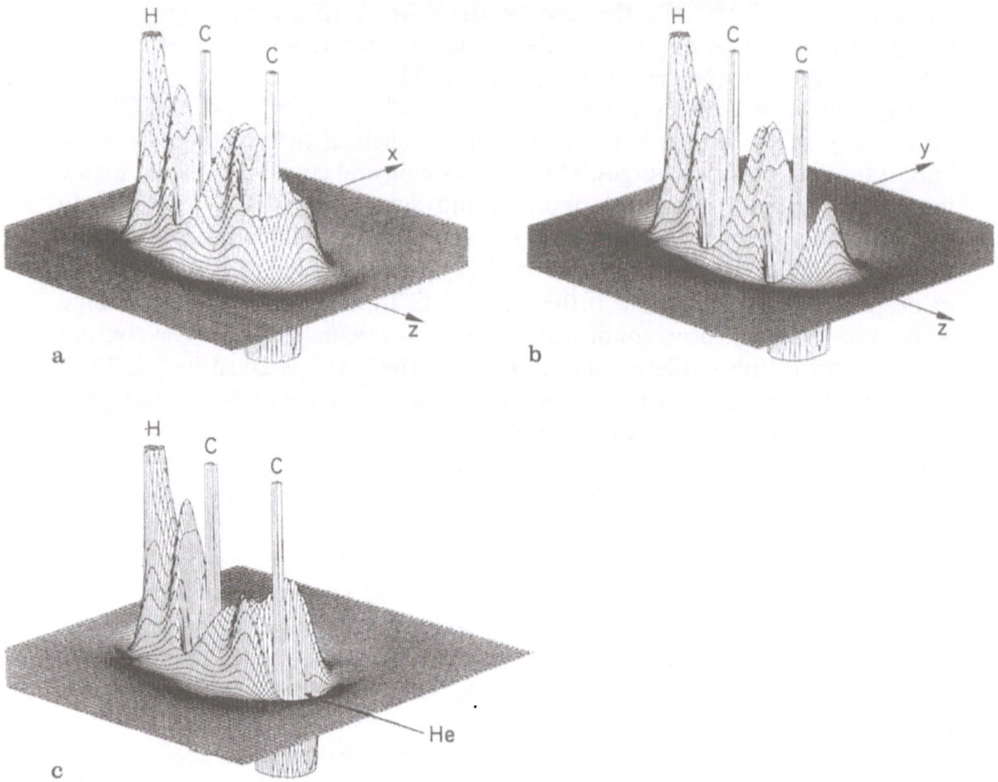

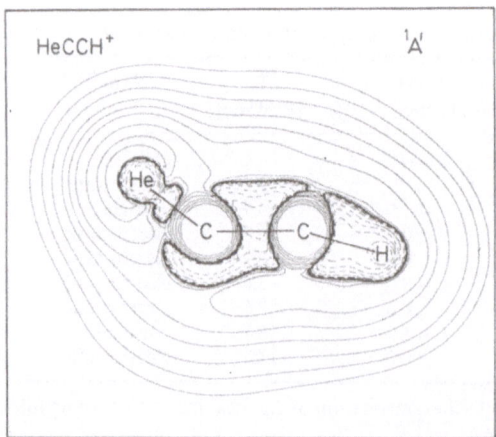

Fig. 21a–c. Perspective drawings of the Laplace concentration, $-\nabla^2\rho(\mathbf{r})$, of **(a, b)** HCC$^+$($^1\Pi$) shown for both the plane with one (a) and with the two paired π-electrons (b), and **(c)** HCC$^+$($^1\Sigma^+$, 4π). Inner shell and valence shell concentrations are indicated. (HF/6-31G(d, p) calculations)

HeCCH$^+$ $^1A'$

Fig. 22. Contour line diagram of the Laplace concentrations, $-\nabla^2\rho(\mathbf{r})$, of HeCCH$^+$. The inner shell concentrations are no longer shown. (HF/6-31G(d, p) calculations). Ref. [7]

energy density H_b reveals that the bond possesses only weak covalent character. Electrostatic interactions between Ne and CCH^+ should play a more important role, which is in line with the relatively low dissociation energy of 7 kcal/mol (Table 15).

It is likely that a replacement of H in $NgCCH^+$ ions by alkyl, aryl or other organic groups does not lead to a significant change in Ng,C bonding and, therefore, monohelium or monoargon alkyne cations should be stable possessing semipolar covalent Ng,C bonds. This is one of the important conclusions of the bond analysis of Ng compounds [7, 13].

$NgCN^+$ ions: The $CN^+(^1\Sigma^+)$ ion is another suitable acceptor for Ng electrons. This is clearly reflected by the Laplace concentration of $CN^+(^1\Sigma^+)$ shown in Fig. 23. There is a σ-hole at the C atom, which is not as deep as the σ-hole at the terminal C atom of $HCC^+(^1\Sigma^+)$ shown in Fig. 21. But it is much deeper than the σ-hole at the N atom of $CN^+(^1\Sigma^+)$ (Fig. 23b). This is indicative of the fact that the $3\sigma_g$ LUMO of $CN^+(^1\Sigma^+)$ has a larger amplitude at the C atom. The $CN^+(^1\Sigma^+)$ ion is an electron acceptor that can pull electrons into its valence shell holes both at the C and the N atom where binding should be stronger via the C atom than the N atom.

He and Ne form only weakly stable van der Waals complexes with CN^+ (Sec. 5.3), which is confirmed by the analysis of $\rho(r)$ and $H(r)$ (Table 15). Obviously, the acceptor ability of $CN^+(^1\Sigma^+)$ does not suffice for the formation of covalently bonded molecules. The properties of $\rho(r)$ and $H(r)$ indicate just weak dipole-induced dipole interactions.

Contrary to He and Ne, Ar forms a covalently bound $ArCN^+(^1\Sigma^+)$ ion with $CN^+(^1\Sigma^+)$ (see Ar, C in Table 15). This is in line with the enhanced donor ability of Ar and the distinct acceptor ability of $CN^+(^1\Sigma^+)$ at its carbon end. Figure 24, which contains contour line diagrams and perspective drawings of calculated Laplace-concentration of $ArCN^+$ and $ArNC^+$, reveals pronounced $3p\sigma$-donation of the Ar atom and concentration of negative charge in the ArC bonding

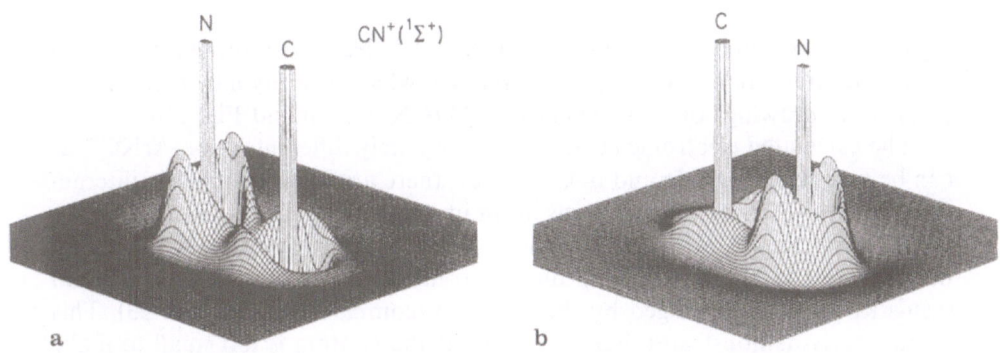

Fig. 23a and b. Perspective drawings of the Laplace concentration, $-\nabla^2\rho(r)$, of the $CN^+(^1\Sigma^+)$ shown (**a**) from the side of the carbon atom, (**b**) from the side of the N atom. (HF/6-31G(d) calculations)

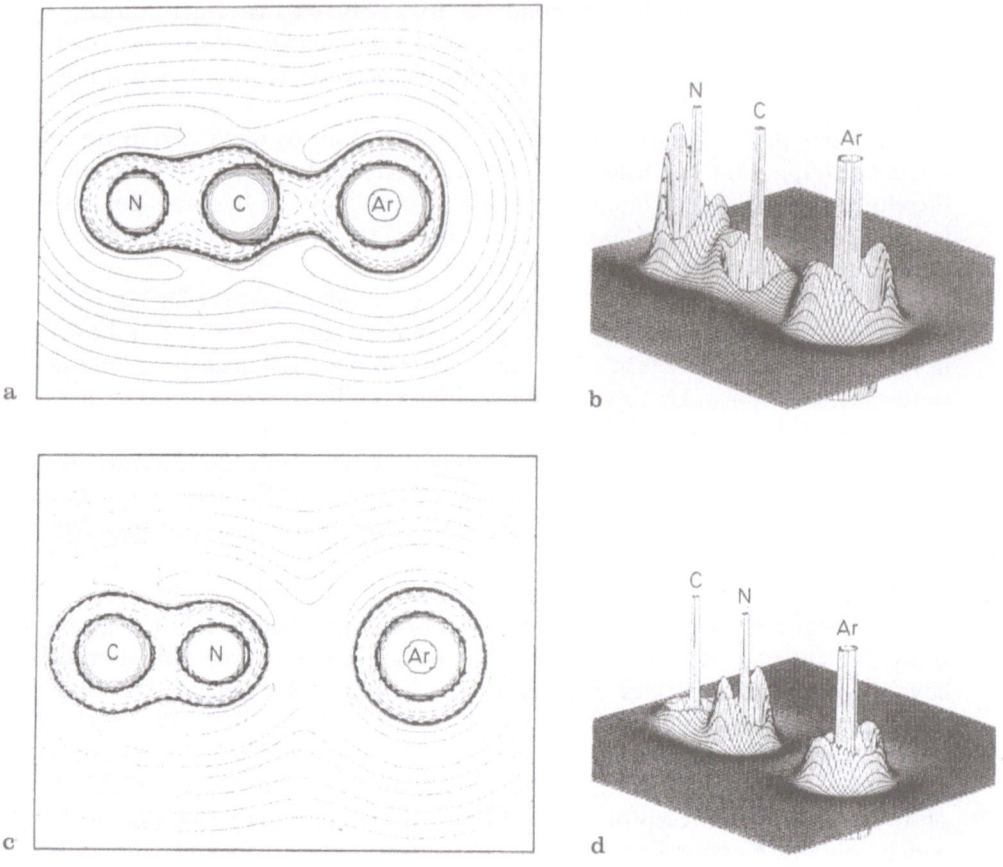

Fig. 24a–d. Contour line diagrams and perspective drawings of the Laplace concentration, $-\nabla^2\rho(\mathbf{r})$, of (**a, b**) $ArCN^+$ and (**c, d**) $ArNC^+$. In the contour line diagrams inner shell concentrations are no longer shown. (HF/6-31G(d) calculations). Ref. [13]

region. Electron donation from Ar to $CN^+(^1\Sigma^+)$ leads to a reorganization of electronic structure in the CN part of $ArCN^+$, which is nicely illustrated by the perspective drawings of $-\nabla^2\rho(\mathbf{r})$ in Fig. 23 ($CN^+(^1\Sigma^+)$) and Fig. 24b.

The calculated electronic structure is completely different in ion $ArNC^+$ as can be seen from Fig. 24c and d. Obviously, there are just electrostatic interactions between Ar and the noble gas atom in line with what has been found for $HeNC^+$ and $NeNC^+$. There is no concentration of negative charge in the region between Ar and N (Fig. 24d) and the electronic structure of the $CN^+(^1\Sigma^+)$ ion seems to be hardly changed by the Ar atom (compare Fig. 24d and 23). This result clearly demonstrates that the σ-hole at the N atom is too small to make $CN^+(^1\Sigma^+)$ an ambident noble gas acceptor with equal strength at the C and N end. The D_e value for $ArNC^+$ is relatively large in view of the electrostatic nature of the interactions (Table 15). However, investigation of ArX^+ ions has shown that ion-induced dipole forces can lead to as much as 25 kcal/mol attraction between Ar and ions such as N^+ [5]. Calculations show that despite the fact that CN^+ is isoelectronic with HCC^+, only $ArCN^+$ of the six possible $NgCN^+$ and $NgNC^+$ systems represents a covalently bonded ion.

6 Neutral Compounds of Helium, Neon and Argon

Soon after the discovery of the noble gas elements, many laboratories made great efforts to induce chemical reactions with Ng. Unsuccessful attempts were made to enforce reactions with compounds known to be very reactive, such as Cl_2, $KMnO_4$, red-hot Mg, alkaline metals, and many others. A survey of unsuccessful attempts to induce chemical reactions with Ng is given in Ref. 122. By 1916 it was established that He, Ne, Ar as well as the other noble gases were chemically inert.

An early report of weakly bound noble gas species came from Druyvesteyn [123] in 1931. He observed two regions of band structure in the violet region for a mixture of helium and neon. Because the resolution of his spectrometer was low, it is difficult to know whether such spectra originate from neutral molecules such as HeNe or from molecular ions such as $HeNe^+$. Reliable information of neutral diatomics NgX was available for the mercury species NgHg as early as 1941, when Heller [124] reported the vibrational wave numbers and dissociation energies for NgHg with Ng = He, Ne, Ar, Kr, Xe. HeHg was reported by Manley as early as 1924 [125].

In the following, we will shortly summarize the knowledge about neutral compounds of the light noble gas elements bound by van der Waals interaction including the "border-line" structures NgBeO. Although no chemically bound molecule of He, Ne, and Ar is known as yet, there is reason to believe that argon compounds are feasible. This is discussed at the end of this section.

6.1 Clathrates and van der Waals Complexes

Clathrates are host-guest complexes in which a crystalline cage of the host compound holds the guest molecule by weak intermolecular forces. Often, the cavities of the guest molecule are formed by a network of hydrogen bonds between covalently bound compounds. Powell [126] names them 'clathrates' from the Latin word 'clathratus', which means enclosed. It is interesting to note that the lattice structure of the host in the clathrate is not its normal crystalline form; the former becomes thermodynamically more stable than the latter only by the formation of the host-guest complex.

The first known clathrate was a gas hydrate of chlorine, discovered by Davy in 1811 [127]. By 1900, many clathrate compounds had been prepared [128], but the nature of these substances was not understood until the X-ray work of Powell and coworkers [126, 129] provided detailed structural information. Although the composition of the clathrate nH.mG (H = host, G = guest) can vary over a long range, the maximum composition formula, which corresponds to completely filled cavities, is stoichiometric. For example, hydroquinone forms a clathrate with many guest molecules with the ratio 3H:1G [130]. This corresponds to 69 ml gas per gramm clathrate.

The best-known noble gas clathrates are hydrates, hydroquinone and phenol clathrates, which have found an increasing number of uses [131]. Clathrates may serve as convenient "storage" for noble gases. Because of the different affinity hydroquinone clathrate prepared from an equal mixture of krypton and xenon liberates 3 times the amount of Xe than Kr [132]. Clathrates are also of interest for nuclear technology. Radioactive isotopes of argon, xenon and krypton can more easily be handled in the compact form of a solid rather than in gas form [133–136].

Hydroquinone clathrates of the nobles gases with the maximum composition formula $(C_6H_4(OH)_2)_3Ng$ are known for Ng = Ar [137], Ng = Kr [138], and Ng = Xe [139]. Attempts to form He or Ne clathrates were unsuccessful. The molecular structures are well known [130, 140]. Each Ng atom is enclosed in a cubic cage formed by 6 benzene rings as shown in Fig. 25 [130]. The stability of the clathrates shows the order Ar < Kr < Xe. The stability order is readily explained by the increase in electric polarizability of the noble gas atoms. One reason that clathrates of He and Ne are not formed is the lower polarizability. But perhaps more important is the size of the cage; the cavity is too large, the phenyl rings are too far away to allow for stabilizing interactions. In fact, up till now there are no clathrates of He and Ne known with any host molecule. A suitable host compound for the two lightest noble gases should provide small cavities; but this yields repulsive intramolecular interactions within the host compound which apparently can not be compensated for by the weak stabilization energy of a He or Ne clathrate.

Hydrates of Ar, Kr, and Xe were first synthesized by Villard in 1896 [141]. They were further studied, as well as hydrates of krypton and xenon, by de Forcrand [142]. Several structures for noble gas hydrates are known [143–146]. All the hydrate structures are different from that of ordinary hexagonal ice. In the two fundamental structures, the water molecules form pentagonal dodecahedra which are stacked with different degrees of distortion from their ideally regular forms [146]. The two types of structures are shown in Fig. 26a and 26b [140]. One structure contains 46 water molecules in the unit cell with 2 small and 6 larger cavities. The other structure has 136 water molecules in the unit cell with 16 small and 8 larger cavities. The formation of the two fundamental types of hydrates depends mainly on the size of the guest species. More detailed data for the two principal clathrate hydrate structures are available from the literature [147].

The phenol clathrates of argon, krypton, and xenon have been prepared by Lahr and Williams [148] by direct combination of the gas with crystalline

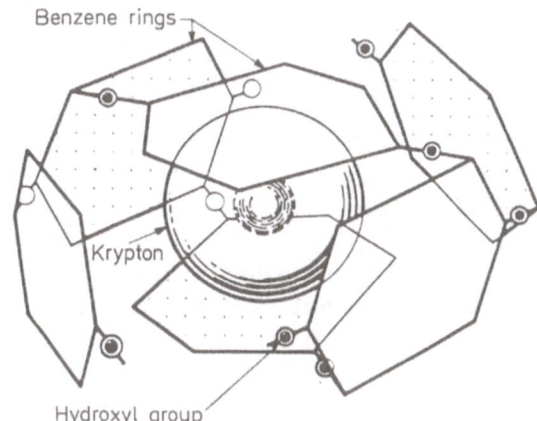

Fig. 25. Schematic display of krypton hydroquinone clathrate [130]

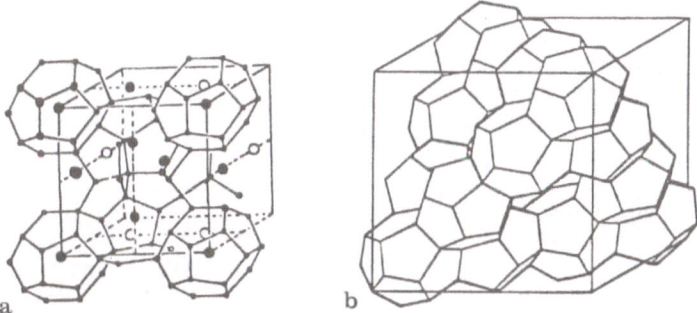

Fig. 26a and b. Schematic representation of the dodecahedra of noble gas hydrates of (a) type I and
(b) type II. [140]

phenol. In the cases of krypton and xenon, clathrates were also formed with
molten phenol. The gas/phenol mole ratio was 1:4 for argon and krypton, and
1:3 for xenon [148]. Also substituted phenol form clathrates with the three
noble gas elements [149]. Details on other substances which form clathrates
with argon, neon, and xenon may be found in the literature [139]. As mentioned
above, no stable clathrates of helium and neon have been prepared.
A semi-empirical quantum mechanical calculation [150] of the stability of
cubane clusters with He and Ne came to the conclusion that there is a lesser
chance of retaining a Ne atom in comparison with a He atom in a cubane cage.
    A different class of noble gas compounds bound by van der Waals interac-
tions are small molecules NgX, where X is either an atom or a diatomic or
triatomic species such as HF, HCl, HBr, ClF, NO, HCN, Cl_2, Br_2. The experimental and theoretical techniques for studying such weakly bound complexes
and the concepts and theoretical models to understand them have recently been
reviewd in several articles and monographs [151–154] and will not be discussed
here. It should be mentioned, however, that detailed structural and dynamical

information of small van der Waals complexes in the gas phase is available from molecular beam experiments using methods such as electric resonance spectroscopy [155] and Fourier transform microwave spectroscopy [156]. The spectroscopic study of these species has accelerated enormously since the advent of the supersonic nozzle [157], which allows the investigation of van der Waals complexes at an extremely low translational temperature. Table 16 shows the dissociation energies of some well studied examples of NgX compounds with X = HF, HCl, HBr, Cl_2, Br_2, J_2, NO. Attractive interactions in these compounds is mainly due to dipole-induced dipole interactions (X = HF, HCl, HBr, NO) or dispersion forces between two induced dipoles (X = Cl_2, J_2). Unlike the clathrates, van der Waals complexes are known for neon and helium.

The data in Table 16 indicate typical dissociation energies D_o for HeX in the range 10–20 cm^{-1}, for NeX they are 20–80 cm^{-1}, and for ArX 80–240 cm^{-1}. As expected, the dissociation energy in NgX increases for a given X with the size of Ng (see the discussion in Sect. 4.5). With 350 cm^{-1} being equal to 1 kcal/mol,

Table 16. Dissociation enrgies D_o (in cm^{-1}) and equilibrium distances r_{NgX}[a] (in Å)

X		He	Ne	Ar	Kr	Xe
HF	D_o			102[f]	133[f]	181[f]
	r_{NgX}			2.63[g]	2.75[h]	2.92[h]
	Structure			lin.[f]	lin.[f]	lin.[f]
HCl	D_o		68[i]	115[f]	154[f]	206[f]
	r_{NgX}		2.41[m]	2.60[i]	2.71[k]	
	Structure		lin.[m]	lin.[f]	lin.[f]	lin.[f]
HBr	D_o			206[j]	247[j]	
	r_{NgX}			2.72[j]	2.83[j]	
	Structure			lin.[j]	lin.[j]	
Cl_2	D_o	9[b]	55[b]	175[b]		
	r_{NgX}	3.8[b]	3.57[b]	4.0[b]		
	Structure	T[b]	T[b]	T[b]		
Br_2	D_o		70[b]			
	r_{NgX}	3.7[b]	3.67[b]			
	Structure	T[b]	T[b]			
J_2	D_o	18[b]	75[b]	235[b]		
	r_{NgX}	4.5[b]				
	Structure	T[b]	T[b]			
NO	D_o	8.5[d]	20[d]	87[d]	130[d]	210[d]
	r_{NgX}	3.63[c]		3.77[e]	3.88[e]	
	Structure	T[c]		T[e]	T[e]	

[a] For linear structures, r_{NgX} is the distance between Ng and the nearest atom; for T-shaped complexes, r_{NgX} is the distance between Ng and center of mass of X; [b] Ref. 182; [c] Ref. 183; [d] Ref. 184; [e] Ref. 185; [f] Ref. 186; [g] Ref. 154b. [h] Ref. 187; [i] Ref. 188; [j] Ref. 189. [k] Ref. 190; [l] D_e value taken from Ref. 191; [m] Ref. 191

binding in van der Waals complexes NgX is very weak with typical dissociation energies $D_o < 1$ kcal/mol. Especially the He and Ne complexes exhibit very weak binding. In agreement with the weak attractive interactions, interatomic distances Ng, X in van der Waals complexes are very large as shown in Table 16. The equilibrium distance between Ng and the nearest atom is always larger than 2.5 Å. van der Waals complexes show distinctively smaller dissociation energies and larger interatomic distances for the 'van der Waals-bond' than typical covalent compounds.

In the course of their systematic theoretical investigation of positively charged helium compounds, Frenking, Cremer, and coworkers found [4, 7, 11, 12] that the electronic structure rather than the charge of the binding fragment X is crucially important for the strength of the interactions in HeX. By extending this concept to neutral molecules, they detected computationally the stable molecule HeBeO [7, 14, 15]. The calculated dissociation energy into He + BeO at the MP4(SDTQ)/6-311G(2df, 2pd)//MP2/6-31G(d, p) level of theory is $D_o = 3.0$ kcal/mol. The equilibrium distance for the linear molecule in the $^1\Sigma^+$ ground state is predicted as $r_{HeBe} = 1.538$ Å [7]. They also performed full-valence CASSCF calculations which give similar results, i.e. $D_e = 3.5$ kcal/mol at $r_{HeBe} = 1.577$ Å [7]. Later, they studied the analogs NeBeO and ArBeO and found [15] for the neon structure $D_o = 2.2$ kcal/mol at $r_{NeBe} = 1.758$ Å, and for ArBeO $D_o = 7.0$ kcal/mol at $r_{ArBe} = 2.036$ Å. A critical evaluation of the theoretical results obtained at several computational levels suggests [15] that the dissociation energies are rather lower bounds to the correct values, and that HeBeO is probably bound by 3–4 kcal/mol. The theoretically predicted results for NgBeO shown in Table 17 differ significantly from the typical data found for van der Waals complexes as shown in Table 16. In particular, the D_o value for HeBeO of 3–4 kcal/mol, which is much higher than the 10–20 cm^{-1} (0.03–0.06 kcal/mol) found for weakly bound helium structures (Table 16), points towards chemical bonding, albeit very weak. In fact, a subsequent natural bond orbital analysis of the electronic structure of HeBeO, NeBeO, and ArBeO by Hobza and Schleyer [158] came to the conclusion that the essential nature of the interactions is charge transfer. For HeBeO they found that the overlap between the donor $1s$ orbital of He with the acceptor σ^* orbital of BeO is quite large (0.42).

In contrast to the natural bond orbital [159] picture, a detailed analysis of the electron density distribution and its associated Laplace field suggests that bonding in HeBeO, NeBeO and ArBeO is caused by strong charge induced

Table 17. Calculated dissociation energies D_o (kcal/mol) and interatomic distances r_{NgBe} (in Å) [15]

	HeBeO	NeBeO	ArBeO
D_o	3.1	2.2	7.0
r_{NgBe}	1.538	1.758	2.036

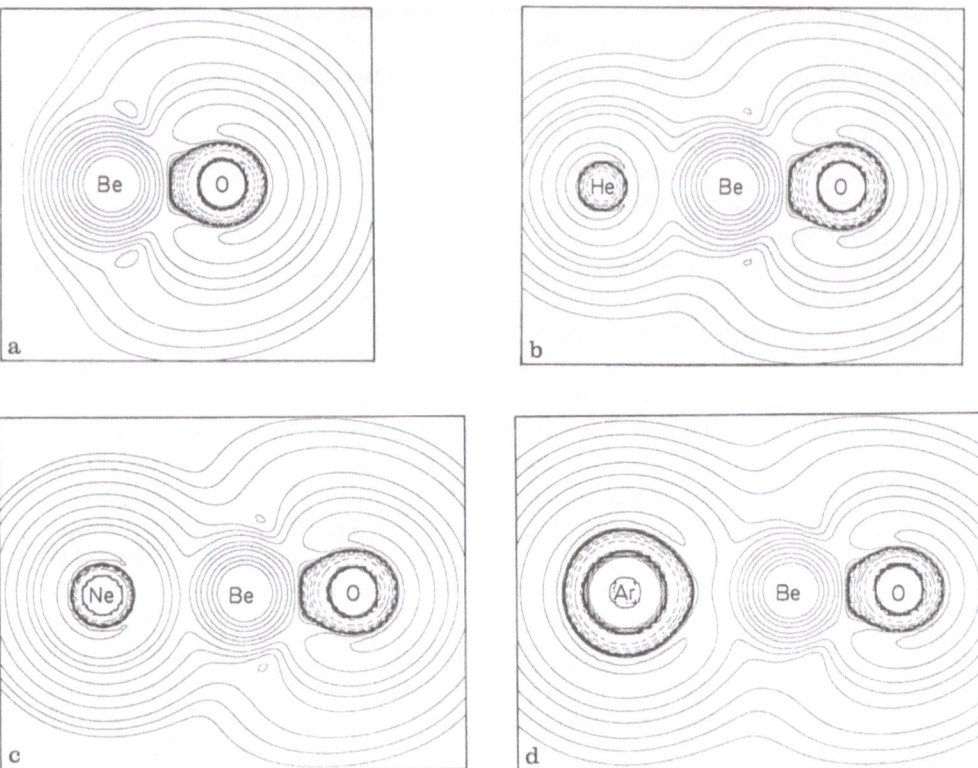

Fig. 27a–d. Contour line diagrams of the Laplace concentration, $-\nabla^2\rho(r)$, of (a) BeO, (b) HeBeO, (c) NeBeO, (d) ArBeO. Inner shell concentrations are not shown. (HF/6-31G(d, p) calculations). Ref. [15]

dipole interactions without covalent contributions [15]. Figure 27 shows the contour line diagrams of the Laplace concentrations of BeO, HeBeO, NeBeO, and ArBeO. The Laplace fields indicate (i) a strong charge transfer from Be to O in BeO as well as in NgBeO, (ii) nearly identical contour line diagrams for BeO as isolated species and for NgBeO, and (iii) hardly any distortion for Ng atoms in NgBeO compared with isolated Ng (Fig. 3). Only for ArBeO there is a small distortion of the Laplace field of the Ar atom. The energy density H_b at the bond critical point for Ng, Be is $+0.120$ (HeBeO), $+0.102$ (NeBeO), -0.005 (ArBeO) [15]. Thus, for ArBeO there is a tendency toward beginning covalency.

The natural bond orbital analysis has been claimed [159] to correct many of the deficiencies [160] of the Mulliken population analysis [161]. But it seems to favor orbital interactions compared with electrostatic interactions as is evident by the results for hydrogen bonding which are also interpreted [159] as resulting from charge-transfer rather than electrostatic interactions. Using a multipole expansion scheme, Dykstra and coworkers [162] concluded that electrostatic forces are the dominant factor in hydrogen bonding. In any case, the light noble gas species HeBeO, NeBeO and ArBeO seem to be examples of molecules in the 'border-line' between typical van der Waals complexes such as shown in Table 16 and covalent compounds.

BeO seems to be a very peculiar acceptor species mainly because of the beryllium atom. Calculations show that, while the oxygen atom may be replaced by isoelectronic NH or CH_2, substitution of Be by any other atom such as Mg, Al, B, does not yield a NgX species which is comparable to NgBeO [163]. The very unique acceptor ability of BeO is demonstrated by the strong complexes formed with CO and N_2, which are calculated with a dissociation energy of $D_0 = 30.0$ kcal/mol (NNBeO) and 40.8 kcal/mol (OCBeO) [164]. Also, neutral BeO is a stronger acceptor for He and Ne than cationic CN^+, as is evident by a comparison of the calculated binding energies for $NgCN^+$ (Table 12) and NgBeO (Table 17).

The idea of using a strong Lewis acid rather than a strong oxidizing agent for binding Ng atoms is not new. In the 1930s, much effort was devoted especially to the investigation of the $Ng + BF_3$ systems, and for some time it was believed that a helium compound was formed. For a survey of the literature see Ref. 165. For example, a thermal analysis of the system $Ar + BF_3$ was reported by Booth and Wilson [166] in 1935. A detailed study of the van der Waals molecule $ArBF_3$ was first made by Janda et al. [167] in 1978. Kaufman and Sachs [168] in 1969 investigated theoretically the systems HeLiH and $HeBeH_2$. They found that $HeBeH_2$ was marginally unstable with respect to $He + BeH_2$ [168b], but HeLiH was bound by 1.8 kcal/mol [168a]. Although more recent highly accurate calculations [15] predict that HeLiH is bound by only ca. 0.1 kcal/mol, the basic idea of Kaufmann and Sachs pointed toward the right direction. The systematic study of binding in light noble gas compounds [7, 13–15] show that Lewis acids exhibit the strongest attracting force toward Ng.

6.2 Chemical Compounds of Helium, Neon and Argon

Soon after the preparation of the first stable noble gas compound by N. Bartlett [1], the isolation of many other noble gas molecules followed [169], it was quickly recognized that many noble gas salts can be classified as derivatives of binary fluorides or oxide fluorides such as $(NgF)^+(A)^-$, $(NgF_3)^+(A)^-$, $(NgOF_3)^+(A)^-$, and many others where Ng = Xe, Kr, and sometimes Rn [170]. Although all attempts to synthesize argon analogs were unsuccessful, stability considerations pointed toward ArF^+ salts as the most promising candidate. Suggestions for possible argon salts are $(ArF)(PtF_6)$ [87] and $(ArF)(BF_4)$ [86]. As discussed in Sect. 4.3, the binding energy of ArF^+ in its $^1\Sigma^+$ ground state has recently been calculated very accurately as $D_0 = 49 \pm 3$ kcal/mol [6]. This value and an estimate of the lattice energies of the possible salt compounds was used by Frenking et al. [6] to predict that $(ArF)(SbF_6)$ and $(ArF)(AuF_6)$ could be stable. However, experimental obstacles to synthesizing these compounds are enormous, because neither AuF_6 nor SbF_6 would probably oxidize the Ar atom, and the need to generate ArF^+ or F^+ ions as precursors for the salts is experimentally a formidable task. Nevertheless, the theoretical study encourages further experimental attempts to synthesize argon salt compounds, which apparently represent the limit of possible neutral, chemical compounds of the noble gas atoms. Unless one includes the (experimentally not yet found) NgBeO structures in the chemical realm!

7 Summary

The research on light noble gas chemistry may be summarized in the following way. For many years chemists tried to induce chemical reactions with noble gas elements using the most reactive agents known. Until 1962, the only success was the observation of various gas phase cations and weakly bound van der Waals complexes and clathrates. Because of the little information on binding interactions in these compounds, the knowledge on the principles of chemical interactions involving noble gas atoms was scarce and not very systematic. It was only known that attractive interactions between Ng and other atoms or molecules increase with the size of Ng. The Lewis paradigm served as a theoretical explanation for the apparent inability of Ng to form any neutral chemical compound. It was so effective that experimental efforts to disprove the strict validity of the octet rule were for some time not very strong.

With Bartlett's landmark experiment in 1962 [1], noble gas chemistry became synonymous with the chemistry of Xe and Kr and very soon became a standard field of inorganic research. Since the obstacles to synthesizing Xe and Kr compounds were shown to be only technical, a renewed search for compounds of the lighter homologues He, Ne, and Ar was started. It was soon realized that experiments, that were successful for Xe and Kr, did not materialize for helium, neon and argon. The explanation for this was the increase in ionization energy with decreasing size of Ng. The ionization energy of Ng replaced the Lewis paradigm as a rationalization for the unreactivity of Ng = He, Ne, Ar. Experimentally obtained information about bonding of the light noble gas elements is still very scarce.

The situation has changed in the last few years by theoretical studies which culminated in a model for chemical bonding in light noble gas compounds, i.e. donor-acceptor interactions [4, 5, 7, 13, 15]. Earlier attempts at a systematic theoretical study of bonding in light noble gas compounds was performed by Liebman and Allen [82, 87]. Because of the crude methods and limited computer facilities available at that time, the results were quantitatively not very reliable. Modern methods of molecular quantum mechanics together with high-speed computers allow theoretical predictions for small molecules with an accuracy that is competitive with high-quality experiments. Using the tools of modern theoretical chemistry, Frenking, Cremer and coworkers [4–15] re-

Table 18. Potentially stable or metastable He, Ne, Ar containing compounds for the thermodynamically stable molecules, the calculated reaction energy ΔE_o (in kcal/mol) for the least endothermic reaction is given. Unless otherwise specified, the results are taken from Ref. 13

Thermodynamically stable molecule	ΔE_o	Metastable
A. Dications		
$He_2C^{2+}(^1A_1)$	13.6	$He_2C^{2+}(^3B_1)$; $He_2N^{2+}(^2B_1)$; $He_2O^{2+}(^1A_1)$; $Ne_2C^{2+}(^1A_1)$; $Ne_2C^{2+}(^3B_1)$; $Ne_2N^{2+}(^2B_1)$; $Ne_2O^{2+}(^1A_1)$; $Ar_2C^{2+}(^1A_1)$; $Ar_2C^{2+}(^3B_1)$; $Ar_2N^{2+}(^2B_1)$; $Ar_2O^{2+}(^1A_1)$;
$ArCCAr^{2+}$	9.4	$HeCCHe^{2+}$; $NeCCNe^{2+}$;
$HeCO^{2+}$	17.7[b]	$HeCF^{3+}$ [b]
B. Monocations		
$HeCCH^+$	12.5	
$NeCCH^+$	5.9	
$ArCCH^+$	54.3	
$ArCN^+$	34.6	
$ArNC^+$	17.5 (ΔE_e)	
C. Neutral Compounds		
HeBeO	3.1[a]	
NeBeO	2.2[a]	
ArBeO	7.0[a]	
$ArFSbF_6$	49[c]	

[a] Ref. 15; [b] Ref. 95; [c] $D_o(Ar, F^+)$, Ref. 6

ported the results of very accurate calculations and extracted a binding model that explains the binding features for light noble gas compounds. The scattered experimental results for NgX molecules could be shown to be examples of molecular species whose bond strength Ng,X should be discussed in terms of donor-acceptor interactions between Ng and X. It is the Lewis acidity (acceptor strength) of X, rather than the electronegativity or electronic charge, that determines the strength of the binding interactions in neutral and ionic NgX molecules. Based on this model, the search for new and strongly bound molecules was focussed on binding partners that were strong σ-acceptors. The surprising results were the predictions of strong covalent Ng,C bonds in molecular cations and the unusually strong interaction in NgBeO. Table 18 shows a list of theoretically predicted stable or metastable Ng compounds [6, 7, 13, 14].

It should be pointed out that the theoretical predictions listed in Table 18 are results *a priori* and not *a posteriori*. The experimental obstacles to verify the theoretical predictions are substantial, however. Light noble gas chemistry has become an area of chemical research where theory and experiment combine and challenge each other.

Acknowledgement. The authors would like to thank the following colleagues and coworkers for stimulating discussions and dedicated cooperation: Prof. C. K. Jørgensen, Prof. N. Bartlett, Prof. J.

F. Liebman, Dr. W. Koch, Dr. J. Gauss, Dr. J. R. Collins, Dr. C. Deakyne, Dr. B. T. Luke, Dipl. Chem. F. Reichel. Part of this research has been supported by the Fonds der Chemischen Industrie.

8 References

1. Bartlett N: Proc. Chem. Soc. 1962: 218
2. For a recent review on the history of the discovery of noble gas compounds see: Laszlo P, Schrobilgen G (1988) Angew. Chem. 100: 495; (1988) Angew. Chem. Int. Ed. Engl. 27: 479
3. Hogness TR, Lunn EG (1925) Phy. Rev. 26: 50
4. Frenking G, Koch W, Gauss J, Cremer D, Liebmann JF (1989) J. Phys. Chem. 93: 3397
5. Frenking G, Koch W, Gauss J, Cremer D, Liebman JF (1989) J. Phys. Chem. 93: 3410
6. Frenking G, Koch W, Deakyne C, Liebmann JF, Bartlett N (1989) J. Am. Chem. Soc. 111: 31
7. Koch W, Frenking G, Gauss J, Cremer, D, Collins JR (1987) J. Am. Chem. Soc. 109: 5917
8. Koch W, Frenking G, Luke BT (1987) Chem. Phys. Lett. 139: 149
9. Koch W, Frenking G (1987) J. Chem. Phys. 86: 5617
10. Koch W, Liu B, Frenking G (in print) J. Chem. Phys.
11. Koch W, Frenking G: J. Chem. Soc. Chem. Commun. 1986: 1095
12. Koch W, Frenking G (1986) Int. J. Mass Spectrom. Ion Proc. 74: 133
13. Frenking G, Koch W, Reichel F, Cremer D (in print) J. Am. Chem. Soc.
14. Koch W, Collins JR, Frenking G (1986) Chem. Phys. Lett. 132: 330
15. Frenking G, Koch W, Gauss J, Cremer D (1988) J. Am. Chem. Soc. 110: 8007
16. See e.g., Hehre WJ, Radom L, Schleyer PVR, Pople JA (1985) Ab initio molecular orbital theory, Wiley, New York
17. (a) Cremer D (1988) In: Maksić ZB (ed) Modelling of structure and properties of molecules. Ellis Horwood, Chichester, England, p 125; (b) Kraka E, Cremer D (in press) In: Maksić ZB (ed) Theoretical models of chemical bonding, vol 2, Springer Verlag, Berlin Heidelberg New York
18. (a) Bader RFW, Preston HJT (1969) Int. J. Quant. Chem. 3: 327; (b) Feinberg MJ, Ruedenberg K (1971) J. Chem. Phys. 54: 1495; (c) Kutzelnigg W (1973) Angew. Chem. 13: 551
19. Cremer D, Kraka E (1984) Croat. Chem. Acta 57: 1259
20. Bader RFW, Essen H (1984) J. Chem. Phys. 80: 1943
21. Frenking G, Koch W (1988) Int. J. Mass Spectrom. Ion Proc. 82: 335
22. (a) Coppens P, Hall MB (eds) (1982) Electron distributions and the chemical bond, Plenum, New York (b) Becker P (ed) (1978) Electron and magnetization densities in molecules and crystals, Plenum, New York
23. Hohenberg H, Kohn W (1964) Phys. Rev. 136B: 864
24. (a) Bader RFW, Nguyen-Dank TT, Tal Y (1981) Rep. Prog. Phys. 44: 893; (b) Bader RFW, Nguyen-Dang TT (1981) Adv. Quantum Chem. 14: 63
25. For a more detailed review, see Ref. 17b
26. Bader RFW, Slee TS, Cremer D, Kraka E (1983) J. Amer. Chem. Soc. 105: 5061
27. Cremer D, Kraka E (1984) Angew. Chem. 96: 612; (1984) Int. Ed. Engl. 23: 627
28. Cremer D, Kraka E (1985) J. Amer. Chem. Soc. 107: 3800, 3811
29. Cremer D, Kraka E, Slee TS, Bader RFW, Lau CDH, Nguyen-Dang TT (1983) J. Amer. Chem. Soc. 105: 5069
30. (a) Cremer D, Gauss J (1986) J. Amer. Chem. Soc. 108: 7467; (b) Cremer D, Kraka E (1988) In: Liebman JF, Greenberg A (eds) Structure and reactivity, VCH Publishers, Deerfield Beach, USA, p 65
31. Cremer D, Gauss J, Kraka E (1988) J. Mol. Struct. (THEOCHEM) 169: 531
32. Note that Bader [24] uses these terms differently by calling all MED paths, also those between closed shell systems, bond paths. This, however, is contrary to general chemical understanding.
33. Morse PM, Feshbach H (1953) Methods of theoretical physics vol 1, McGraw-Hill, New York, p 6
34. Bader RFW, MacDougall PJ, Lau CDH (1984) J. Am. Chem. Soc. 106: 1594 See also Ref. 19

35. Bader RFW (1980) J. Chem. Phys. 73: 2781. G(r) is always positive while V(r) is always negative. When integrated over total molecular space, they yield the kinetic and potential energy of a molecule: $E = \int H(r)\, dr = \int G(r)\, dr + \int V(r)\, dr$
36. Tüxen O (1936) Z. Physik 103: 463
37. Hornbeck JA, Molnar JP (1951) Phys. Rev. 84: 621
38. Inghram MG (1953) Natl. Bur. Standards (U.S.) Circ. No. 522: 204
39. (a) Fukui K (1971) Accts. Chem. Res. 4: 57; (b) Fleming I (1976) Frontier orbitals and organic chemical reactions, Wiley, Chichester
40. Munson MSB, Franklin JL, Field FH (1963) J. Phys. Chem. 67: 1542
41. Weise H-P, Mittmann H-U (1973) Z. Naturforsch. 28a: 714
42. Wadt WR (1978) J. Chem. Phys. 68: 402
43. Black JH: cited in Ref. 44
44. Yu N, Wing WH (1987) Phys. Rev. Lett. 59: 2055
45. Huber KP, Herzberg G (1979) Molecular spectra and molecular structure, vol IV, Constants of diatomic molecules. Van Nostrand Reinhold, New York
46. Christodoulides AA, McCorkle DL, Christophoru LG (1984) In: Christophorou LG (ed) Electron-molecule interactions and their applications, vol 2, Academic, London
47. Bondyby V, Pearson PK, Schaefer HF (1972) J. Chem. Phys. 57: 1123
48. Pauling L (1933) J. Chem. Phys. 1: 56
49. Guilhaus M, Brenton AG, Beynon JH, Rabenovic M, Schleyer PVR (1984) J. Phys. B17: L605
50. Frenking G, Koch W, Liebmann JF (1989) In: Liebmann JF, Greenberg A (eds) Molecular structure and energetics: From atoms to polymers. Isoelectronic analogies, VCH Publishers, p 169
51. Gill PMW, Radom L (1987) Chem. Phys. Lett. 136: 294
52. (a) Yagisawa H, Sato H, Watanabe, T (1977) Phys. Rev. A16: 1352; (b) Cohen JS, Bardsley JN (1978) Phys. Rev. A18: 104
53. Mercier E, Chambaud G, Lewy B (1985) J. Phys. B18: 3591
54. Montanbonel M-CB; Cimiraglia R, Persico M (1984) J. Phys. B17: 1931
55. The ground states on Ne^{2+} and Ar^{2+} have ^3P symmetry. The lowest lying electronic state of $Ng^{2+} + He(^1S)$ (Ng = Ne, Ar) which can mix with the (1) $^1\Sigma^+$ state is the ^1D state of Ng^{2+}. This state is 3.204 eV (Ne^{2+}) and 1.737 eV (Ar^{2+}) higher in energy than the ^3P ground state [62]
56. The situation is actually more complicated because more than two states are sometimes involved in the mixing of the ground state curves: Stärk D, Peyerimhoff SD (1986) Mol. Phys. 59: 1241
57. Stärk D (Personal communication)
58. Helm H, Stephan K, Märk TD, Huestis DL (1981) J. Chem. Phys. 74: 3844
59. Stephan K, Märk TD, Helm H (1982) Phys. Rev. A26: 2981
60. Carrington A, Softley TP (1983) In: Miller TA, Bondybey VE (eds) Molecular ions: Spectroscopy, structure and chemistry, North-Holland, Amsterdam
61. Beach JY (1936) J. Chem. Phys. 4: 353
62. Moore CE (1971) Atom energy levels as derived from the analyses of optical spectra, Nat. Stand Ref. Data Ser. Nat. Bur. Stand. (U.S.) NSRDS-NBS
63. (a) Kolos W, Peek JM (1976) Chem. Phys. 12: 381; (b) Kolos W (1976) Int. J. Quantum Chem. 10: 217
64. (a) Rosmus P (1979) Theoret. Chim. Acta 51: 359; (b) Rosmus P, Reinsch EA (1980) Z. Naturforsch. 35a: 1066; (c) Klein R, Rosmus P, (1984) Z. Naturforsch. 39a: 349; (d) Rosmus P, Reinsch EA, Werner HJ (1983) In: Berkowitz J, Groeneveld KO (eds) Molecular ions: geometric and electronic structures, Plenum, New York
65. Bernath P, Amano T (1982) Phys. Rev. Lett. 48: 20
66. Wong M, Bernath P, Amano T (1982) J. Chem. Phys. 77: 693
67. Lorenzen J, Hotop H, Ruf MW, Morgner H (1980) Z. Physik A, 297: 19
68. Ram RS, Bernath PF, Brault JW (1985) J. Mol. Spectrosc. 113: 451
69. Brault JW (1978) Proceeding of the Workshop on Future Solar Observations, Needs and Constraints, Florence
70. Johns JWC (1984) J. Mol. Spectrosc. 106: 124
71. Wells BH, Wilson S (1986) J. Phys. B19: 17
72. Butler SE, Bender CF, Dalgarno A (1979) Astrophys. J. Lett. 230: 59
73. Chambaud G, Levy B (1986) Ann. Phys. 11: Suppl. 3, 107

74. (a) Dalgano A, McDowell MRC, Williams A (1958) Phil. Trans. R. Soc. A250: 411; (b) Mason EA, Schamp HW Jr, (1958) Ann. Phys. 4: 233; (c) Polark-Dingels P, Rajan MS, Gislason EA (1982) J. Chem. Phys. 77: 3983
75. Böttner E, Dimpfl WL, Ross U, Toennies JP (1975) Chem. Phys. Lett. 32: 197
76. (a) Subbaram KV, Coxon JA, Jones WE (1976) Can. J. Phys. 54: 1535; based on the experimental data, slightly different dissociation energies have been predicted by: (b) Goble JH, Winn JS (1977) J. Chem. Phys. 67: 4206; (c) Le Roy RJ, Lam W-H (1980) Chem. Phys. Lett. 71: 544; (d) Vahala L, Havey M (1984) J. Chem. Phys. 81: 4867
77. Ding A, Karlau J, Weise J, Kendrick J, Kuntz PJ, Hillier IH, Guest MF (1978) J. Chem. Phys. 68: 2206
78. Hillier IH, Guest MF, Ding A, Karlau J, Weise J (1979) J. Chem. Phys. 70: 864
79. Estimated value in Ref. 82c.
80. Ding A, Karlau J, Weise J (1977) Chem. Phys. Lett. 45: 92
81. Berkowitz J, Chupka W (1970) Chem. Phys. Lett. 7: 447
82. (a) Liebmann JF, Allen LC (1972) Inorg. Chem. 11: 1143; (b) Liebmann JF, Allen LC (1970) J. Am. Chem. Soc. 92: 3539; (c) Liebmann JF, Allen LC (1971) Int. J. Mass Spectrom. Ion Phys. 7: 27
83. Cooper DL, Wilson S (1981) Mol. Phys. 44: 161
84. Bernardi F, Epiotis ND, Cherry W, Schlegel HB, Whangbo M-H, Wolfe S (1976) J. Am. Chem. Soc. 98: 469
85. Moore CE (1970) Ionization potentials and ionization limits derived from the analyses of optical spectra nar. Stand Ref. Data Ser. Nat. Bur. Stand. (U.S.) NSRDS-NBS 34
86. Jørgensen CK (1986) Z. Anorg. Allg. Chem. 540: 91
87. Liebmann JF, Allen JF: J. Chem. Soc. Chem. Commun. 1969: 1355
88. Jonkman HT, Michl J (1981) J. Am. Chem. Soc. 103: 733
89. Jonathan P, Boyd RK, Brenton AG, Beynon JH (1986) Chem. Phys. 110: 239
90. (a) Hayes EF, Gole JL (1971) J. Chem. Phys. 55: 5132; (b) Harrison SW, Massa LJ, Solomon P (1972) Chem. Phys. Lett. 16: 57; (c) Alvarez-Rizzatti M, Mason EA (1975) J. Chem. Phys. 63: 5290; (d) Schleyer PvR (1985) Adv. Mass Spectrom. 10a: 287
91. Harrison SW, Henderson GA, Masson LJ, Salomon P (1974) Astrophys. J. 189: 605
92. (a) Müller EW, McLane SB, Panitz JA (1969) Surf. Sci. 17: 430; (b) Tsong TT, Müller EW (1970) Phys. Rev. Lett. 25: 911; (c) Tsong TT, Müller EW (1971) J. Chem. Phys. 55: 2884; (d) Müller EW, Tsong TT (1973) Prog. Surf. Sci., 4: 1; (e) Müller EW, Krishnaswamy (1973) Phys. Rev. Lett. 31: 1282; (f) Tsong TT, Kinkus TJ (1983) Physica scripta, T4: 201; (g) Tsong TT, Kinkus TJ (1984) Phys. Rev. B29: 529; (h) Tsong TT (1984) Phys. Rev. B30: 4946; (i) Tsong TT, Liou Y (1985) Phys. Rev. Lett. 55: 2180
93. Hotokka M, Kindstedt T, Pyykkö P, Roos B (1984) Mol. Phys. 52: 23
94. Wong MW, Radom L (1988) J. Am. Chem. Soc. 110: 2375
95. (a) Wong MW, Nobes RH, Radom L: J. Chem. Soc., Chem. Commun. 1987: 233; (b) Wong MW, Nobes RH, Radom L (1987) Rapid Commun. Mass Spectrom. 1: 3; (c) Radom L, Gill PMW, Wong MW, Nobes RH (1988) Pure Appl. Chem. 60: 183
96. Hirschfelder JO, Curtiss CF, Bird RB (1986) Molecular theory of gases and liquids, Wiley, New York
97. Wilson S, Green S (1980) J. Chem. Phys. 73: 419
98. Bohme DK, Adams NG, Mosesman M, Dunkin DB, Ferguson EE (1970) J. Chem. Phys. 52: 5094
99. Munson MSB, Field FH, Franklin JL (1962) J. Chem. Phys. 37: 1790
100. Adams NG, Bohme DK, Ferguson EE (1970) J. Chem. Phys. 52: 5101
101. Milleur MB, Matcha RL, Hayes EF (1974) J. Chem. Phys. 60: 674
102. Matcha RL, Milleur MB, Meier PF (1978) J. Chem. Phys. 68: 4748
103. Matcha RL, Milleur MB (1978) J. Chem. Phys. 69: 3016
104. Dykstra CE (1983) J. Mol. Struct. THEOCHEM 103: 131
105. Matcha RL, Pettitt BM, Meier PF, Pendergast P (1978) J. Chem. Phys. 69: 2264
106. Clampiti R, Jefferies DK (1970) Nature 226: 142
107. (a) Leopold DG, Murray KK, Stevens Miller AE, Lineberger WC (1985) J. Chem. Phys. 83: 4849; (b) Bunker PR, Sears TJ (1985) J. Chem. Phys. 83: 4866
108. For the problem of graphically illustrating Δ states see: Salem S (1982) Electrons in chemical reactions: First principles, Wiley, New York
109. The orbital terms in Fig. 15 correspond to heteronuclear diatomics XY. For homonuclear

species XX, the 3σMO becomes 2σg orbital, 4σ becomes $2\sigma_u$, etc. For simplicity reasons, we have used the heteronuclear notations throughout the text.

110. Kaul W, Fuchs R (1960) Z. Naturforschg. 15a: 326
111. Teng HH, Conway DC (1973) J. Chem. Phys. 59: 2316
112. The ground state of CCH$^+$ is a triplet: (a) Krishnan R, Frisch M, Whiteside RA, Pople JA, Schleyer PVR (1981) J. Chem. Phys. 74: 4213; (b) Glaser R (1987) J. Am. Chem. Soc. 109: 4237
113. The $^1\Pi$ state is unstable, however, towards Renner distortion. The lowest lying singlet state of CCH$^+$ has $^1A'$ symmetry and is 4.8 kcal/mol lower than $^1\Pi$: Koch W, Frenking G: J. Chem. Phys. (in print)
114. In Ref. 7, the binding energy of the He–C bond in HeCCH$^+$ has been calculated using the $^1\Delta$ state of CCH$^+$. However, the $^1\Pi$ state of CCH$^+$ was later [113] calculated to be lower-lying than the $^1\Delta$ state. The dissociation energy of HeCCH$^+$ yielding He + ($^1\Pi$) CCH$^+$ is reported in Ref. 13
115. Ikuta S, Iwata S, Imamura M (1977) J. Chem. Phys. 66: 4671
116. Snell AH, Pleasonton F (1958) J. Phys. Chem. 62: 1377
117. Cacace F (1970) Adv. Phys. Org. Chem. 8: 79
118. Wexler S, Anderson GR, Singer LA (1960) J. Chem. Phys. 32: 417
119. (a) Carlson TA, White RM (1962) J. Chem. Phys. 36: 2883; (b) Carlson TA, White RM (1962) J. Chem. Phys. 39: 1748; (c) Carlson RT, White RM (1962) J. Chem. Phys. 38
120. Holtz D, Beauchamp JL (1971) Science 173: 1237
121. See e.g., Gimarc BM (1979) Molecular structure and bonding, the qualitative molecular orbital approach, Academic Press, New York
122. Holloway JH (1968) Noble gas chemistry, Methuen, London
123. Druyvesteyn MJ (1931) Nature 128: 1076
124. Heller R (1941) J. Chem. Phys. 9: 154
125. (a) Manley JJ (1924) Nature 114: 861; (b) Manley JJ (1925) Nature 115: 337
126. Powell HM: J. Chem. Soc. 1948: 61
127. Davy H (1811) Phil. Trans. 101: 1
128. Mylius F: J. Chem. Soc. Abstracts 1886: 50
129. (a) Palin DE, Powell HM: J. Chem. Soc. 1947: 208; (b) Palin DE, Powell HM: J. Chem. Soc. 1948: 571; (c) Palin DE, Powell HM: J. Chem. Soc. 1948: 815; (d) Rayner JH, Powell HM: J. Chem. Soc. 1952: 319; (e) Wallwork SC, Powell HM: J. Chem. Soc. 1956: 4855; (f) Lawton D, Powell HM: J. Chem. Soc. 1958: 471
130. Balek V (1970) Anal. Chem. 42: (9) 16A
131. (a) Hagan SMM (1962) Clathrate inclusion compounds, Reinhold, New York (b) Gawalek G (1969) Einschlußverbindungen, Additionsverbindungen, Clathrate. Deutscher Verlag der Wissenschaften, Berlin
132. Hagan SMM (1963) J. Chem. Educ. 40: 643
133. (a) Adams RE, Browning WE, Ackley RD (1959) Ind. Eng. Chem. 51: 1467; (b) Steinberg M, Manowitz B (1959) Ind. Eng. Chem. 51: 47
134. (a) Swift WE (1957) Nucleonis 15: 66; (b) Wilson EJ, Dibbs HP, Richards S, Eakins JD (1958) Nucleonics 16: 110
135. Mock JE (private communication) cited in: Mock JE, Myers JE, Trabant EA (1961) Ind. Eng. Chem. 53: 1007
136. McClain JW, Diethorn WS (1963) J. Appl. Radiation and Isotopes 14: 527
137. (a) Powell HM, Guter M (1949) Nature 164: 240; (b) Powell HM: J. Chem. Soc. 1950: 298
138. Powell HM: J. Chem. Soc. 1950: 300
139. Powell HM: J. Chem. Soc. 1950: 468
140. Mandelcorn L (ed) Non-stoichiometric compounds, Academic, New York, p 438
141. Villard P (1896) Compt. Rend 123: 377
142. (a) De Forcrand R (1923) Compt. Rend 176: 335; (b) De Forcrand R (1925) Compt. Rend 181: 15
143. van der Waals JH, Platteeuw JC (1959) Adv. Chem. Phys. 2: 1
144. Beurskens PT, Jeffrey GA (1964) J. Chem. Phys. 40: 906; and earlier papers cited therein
145. Barrer RM (1964) In: Mandelcoin L (ed) Non-stoichiometric compounds, Academic, New York
146. Von Stackelberg M, Müller HR (1954) Z. Elektrochem. 58: 25
147. Davidson DW, Garg SK, Gough SR, Handa YP, Ratcliffe CI, Tse JS, Ripmeester JA (1984) J. Inclus. Phenom. 2: 231

148. Lahr PH, Williams HL (1959) J. Phys. Chem. 63: 1432
149. Barrer RM, Shanson VH: J. Chem. Soc. Chem. Commun. 1976: 333
150. Bolotin AB, Bolotin VA, Balyavichus M-LZ, Gantmakher BF, Stumbris EP, Tatevskii VM, Shapiro EI, Yaravoi SS (1982) Teor. Eksp. Khim 18: 212
151. Weber A (ed) (1987) Structure and dynamics of weakly bound molecular complexes, Reidel, Dordrecht
152. See the review articles in: (1988) Chemical Review 88
153. Van Lenthe JH, van Duijneveldt-van der Rijdt JGCM, van Duijneveldt FB (1987) Adv. Chem. Phys. 69: 521
154. (a) Legon AC, Millen DJ (1986) Chem. Rev. 86: 635; (b) Peterson KI, Fraser GT, Nelson DD Jr, Klemperer W (1985) In: Bartlett RJ (ed) Comparison of ab initio quantum chemistry with experiment for small molecules, Reidel, Dordrecht
155. English TC, Zorn JC, (1972) In: Williams D (ed) Methods of experimental physics, vol 3 (2nd edn), Academic, New York
156. Balle TJ, Campbell EJ, Keenam MR, Flygare WH (1979) J. Chem. Phys. 71: 2723
157. (a) Levy DH (1981) In: Jortner J, Levine RD, Rice SA (eds) Photoselective chemistry, Wiley, New York; (b) Levy DH (1980) Annu. Rev. Phys. Chem. 31: 197
158. Hobza P, Schleyer PvR: Coll. Czech. Chem. Commun. (in print)
159. Reed AE, Curtiss LA, Weinhold F (1988) Chem. Rev. 88: 899
160. Bachrach SM, Streitwieser A Jr (1984) J. Am. Chem. Soc. 106: 2283
161. Mulliken RS (1955) J. Chem. Phys. 23: 1833
162. Dykstra CE, Liu S-Y (1987) In: Weber A (ed) Structure and dynamics of weakly bound molecular complexes, Reidel, Dordrecht
163. Koch W, Frenking G (unpublished results)
164. (a) Frenking G, Koch W, Collins JR: J. Chem. Soc., Chem. Commun. 1988: 1147; (b) Frenking G, Koch W, Cremer D, Gauss J, Collins JR (manuscript in preparation)
165. Hawkins DT, Falconer WE, Bartlett N (1978) Noble gas compounds, Plenum, New York
166. Booth HS, Willson KS (1935) J. Am. Chem. Soc. 57: 2273
167. Janda KC, Bernstein LS, Steed JM, Novick SE, Klemperer W (1978) J. Am. Chem. Soc. 100: 8074
168. (a) Kaufman JJ, Sachs LM (1970) J. Chem. Phys. 52: 3534; (b) Kaufman JJ, Sachs LM (1969) J. Chem. Phys. 51: 2992
169. Hyman HH (ed) (1963) Noble gas compounds, University of Chicago press, Chicago
170. Selig H, Holloway JH (1984) Topics Current Chem. 124: 33
171. Miller TM, Bederson B (1977) Adv. At. Mol. Phys. 13: 1
172. Dabrowski L, Herzberg G (1978) J. Mol. Spectrosc. 73: 183
173. Liao MZ, Balasubramanian K, Chapman D, Lin SH (1987) Chem. Phys. 111: 423
174. Dabrowski L, Herzberg G, Yoshino K (1981) J. Mol. Spectrosc. 89: 491
175. Trevor DJ (1980) Ph. D. Thesis, University of California, Berkeley
176. Pratt ST, Dehmer PM (1982) J. Chem. Phys. 76: 3433
177. Dehmer PM, Pratt ST (1982) J. Chem. Phys. 76: 843
178. Dehmer PM, Pratt ST (1982) J. Chem. Phys. 77: 4804
179. Pratt ST, Dehmer PM (1982) Chem. Phys. Lett. 87: 533
180. Ng CY, Trevor DJ, Mahan BH, Lee YT (1976) J. Chem. Phys. 65: 4327
181. Balasubramanian K, Liao MZ, Lin SH (1987) Chem. Phys. Lett. 138: 49
182. Cline JI, Edvard DD, Reid BP, Sivakumar N, Thommen F, Janda KC (1987) In: Weber A (ed) Structure and dynamics of weakly bound molecular complexes, Reidel, Dordrecht
183. Beneventi L, Casacecchia P, Volpy GG (1987) In: Weber A (ed) Structure and dynamics of weakly bound molecular complexes, Reidel, Dordrecht
184. Estimated value by Miller JC (1987) J. Chem. Phys. 86: 3166
185. Casavecchia P, Lagana A, Volpi GG (1984) Chem. Phys. Lett. 112: 445
186. Pine AS (1987) In: Weber A (ed) Structure and dynamics of weakly bound molecular complexes, Reidel, Dordrecht
187. Fraser GT, Pine AS (1986) J. Chem. Phys. 85: 2502
188. Howard BJ, Pine AS (1985) Chem. Phys. Lett. 122: 1
189. Keenan, MR, Campbell EJ, Balle TJ, Buxton LW, Minton TK, Soper PD, Flygare WH (1980) J. Chem. Phys. 72: 3070
190. Balle TJ, Campbell EJ, Keenan MR, Flygare WH (1979) J. Chem. Phys. 71: 2723
191. Hutson JM, Howard BJ (1982) Molec. Phys. 45: 769
192. A noble gas atom may become an electron acceptor (Lewis acid) in molecules if electronic

charge is withdrawn by other atoms or ions. An example is provided by the recently synthesized first compound with a krypton-nitrogen bond $HCN-KrF^+AsF_6^-$. In KrF^+, electronic charge is withdrawn from Kr and a donor-acceptor bond between Kr and HCN is formed, but here the (partially positively charged) Kr atom serves as electron acceptor and HCN (via the nitrogen lone-pair electrons) is the electron donor: Schrobilgen GJ: J. Chem. Soc. Chem. Commun. 1988: 863; MacDougall PJ, Schrobilgen GJ, Bader RFW (1989) Inorg. Chem. 28: 763

193. For a survey of the techniques and the method see: Bierbaum VM, Ellison GB, Leone SR (1984) In: Bowers MT (ed) Gas phase ion chemistry, vol 3, Academic, Orlando, p 2

Chemistry of Inorganic Vapors

Klaus Hilpert

Institute for Reactor Materials, Nuclear Research Centre (KFA) Jülich, P.O. Box 1913, D-5170 Jülich, FRG

The chemistry of inorganic vapors is for the most part the chemistry of high-temperature vapors. This article essentially reports on the gaseous species observed in the equilibrium vapors and their dissociation or atomization energies. Most of the results were obtained by Knudsen effusion mass spectrometry. The fundamentals of this method are described with respect to the recent methodic developments. Accounts of the investigations of metals, alloys, oxides, metal halides, and technical materials are given and the results are reviewed. Vaporization processes and thermochemical properties of condensed phases evaluated from gas phase data are also considered. The results are for example of interest for the active fields of energy, high-temperature materials, and cluster research. An extended summary with outlook is given in Sect. 7—"Summary and conclusions"—of this article.

Structure and Bonding 73
© Springer-Verlag Berlin Heidelberg 1990

1 Introduction

The chemistry of inorganic vapors is for the most part the chemistry of high-temperature vapors, since high temperatures are necessary to evaporate most of the elements or inorganic phases. The first step in the study of gas phase chemistry is the identification of the gaseous species and the determination of their concentration. Moreover, the molecular parameters of the gaseous species, such as geometry and electronic structure, as well as their thermochemical properties, such as bond energies, are of interest. A variety of different methods is applied in such studies.

Vibrational frequencies for example result from the application of infrared, laser-induced fluorescence, Raman, and Raman resonance spectroscopy. Spectroscopy in the visible and near-UV regions yields information on electronic transitions. Electron spin resonance spectroscopy is used in determining the geometric and electronic structure. These methods were applied to study the gaseous species trapped at low temperatures in a solid inert rare gas matrix (matrix isolation technique) as well as in the free state.

Molecular structures and rotational constants are obtained by electron diffraction and microwave spectroscopy, respectively.

Photoionization mass spectrometry is the most accurate method for the determination of dissociation energies of molecular species. Photoelectron spectroscopy yields for example valuable information on the nature of bonding of the molecule.

Review articles on the aforementioned methods and others, as well as applications, are for example given in Margrave's book "The Characterization of High-Temperature Vapors" [1], the NBS special publication "Characterization of High Temperature Vapors and Gases", by Hastie [2], and in the Series "Advances in High-Temperature Chemistry" by Eyring [3].

The comparatively new field of the spectroscopy of small free metal atom clusters is described in the recent review by Gole [4]; the application of the matrix isolation technique to the study of such species is reported by Moskovits [5] in the same book.

Most of the gaseous species have been identified by the use of mass spectrometry. A variant of this technique, frequently used in inorganic gas phase chemistry, is Knudsen effusion mass spectrometry. Vaporization studies up to temperatures above 2500 K were carried out by this method and thermodynamic properties determined. The dissociation energies of numerous gaseous species were obtained in this way.

The emphasis of this article is on the gaseous species observed in equilibrium high-temperature vapors and their dissociation energies. The fundamentals of Knudsen effusion mass spectrometry and recent methodic developments are described in Sect. 2. The results obtained by this method for metals and alloys, oxides, metal halides, and technical materials are reviewed in the following four sections. The results are discussed considering, in addition, relevant studies by

other methods. Vaporization processes and thermochemical properties of condensed phases evaluated from gas phase data are also considered. Investigations
by Knudsen effusion mass spectrometry generally mean the study of equilibrium
reactions involving neutral molecules and/or atoms. This meaning has also been
adopted for this article. The reader is referred to Sect. 5.5 for the study of ion-ion
and ion-molecule equilibrium reactions by Knudsen effusion mass spectrometry.

2 Knudsen Effusion Mass Spectrometry

2.1 Principles of the Method

Figure 1 shows a magnetic-type sector field mass spectrometer coupled with
a Knudsen cell. The most important part of the instrument is the Knudsen cell.
It can be heated up to temperatures above 2500 K. The temperatures are
measured with an optical pyrometer or a thermocouple. There would be
thermodynamic equilibrium in the Knudsen cell if it were closed. However, real
Knudsen cells have an effusion orifice (typical diameter: 0.1 to 1 mm) through
which a small fraction of the molecules effuse without practically disturbing the
equilibrium in the cell. A molecular beam representing the equilibrium vapor in

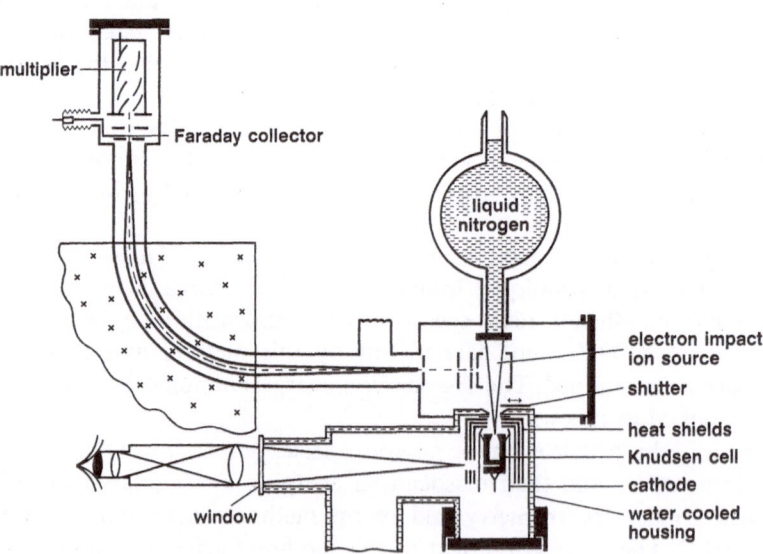

Fig. 1. Schematic representation of a Knudsen cell-magnetic sector field mass spectrometer system

the cell is formed by the effusing species. This molecular beam crosses the electron impact ion source and is analyzed mass spectrometrically.

Ion intensities and Knudsen cell temperatures are the quantities measured in the course of an investigation by the Knudsen effusion mass spectrometric method. Partial pressures are computed from these quantities for the vapor species identified.

The potential of Knudsen effusion mass spectrometry can be summarized as follows:

1. identification of the gaseous species forming the equilibrium vapor in the Knudsen cell,
2. determination of their partial pressures in the range from generally between 10^{-5} and 10 Pa and up to temperatures above 2500 K, and
3. computation of thermodynamic properties from the measured partial pressures and their temperature dependences, as, for example, enthalpies and entropies of vaporization, of dissociation and of formation.

Points 1 and 2 mean a qualitative and quantitative analysis of equilibrium vapors. Point 3 mentions some of the thermochemical properties which can be evaluated from the analytical data.

Details on points 1, 2, and 3 are given in Sects. 2.3, 2.4, and 2.5, respectively. The given lower limit of the dynamic range of the vapor pressure measurement is determined by the sensitivity of the instrument (upper limit see Sect. 2.4). Recent developments to improve the sensitivity are mentioned in Sect. 2.6.1.

2.2 General Aspects and General Review

Mass spectrometry for the determination of physico-chemical data was first used by Chupka and Inghram [6] as well as Honig [7] to study the free evaporation of carbon. These authors identified different polymeric carbon species at high temperatures and they concluded that the trimer C_3 is the most abundant vapor species at temperatures of 2450 and 2600 K. The numerous results obtained thereafter refer essentially to equilibrium vaporization investigated by Knudsen effusion mass spectrometry and are summarized in the first comprehensive review by Inghram and Drowart [8] in 1960. The systems studied and the results obtained up to 1964 are tabulated by Drowart [9]. An extensive compilation of literature is also given by Grimley [10] and in the recent review by Drowart [11].

Thermodynamic properties of condensed phases, such as mixing properties, can be additionally obtained by the investigation of their equilibrium vapor. An account of the investigations by Knudsen effusion mass spectrometry and of the evaluation methods used is given by Chatillon et al. [12] as well as Raychaudhuri and Stafford [13]. Gingerich [14] reports on the potential of the method for the determination of the homogeneity range of binary compounds. Gilles et al. [15] described the measurement of the extent of non-stoichiometry in oxide phases. Tomiska [16] discusses approximative equations for thermodynamic excess functions with regard to best fit of mass spectrometric data on alloys.

The high potential of Knudsen effusion mass spectrometry for the analysis of salt vapors is pointed out in the early paper of Berkowitz and Chupka [17], thereby also resuming the pioneering work of Ionov [18]. Accounts of inter-metallic and non-metallic molecules as well as their enthalpy of dissociation are given by Gingerich in Refs. 19, 20 and particularly in Ref. 21. Sublimation processes are additionally described in Ref. 20. General aspects on the relative concentrations of the high-temperature molecules and their temperature de-pendence are discussed by Goldfinger [22] as well as Drowart and Goldfinger [23]. The fundamentals of Knudsen effusion mass spectrometry are described extensively in these two references. In addition, these references contain com-pilations of molecules observed in high-temperature vapors together with their enthalpies of dissociation.

A particularly extensive general description of the Knudsen effusion mass spectrometric method is given by Drowart [24] and Grimley [10]. Drowart et al. [25] describe the numerous methods for the determination of thermo-dynamic data of gaseous molecules and condensed phases. They also report on the determination of evaporation and reflection coefficients. A Knudsen cell set up with cooled aperture for the molecular beam is additionally described. Questions of the adjustment of equilibrium in Knudsen cells and the influence of residual gases on this adjustment are discussed by Babeliowsky [26]. Further reviews are given by Refs. 27–31. The limitations in applying Knudsen effusion mass spectrometry to high temperature equilibrium studies are discussed by Stafford [32]. Finally, the brief general articles of Drowart [33], DeMaria [34], Clark [35], and Avery et al. [36] should be mentioned.

Hastie [37] describes recent developments in Knudsen effusion mass spec-trometry and in the application of mass spectrometry to high-temperature materials chemistry.

The use of Knudsen effusion mass spectrometry for the study of non-equilibrium processes, such as catalytic decomposition or recrystallization of amorphous films, is described in Refs. 38–42.

2.3 Identification of Gaseous Species

The identification of the ions in the mass spectrum originating in the vapor species over the sample can be easily carried out by their mass, their isotopic distribution, and by interrupting the molecular beam from the Knudsen cell with the shutter. The mass spectra for molecular ions can be computed from those of the component elements. The essential step for the identification of the gaseous species in the equilibrium vapor is the assignment of the ions observed in the mass spectrum to their neutral precursors. This can be difficult due to fragmentation. The following rules are useful for the assignment:

a) the intensities of ions generated from the same neutral molecule often show the same temperature dependences,

b) the appearance potentials of molecular ions formed by simple ionization are generally smaller than those of fragments generated from the same neutral precursor, they increase with increasing degree of fragmentation,

c) fragmentation can be indicated by the shape of ionization efficiency curves,

d) the angular distributions of vapor species effusing from a non-ideal orifice deviate from each other for different species [43], and

e) homologous gaseous species show similar fragmentation patterns.

Deviations from rules (a) and (b) are possible by temperature dependent fragmentation [11, 44, 45, 496] and by the formation of negative ions [46], respectively. Temperature dependent fragmentation of small inorganic molecules was first reported by Gräber and Weil [496]. Some further remarks on points a), b), c), and e) are made in the following.

The temperature dependences of the ion intensities, I, are represented as apparent enthalpies of sublimation "$\Delta_s H_T$" and result from the slopes of the log(IT) versus 1/T plots (see e.g. Fig. 2) by the equation

$$\text{"}\Delta_s H_T\text{"} = -R\frac{d \ln(IT)}{d(1/T)}. \tag{1}$$

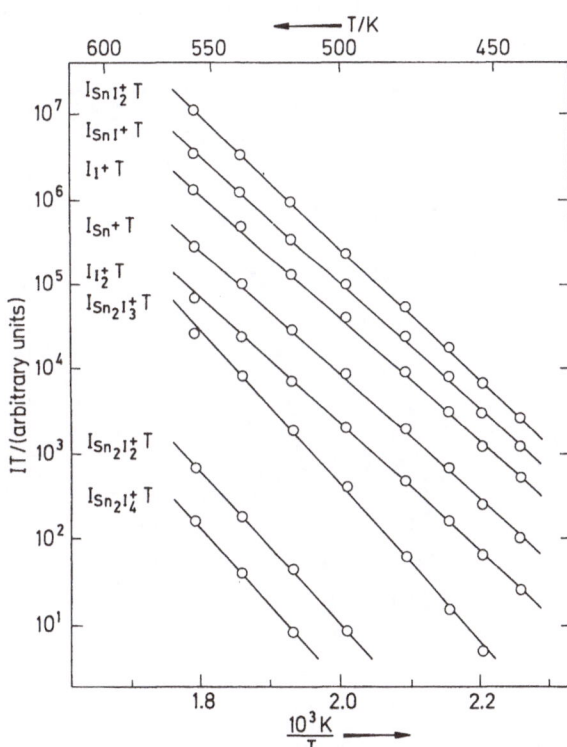

Fig. 2. Temperature dependence of the intensities of the ions observed on vaporizing SnI_2 [52]

Figure 3 shows a typical plot of ionization efficiency curves obtained by electron impact ionization upon vaporizing PbI_2 in the Knudsen cell. Appearance potentials can be obtained from such plots. The different methods used for their determination are given by Franklin et al. [47] and references quoted therein. The electron impact energy has to be calibrated for the determination of appearance potentials. Hilpert and Gingerich [48] describe a calibration technique by vaporizing Hg, Ag, and Dy and using the known ionization potentials of Hg^+, Ag^+, and Dy^+ determined spectroscopically [49]. The appearance potentials obtained in this way for Cs^+, CsI^+ [50]; $Na_2I_2^+$ [51]; SnI_2^+, PbI_2^+, and SnI_4^+ [52] deviate by less than 0.2 eV from the ionization potentials obtained by photoionization and photoelectron spectroscopy as shown in the references given in parenthesis.

However, the deviations may be larger than 0.2 eV due to low lying electronic states or the energy spread of the ionizing electrons [53]. An account of ionization and appearance potentials for different ions is given in Refs. 54–56.

The break of the ionization efficiency curves of PbI^+, PbI_2^+ at 15 to 25 V and the strong increase of the I^+ ion intensity indicate the formation of the I^+ ions by fragmentation from the molecular ions (cf. Fig. 3). Further indications of fragmentation from the shape of ionization efficiency curves are for example given by Scheuring and Weil [57] as well as Odoj and Hilpert [58].

Homologous gaseous species showing similar fragmentation patterns are for example the metal halides. On ionizing metal halide species such as MX_n or

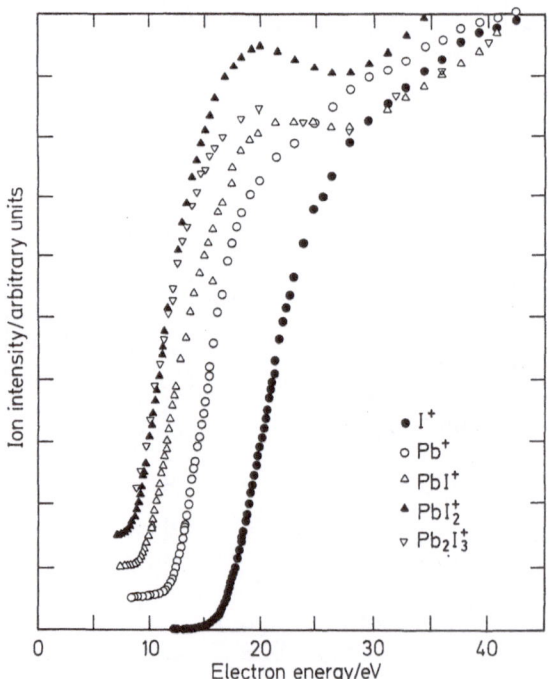

Fig. 3. Ionization efficiency curves of ions observed on vaporizing PbI_2 [485]

$(MX_n)_i$ (n, i = 1, 2, 3, 4) by electron impact, a halide atom is split off preferentially and fragment ions MX_{n-1}^+ or $M_iX_{(nxi)-1}^+$ are formed in the mass spectrum with large relative abundances. This is generally explained by the relative instability of the positively charged molecule-ions generated by simple ionization [59].

An example of the assignment of different ions to neutral precursors on the basis of some of the aforementioned rules is given in Table 1. It shows the results obtained on studying the vaporization of FeI_2 [60]. The assignment is valid for an electron impact energy of 2.3 to 2.6 aJ. Additional information supporting the assignment in Table 1 can be used [60].

If an ion has two different neutral precursors, it is necessary for the partial pressure determination to know the contribution of each precursor to its intensity. The technique used for example to determine the contributions to the Na^+ ion intensity generated by electron impact ionization from the precursors $NaI(g)$ and $NaA(g)$ $(A = SnI_3, PbI_3, FeI_3, DyI_4$ or $ScI_4)$ are described by Hilpert and coworkers [61–64]. The use of a two compartment Knudsen cell can be very useful in connection with this technique [62, 64]. Other techniques used for the quantitative determination of the aforementioned contributions are described by Pittermann and Weil [65] as well as Grimley and Forsman [43]. Techniques for the assignment of ions, of the mass spectra of binary salt systems with a common cation are described by Kapala and Skudlarski [66, 67] and Miller [68].

Modulation of the molecular beam from the Knudsen cell with phase analysis of the ion signal is useful for the assignment of ions to noncondensible gaseous species in the cell [37].

Table 1. Identification of gaseous neutral species ("Δ_sH_T" = Apparent enthalpy of sublimation)

Neutral precursor	Ion	Appearance potential in V	"Δ_sH_T" kJ mol^{-1}	Relative intensity at 725 K
FeI_2	FeI_2^+	9.9	187 ± 5	1
	FeI^+	11.5	183 ± 3	1.6×10^{-1}
	Fe^+	15.2	185 ± 2	2.3×10^{-3}
$(FeI_2)_2$	$Fe_2I_4^+$	9.1	217 ± 8	1.3×10^{-1}
	$Fe_2I_3^+$	10.2	219 ± 4	2.1×10^{-1}
	$Fe_2I_2^+$	12.2	223 ± 2	2.9×10^{-2}
$(FeI_2)_3$	$Fe_3I_6^+$	8.3	258 ± 5	1.4×10^{-3}
	$Fe_3I_5^+$	9.4	263 ± 4	5.0×10^{-3}
	$Fe_3I_4^+$	10.2	257 ± 3	5.2×10^{-3}
FeI_3	FeI_3^+	9.7	212 ± 6	1.6×10^{-2}
$(FeI_3)_2$ or Fe_2I_5	$Fe_2I_5^+$	7.7	224 ± 7	3.2×10^{-3}
I	I^+	10.5	166 ± 4	4.5×10^{-2}
I_2	I_2^+	9.4	175 ± 6	5.2×10^{-2}

2.4 Partial Pressures

The partial pressures p_i of the species i at the temperature T result from the relation

$$p_i = k \frac{1}{\sigma_i} I_i T, \tag{2}$$

where k is the pressure calibration factor and σ_i the ionization cross section of i. The intensity, I_i, representing the species, i, is obtained from the equation

$$I_i = \sum_1 \frac{100}{\gamma_{i,l} A_{i,l}} I_{i,l} \tag{3}$$

$A_{i,l}$, $\gamma_{i,l}$, and $I_{i,l}$ are the isotopic abundances in percent, the multiplier gains, and the intensities of the ions, l, which are generated by ionization from the neutral species, i. If an ion has more than one neutral precursor, the quantitative contribution from each neutral precursor to the measured total intensity of this ion has to be determined (cf. Sect. 2.3); $I_{i,l}$ in Eq. (3) then denotes the fraction of this total ion intensity originating in the neutral precursor, i.

Knowledge of multiplier gains is necessary for the evaluation, if an electron multiplier is used for the ion detection, and if the multiplier output current is measured. Due to the mass discrimination of the multiplier it would be ideal to determine this gain for each ion considered. The determination of multiplier gains by the use of a Faraday cup is, however, often impossible since low ion currents cannot be measured with this device. An ion counting system showing practically no mass discrimination is used in our laboratory. The intensities of all ions of the same measurement are converted to counting rates independent of whether they are determined with the Faraday cup or with the multiplier using the output current. The suitable conversion factors can be easily determined. The (multiplier output current)/(counting rate) ratio is practically proportional to the multiplier gain γ. The values for this ratio obtained so far generally decrease if the mass of the ions increases. The stated decrease for molecular ions is less than that predicted by the law generally used for the estimation of the mass discrimination of a multiplier $\gamma_M \approx M^{-1/2}$ [15], where M is the mass of the ion. A formula for the estimation of the multiplier gain for molecules is given by Goldfinger [22]. Additional information on the mass discrimination of the multiplier is for example given in the review by Gingerich [21] and the references quoted therein.

The maximum ionization cross sections of the atoms were computed by Otvos and Stevenson [69] as well as Mann [70, 71]. The cross sections determined by Mann are available for different ionization energies between 0 and 220 V with steps of 1 eV [72]. Mann's data are generally used in Knudsen effusion mass spectrometry. According to Drowart [11], Mann's maximum values agree within some 20 to 50% with the experimental ones; the agreement is generally improved if the experimental and Mann's cross sections are compared at the same energy. Experimental and theoretical data for the electron impact ionization cross section of the elements are compiled in a recent paper by Tawara and Kato [73]. A comprehensive recent review of Märk and Dann [74] on the subject of electron impact ionization with emphasis on cross sections should also be mentioned.

Ionization cross sections of molecules, σ_M, can be obtained from those of the component atoms, σ_A, by applying the additivity rule, $\sigma_M = \Sigma \sigma_A$, postulated by Otvos and Stevenson [69]. This rule was for example used in Refs. 75, 76, 77 for metal halide dimers and polymers. Other workers [78–80] used a somewhat smaller ionization cross section than that resulting from the additivity rule by employing the equation: $\sigma_M = 0.75 \Sigma \sigma_A$. Further equations for the estimation of ionization cross sections of comparatively large polymers are given in Refs. 81–85. The selection of suitable ionization cross sections for Knudsen effusion mass spectrometry is critically discussed in the recent reviews of Drowart [11] and Hastie [37].

The pressure calibration factor k takes into account losses of species caused for example by the intensity distribution of the molecular beam from the Knudsen cell or the transmission of the ion source and the analyzer. Three different methods are generally used [86]:

1. A substance, A, with known vapor pressure p_A, is vaporized in the cell, k results from the equation:

$$k = \sigma_A\, p_A / (I_A T). \tag{4}$$

2. A pressure dependent equilibrium such as

$$A_2(g) \leftrightarrows 2A(g) \tag{5}$$

present in the vapor to be investigated can be used if the equilibrium constant $K_p = p_A^2 / p_{A_2}$ is known or can be estimated. The constant k is then obtained by the equation

$$k = \frac{\sigma_A^2}{\sigma_{A_2}} \frac{I_{A_2}}{I_A^2} \frac{1}{T} K_p. \tag{6}$$

3. The method of quantitative vaporization is based on the Hertz Knudsen equation [87]. The mass loss rate dm/dt of a substance A, vaporized at the temperature T in the Knudsen cell, is determined by weighing, in addition to the ion intensity measurement. k is obtained by the equation

$$k = \frac{\sigma_A}{I_A T} \frac{1}{qa} \sqrt{\frac{2\pi\,RT}{M_A}} \frac{dm}{dt}. \tag{7}$$

M is the molecular weight and, qa, is the area of the effusion orifice corrected by the Clausing factor, a, [87]. An example of the application of this method is given in Ref. 88.

Method 3 is an absolute method for pressure calibration in contrast to the relative methods 1 and 2. A variant of method 3 is possible by the use of a vacuum microbalance (see e.g. Hilpert [51] and Biefeld [78]).

Ratios of ionization cross sections are necessary for the pressure computation according to Eq. (2) since k also depends on the ionization cross section of the reference species. The substance used for the pressure calibration (method 1) should, therefore, also be selected in a way that the uncertainty arising from estimated ionization cross section ratios is minimized. Hilpert et al. [61] for example used NaI(s) instead of Ag(s) for the determination of partial pressures over phases of the NaI-FeI$_2$ system.

Equation (2) can only be used for the pressure determination if there is molecular flow in the effusion orifice of the Knudsen cell. According to Boerboom [30] molecular flow exists as a coarse approximation for p < 3.6/r (p in Pa, r in mm), where p is the pressure in the Knudsen cell and r is the radius of the effusion orifice. The demand for a molecular flow and the effusion orifice diameter thus determines the upper limit of the range for the vapor pressure measurement. Experimental and theoretical studies on the transition range between molecular and hydrodynamic flow were recently reviewed by Wahlbeck [89] and references quoted therein. Thermodynamic studies at the limit of the Knudsen flow region are discussed by Hilpert and Gingerich [80].

2.5 Thermodynamic Properties

Typical reactions studied by Knudsen effusion mass spectrometry are for example the dissociation, the isomolecular exchange, and sublimation reactions:

$$AB(g) \leftrightarrows A(g) + B(g), \tag{8}$$

$$AB(g) + C(g) \leftrightarrows A(g) + BC(g), \tag{9}$$

$$AB(s) \leftrightarrows A(g) + B(s), \tag{10}$$

and

$$A(s) \leftrightarrows A(g). \tag{11}$$

The equilibrium constants K_p as well as the enthalpy and entropy changes for these reactions at the mean temperature T_m are determined from the partial pressures and their temperature dependences by the use of equations:

$$K_p = \Pi[p(i)/p°]^\nu \tag{12}$$

and

$$\log K_p = A/T + B \tag{13}$$

where

$$p° = 101325 \text{ Pa,}$$

$$A = -\Delta_r H°_{T_m}/(2.303 \text{ R}) \tag{14}$$

and

$$B = \Delta_r S^\circ_{T_m} / (2.303 \ R). \tag{15}$$

The coefficients A and B for this second-law evaluation are obtained by a least squares computation.

Enthalpy changes can additionally be evaluated according to the third-law method

$$\Delta_r H^\circ_{298} = - T \left[R \ln K_p + \Delta \frac{G^\circ_T - H^\circ_{298}}{T} \right]; \tag{16}$$

where

$$\Delta \frac{G^\circ_T - H^\circ_{298}}{T} = \Delta \frac{H^\circ_T - H^\circ_{298}}{T} - \Delta S^\circ_T \tag{17}$$

is the change of the Gibbs energy function for the reaction considered. This change can be obtained from the Gibbs energy functions of the reacting materials. They can be obtained from compilations of thermodynamic data by the JANAF group [90], by Barin and Knacke [91], by Barin et al. [92], Glushko et al. [93], Hultgren et al. [94], Kelley [95], and Kelley and King [96], as well as by Refs. 97–103 for oxides (cf. Sect. 4.1). Rules for their estimation for condensed phases are given by Kubaschewski and Alcock [104] as well as Lewis et al. [105]. Gibbs energy functions for gaseous species can be computed from molecular parameters by the method of statistical thermodynamics. Molecular parameters of numerous molecules are for example tabulated by Refs. 106–108. They can be estimated by the use of the F,G matrix computations (see e.g. Refs. 52, 109) and the use of the rules for force constants or frequencies by Guggenheimer [110] or Gordy [111]. Interatomic distances can be estimated from the atomic radii. Frurip and Blander [112] developed the dimensional model for the estimation of Gibbs energy functions for gaseous molecules.

In addition to the determination of thermodynamic data for homogeneous and heterogeneous equilibria of the type Eqs. (8–11), important thermodynamic properties of condensed phases can be evaluated from gas phase studies.

Thermodynamic data for the formation of solid compounds can for example be evaluated by thermodynamic cycles from the data of reactions such as in Eq. (10) (see for example Refs. 58, 113, 114).

The evaluation of mixing properties of melts and solid solutions from measured ion intensities and temperatures are described in the reviews by Chatillon et al. [12], Sidorov and Korobov [115], as well as Raychaudhuri and Stafford [13] and the references quoted in these articles. The chemical activities, or activity coefficients, can be obtained from the partial pressures of the mixture components. The pressure calibration constant (see Sect. 2.4) has to be determined in this case. The pressure calibration can be avoided by the use of the ion intensity ratio method described by Lyubimov et al. [116], Belton and Fruehan [117], as well as Neckel and Wagner [118, 119]. The Gibbs-Duhem relation is used to obtain the activity coefficient f_A of the component A in the mixture {xA + (1 − x)B} from the ion intensity ratio $I(A^+)/I(B^+)$ by the equation

$$\ln f_A = - \int_{x=1}^{x} (1 - x) \, d\ln[x \, I(B^+)/(1 - x) \, I(A^+)]. \tag{18}$$

Chemical activities are obtained by a similar equation. The ion intensity ratio method has been used successfully for the study of numerous systems (see e.g. Sect. 3.4). However, no consistency check of the activities by the Gibbs-Duhem relation is possible since this relation has to be used to obtain the equations necessary for the evaluation according to the ion intensity ratio method.

Phase boundaries can be determined from the ion intensities by the use of the phase rule as reported for example by Hilpert et al. [120] for the Al-Ni system and Chastel and Bergman [121].

2.6 Instrumental

2.6.1 General Aspects

Different mass spectrometer-Knudsen cell systems are used at various laboratories. Single-focusing magnetic sector field or quadrupole mass spectrometers are used in most cases for the analysis of the molecular beam.

Magnetic sector field mass spectrometers were used first by Chupka and Inghram [6] in Knudsen effusion mass spectrometry. Subsequently many other groups (see e.g. Refs. 122–124) used these instruments. Figure 1 in Sect. 2.1 shows one of the commercially available mass spectrometer – Knudsen cell systems employed at our laboratory. The system containing a single-focusing mass spectrometer of the type CH5 was supplied by Varian MAT, now Finnigan MAT, Bremen, Federal Republic of Germany, and has been modified substantially by us. An ion counting system [125] was mounted and special baking procedures [86] were developed thereby increasing the sensitivity of the instrument. By this means, the determination of partial pressures in the Knudsen cell down to 10^{-7} Pa became possible. The sensitivity could be further improved [126] by the development of a special ion source [127]. Other electron impact ion sources used in Knudsen effusion mass spectrometry are for example described in Refs. 124, 128–130.

A fast and precise mass adjustment is possible by the use of a Hall probe and a mass programmer. This is very useful for the study of metal halides with their complex mass spectra [60] and the study of gas phase equilibria with very different partial pressures between the reacting species [80].

A new, completely computer controlled mass spectrometer – Knudsen cell system has been recently developed by Finnigan MAT, Bremen, Federal Republic of Germany, using a single focusing magnetic sector field instrument. This system combines high resolution with high sensitivity for the vapor pressure measurement due to improvements of the ion optics.

Hastie [131] coupled for the first time a quadrupole mass spectrometer with a Knudsen cell. One of the quadrupole mass spectrometer – Knudsen cell systems used at our laboratory is shown in Fig. 4. The system has been developed to study small alkali metal clusters under equilibrium conditions (see Sect. 3.2). Broad-band photoionization by a 1 kW Hg/Xe lamp is used for the first time in Knudsen effusion mass spectrometry to reduce fragmentation. Other quadrupole mass spectrometer – Knudsen cell systems have for example been developed by Hilpert [132, 133], Fraser and Rammensee [134], Plante [135], Ono et al. [136], Kematick et al. [137], and Edwards et al. [138]. Cryogenic pumping is used in the device by Hilpert to reduce mercury background ion intensities for the study of amalgams [132]. The instruments described in Refs. 134, 135 use a chopper to modulate the molecular beam from the Knudsen cell. Interfering background ion intensities can, thereby, be subtracted. The apparatus developed by the authors of Refs. 137, 138 renders possible the simultaneous application of Knudsen effusion mass spectrometry and the mass-

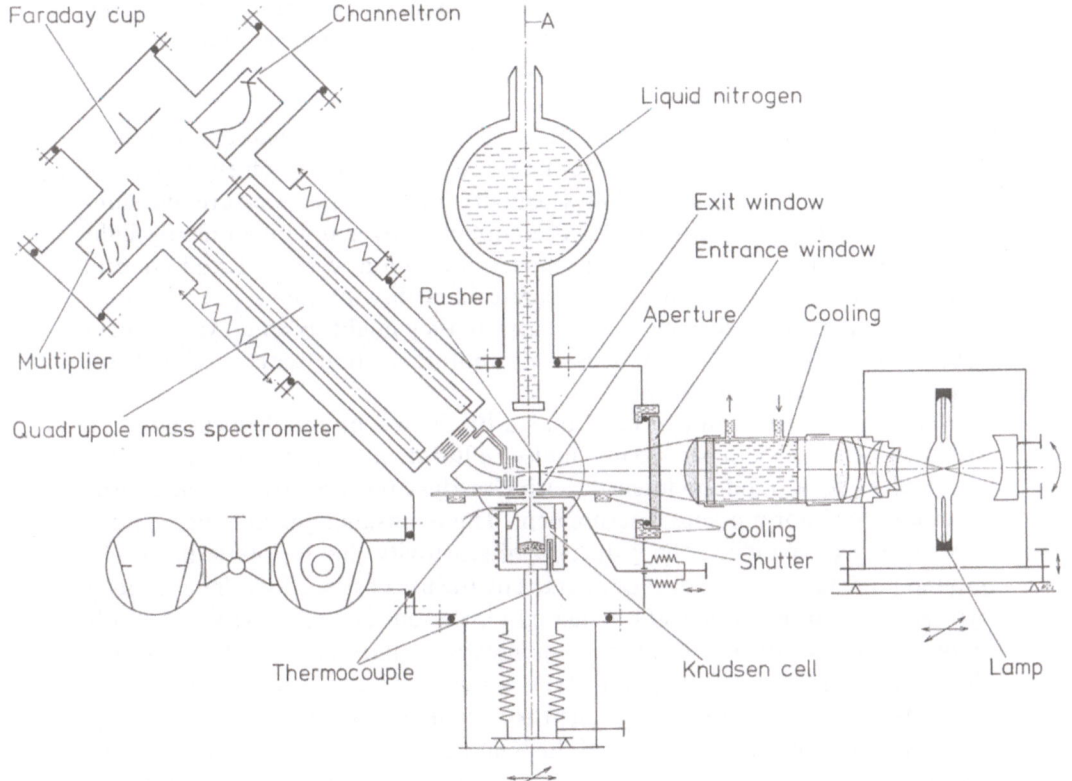

Fig. 4. Schematic representation of a Knudsen cell-quadrupole mass spectrometer system using broad-band photoionization (The entrance window, the optical system, and the lamp has to be rotated on the axis A by 90°)

loss Knudsen effusion method using a microbalance [137] or target collection of the vapor [138].

Quadrupole mass spectrometers generally use a constant radio frequency voltage superimposed onto a dc voltage for the mass separation. The mass range of such instruments increases and its transmission decreases if this frequency is reduced. According to our experience similar sensitivities for the vapor pressure measurement can be obtained by using a quadrupole mass spectrometer with a mass range of up to about 300 and the CH5 mass spectrometer without modification. Quadrupole mass spectrometers are disadvantageous if species with high masses and very low partial pressure have to be investigated, as was for example the case for the study of $(CsI)_4$ [109] or $Au_3(g)$ [80]. Moreover, quadrupole mass spectrometers generally show a mass-dependent transmission, which can, however, be corrected [139]. Quadrupole mass spectrometers are well suited for investigations in the range up to about 200 or 300 amu. In addition to their comparatively low price, it may also be advantageous that, after calibration, masses can be adjusted with high reproducibility and long-time stability without measuring the ion intensities. The long-time stability of the mass adjustment for a magnetic type sector field mass spectrometer with a Hall probe is less than that of a quadrupole. It is, however, sufficient for its use in Knudsen effusion mass spectrometry [86].

In addition to magnetic type sector field and quadrupole mass spectrometers double-focusing instruments (see e.g. Refs. 130, 140, 141) and time-of-flight mass spectrometers (see e.g. Refs. 142, 143) were also used in Knudsen effusion mass spectrometry. A comparatively low sensitivity for the vapor pressure measurement in the Knudsen cell has so far been obtained by coupling a time-of-flight mass spectrometer using electron impact ionization with a Knudsen cell. Time-of-flight mass spectrometers operate in a discontinuous pulsed mode. Therefore, and due to their high transmission, they might by employed in mass spectrometer – Knudsen cell systems using photoionization (see e.g. Ref. 144 and References quoted therein) by lasers.

High mass resolution can be obtained by the use of double-focusing mass spectrometers. This can be used in Knudsen effusion mass spectrometry to avoid interfering background ion intensities which reduce the sensitivity of the instrument for the vapor pressure measurement. Time-consuming baking procedures can thereby be avoided. The reduction of sensitivity due to interfering background ion intensities can also be avoided by the use of single-focusing magnetic type sector field instruments at increased resolution. The sensitivity at a resolution of $m/\Delta m = 1000$ of the instrument integrated in the aforementioned newly developed Knudsen cell-magnetic type sector field mass spectrometer and a standard double-focusing instrument is roughly comparable.

A valve between the chambers of the Knudsen cell and the ion source is very advantageous according to our experience. Special constructions are available (see e.g. Refs. 86, 126), which practically do not lead to a decrease of the sensitivity due to the slight increase of the distance between ion source and Knudsen cell. The valve renders possible a fast change of the material in the cell without change of the pressure calibration constant by more than 5 to 15 percent maximum. No baking procedure is necessary after the change of the sample and all parameters of the ion source can be kept constant during the sample change.

Gomez et al. [145] describe skimmer diaphragms to define the molecular beam originating in the Knudsen cell for the analysis thereby increasing the potential for O_2 and N_2 partial pressure measurements in the cell.

2.6.2 Knudsen Cells

The investigations can be carried out with very different Knudsen cells depending on the particular target of the investigation. Some of the cells are shown in Fig. 5. They are for example made of tungsten or molybdenum. Linings made of ceramics [120] or rhenium [58] can be used to avoid chemical interactions between the sample and the cell material. The generation of a protecting SiC layer on the inner walls of a graphite Knudsen cell is described by Riekert et al. [146]. Two-compartment Knudsen cells [62, 147] are used if vapor species formed from two different materials with very different vapor pressures have to be investigated. They are also very useful for the qualitative and quantitative assignment of ions to their neutral precursors (cf. Sect. 2.3). Moreover, they can be used to superheat vapors from the sample in compartment B. Compartment A contains no sample material in this case. Compartment B is replaced by a gas inlet system if gases are involved in the chemical reactions to be studied. The potential of short-term cooling for comparatively large Knudsen cells [132] speeds up the temperature change at low temperatures. A Knudsen cell completely avoiding chemical interactions with the air or moisture upon changing reactive samples is reported by Hilpert [127].

Multiple Knudsen cells for the determination of chemical activities have been reported by Chatillon and coworkers [148–151], Fraser and Rammensee [134], Zimmermann [152], Paulaitis and Eckert [153], and Hackworth et al. [154]. Multiple cells have practically the same temperature. One cell contains a reference material, the other for instance samples of different compositions. A comparison of the vapor in the different cells is generally rendered possible by their displacement from outside so that the molecular beams originating in the different cells can be analyzed successively. The quadrupole mass spectrometer and not the multiple Knudsen cell is displaced in the Knudsen cell-mass spectrometer system described by Ref. 153. No displacement of the mass spectrometer or the cell is necessary if the triple cell by Hackworth et al. [154] is used. Backdiffusion through the orifice has, however, to be considered for the two cells containing sample and reference.

Gupta and Gingerich [155] describe a comparatively simple, directly resistant heated Knudsen cell. Lukas et al. [156] used a Knudsen cell made of quartz glass with a Ta cover.

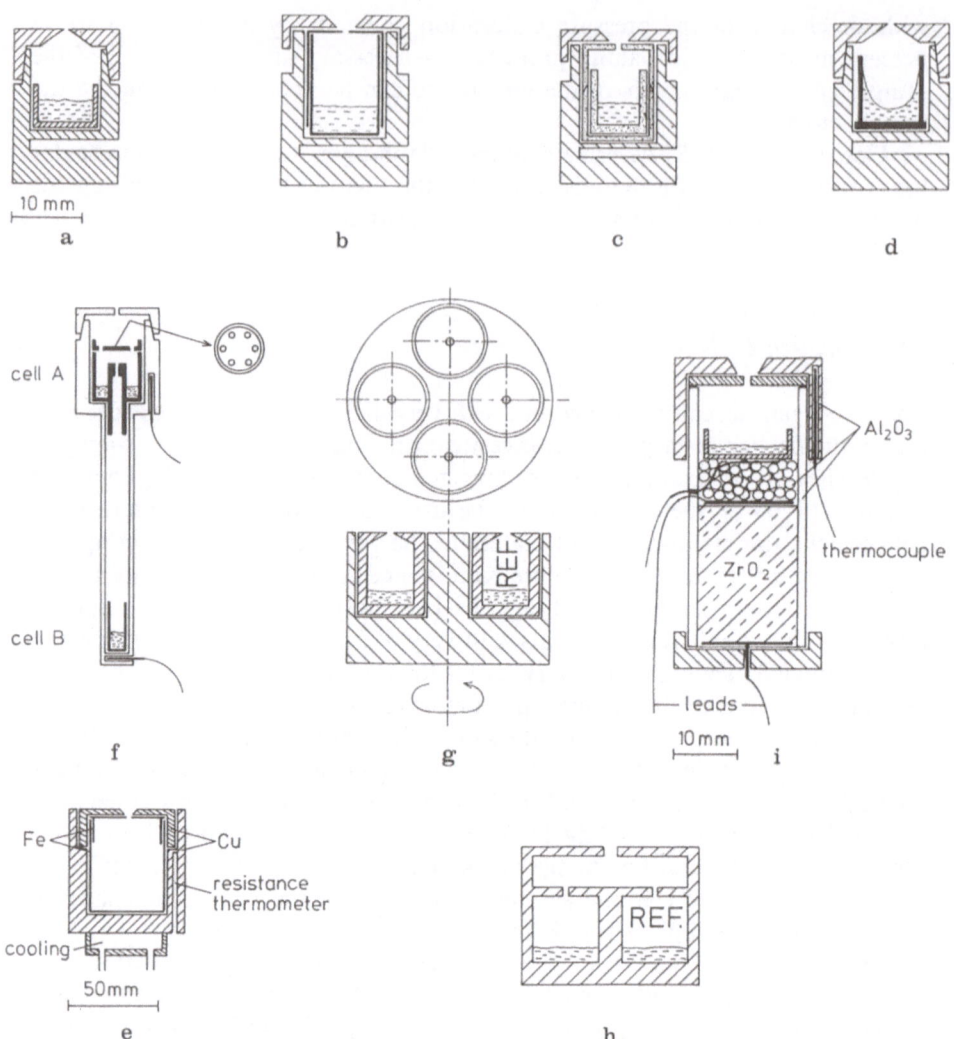

Fig. 5. Standard Knudsen cell (**a**), with lining made of metal (e.g. Re, Ir) (**b**), and of ceramics (e.g. Al_2O_3, ZrO_2) (**c**), "anti-creep" Knudsen cell (**d**), low-temperature Knudsen cell with cooling (**e**), two-compartment cell (**f**), multiple cell (**g**), triple cell (**h**), and electrochemical Knudsen cell for oxygen injection from non-stoichiometric zirconia (**i**)

Electrochemical Knudsen cells were reported first in Refs. 157, 158. An interesting development is Knudsen cells rendering possible electrochemical control of the oxygen pressure in the cell [145, 159, 160]. Knife-edged and cylindrical effusion orifices are generally used in Knudsen effusion mass spectrometry. Cylindrical effusion orifices maintain intensity along the orifice axis, thereby rendering possible a high sensitivity of the system at reduced effusion rate, due to the comparatively low transmission probability of such orifices given by their Clausing factor (for more details see Ref. 89).

The effect of surface diffusion on the walls in and near the effusion orifice of a Knudsen cell and its detection is discussed by Chatillon et al. [161] (see also Ref. 89).

Knudsen cells can be heated by electron bombardment (see e.g. Refs. 124, 126), by radio frequency (see e.g. Refs. 133, 162) or by resistance heating (see e.g. Refs. 135, 140, 163).

3 Metals and Alloys

3.1 Introduction and General Aspects

The most abundant species in the equilibrium vapor over condensed metals are generally atoms as shown by Knudsen effusion mass spectrometry. The relative abundance of the homonuclear diatomic molecules was determined to be between 10 and 10^{-5} percent if they are detectable. Exceptions are bismuth and antimony. The dimer partial pressure over liquid bismuth at temperatures below 1000 K exceeds that of the monomer [83, 164, 165]. The tetramer is the most abundant species over liquid antimony [85]. Other polyatomic homonuclear species have so far been observed under equilibrium conditions for some of the alkali metals, as well as the group Ib, IVb, and Vb metals. Particularly large polymers up to Ge_7 and Sn_7 were detected for germanium [84, 166, 167] and tin [168].

In addition to the homonuclear metallic species numerous heteronuclear intermetallic species have also been detected and their bond energies determined as in the case of homonuclear species. The vaporization of the different metals and the bond energies of many gaseous species are reported in the general review by Gingerich [21] emphasizing the results obtained by Knudsen effusion mass spectrometry. Further reviews on metallic and intermetallic gaseous species with the emphasis on mass spectrometry are in Refs. 9, 11, 20, 169, 170. Drowart [172], in addition to mass spectrometry, also considered the results of spectroscopic investigations and discusses the dissociation energies of homonuclear metallic and heteronuclear intermetallic molecules. A general parallelism between the bonds in the alloy and the intermetallic molecules is suggested.

The vapor pressures and the thermodynamic functions of the gaseous atomic species and condensed phases such as enthalpy and entropy increments as well as enthalpies of evaporation are given in the tables by Hultgren et al. [94] for all metals. Data for the polyatomic homonuclear molecules are only given for a few comparatively abundant species. Vargaftik [173] gives a compilation of the monomer and dimer partial pressures over alkali metals. Thermodynamic data on metals and their vapor species are also given in the general compilations of thermodynamic data in Refs. 90–93, 95, 96. Bond energies of gaseous species and molecular parameters are tabulated by Huber and Herzberg [106], Rosen [108] as well as Mizushima [107].

Empirical models have been developed to predict the bond energies of metallic and intermetallic molecules, such as the following: the Pauling model of a polar single bond [174], the valence bond model for certain multiply bonded metallic molecules by Brewer [175] and Gingerich [176], and the macroscopic atom or atomic cell model by Miedema and Gingerich [177].

These models help us to understand the chemistry of vapors composed of different metals and/or metalloids since they lead to predictions regarding the stability of various possible intermetallic molecules. Gingerich [21, 169, 170] discusses the experimental bond energies of diatomic metallic and intermetallic compounds in terms of the different empirical models of bonding. An insight into the nature of the bond is obtained by complex ab initio computations. The interplay of theoretical and experimental investigations by Knudsen effusion mass spectrometry has recently been reported by Gingerich et al. [178]. Knudsen effusion mass spectrometry is a useful method in cluster research, and the results obtained are for example valuable for the preparation of materials, such as thin films, from the vapor [179]. The merits of Knudsen effusion mass spectrometry in cluster research are also elucidated in the recent articles by Hartmann and Weil [171] as well as Gole [4].

Thermodynamic data of alloys, such as mixing properties, can additionally be evaluated from the data obtained for the gaseous species over these alloys. A comprehensive bibliography of the studies by Knudsen effusion mass spectrometry up to the year 1974 is given by Chatillon et al. [12].

An account of the investigations carried out by Knudsen effusion mass spectrometry since about the year 1979 is given in the following sections. The condensed phase in the Knudsen cell, the temperature range of the measurement, and the vapor species identified are tabulated. The given investigations are commented and discussed. The investigations are grouped under those aimed at the study of gaseous homonuclear and heteronuclear species (Sects. 3.2 and 3.3) and those devoted to the study of condensed phases (Sect. 3.4).

3.2 Homonuclear Gaseous Species

Table 2 shows the enthalpies of dissociation of selected homonuclear gaseous species observed during the investigations listed in Table 3. The ion intensities determined are shown in Table 4 for some species as an example. Obviously, they decrease strongly if the size of the molecules increases.

The polyatomic species Li_4, Na_3, Cu_3, Ag_3, and Au_3 were detected for the first time by Knudsen effusion mass spectrometry, and their enthalpies of dissociation were determined [80, 127, 180]. The study of the molecule Li_4 by Wu [180] complements his earlier investigation of Li_2 and Li_3 [223]. The binding energies of alkali metal dimers and polymers obtained by different

Table 2. Enthalpies of dissociation of homonuclear molecules for the reactions
$M_i(g) \rightleftarrows iM(g)$

Molecule	ΔH_o°	Ref.	Molecule	ΔH_o°	Ref.
	kJ mol^{-1}			kJ mol^{-1}	
Li$_4$	326 ± 13	[180]	Cu$_2$	190 ± 5	[209]
Na$_2$	71.8 ± 4.0	[127]	Cu$_3$	298 ± 13	[80]
			Ag$_3$	254 ± 13	[80]
Mg$_2$	11.3 ± 4.2	[182]	Au$_3$	368 ± 13	[80]
Nb$_2$	503 ± 10	[155]	Hg$_2$	8 ± 2	[213]
Cr$_2$	139 ± 5	[126]	Ge$_2$	260.4 ± 7.0	[219]
Pd$_2$	99 ± 15	[206]	Sn$_2$	182.4 ± 6.5	[179]
Pt$_2$	358 ± 15	[198]	Sn$_3$	404 ± 15	[179]

theoretical computations and further results are given in the recent review by
Koutecký [224] on alkali metal clusters. There is an increase of the enthalpy of
dissociation per atom for the reaction $M_i(g) \rightleftarrows iM(g)$ for the Ib group metals on
going from the dimer to the trimer [80]. The same trend is also observed for the
species Li$_n$ (n = 2, 3, 4) and Sn$_n$ (n = 2, 3).

Hilpert and Kath [225] investigated the vaporization of sodium with
a Knudsen cell – mass spectrometer system using broad-band photoionization
to reduce fragmentation. Relative ion intensities of 1 for Na$^+$ and of 9.7×10^{-2}
for Na$_2^+$ were obtained at 500 K if the intensities are referred to the same rate of
ionizing incident photons. Na$_3^+$ and Na$_4^+$ were observed additionally. The high
relative Na$_2^+$ ion intensity obtained by photoionization in comparison to that by
electron impact might be caused by the very low ionization cross section for Na
near the ionization threshold and/or by autoionizing Rydberg states in Na$_2$
[225]. Results obtained on vaporizing lithium and potassium are reported in
Ref. 188. The relative K$^+$ and K$_2^+$ ion intensities obtained by photoionization
behave similar to those described for Na$^+$ and Na$_2^+$ unlike those of Li$^+$ and Li$_2^+$.

Shim and Gingerich [206] as well as Kingcade et al. [219] report on the
study of the molecules Pd$_2$ and Ge$_2$, respectively. Investigations by Knudsen
effusion mass spectrometry as well as all electron ab initio Hartree-Fock and
configuration interaction calculations of low-lying electronic states of these
molecules were carried out. The results of the theoretical and experimental
studies were combined to obtain new values for the enthalpies of dissociation for
Pd$_2$ and Ge$_2$ integrated in Table 2. The new value determined by Gingerich et al.
[179] for the enthalpy of atomization of Sn$_3$ according to the second- and
third-law methods (Table 2) is by more than 70 kJ mol^{-1} smaller than that
reported earlier [21]. No spectroscopic or theoretical studies on the structure of
Sn$_3$ have been carried out so far. A comparison of the second- and tird-law
values support a bent structure.

The enthalpy of dissociation of the molecule Cr$_2$ is of interest since the large
number of recent theoretical computations lead to a variety of results and
controversies (cf. Refs. 126, 226, 227). The experimental value for this enthalpy
by Kant and Strauss [228] has to be considered as very preliminary [126].

Table 3. Mass spectrometric investigations for the study of gaseous metallic and intermetallic species. (The gaseous species are underlined if their enthalpies of formation or dissociation are given)

System in the cell	Temperature range in K	Gaseous species	Ref./Year
Species with alkali and alkali earth metals			
Li	986–1135	Li, Li_2, Li_3, $\underline{Li_4}$	[180]/1983
Li + Mg	905–1075	Li, Li_2, Mg, $\underline{LiMg}$, $\underline{(LiMg)_2}$	[181]/1980
Li + Ca	728–940	Li, Ca, $\underline{LiCa}$	[182]/1983
LiTl	892–1008	Li, Li_2, Tl, $\underline{LiTl}$	[183]/1983
Li + Ge	1369–1413	Li, Li_2, Ge, Ge_2, $\underline{LiGe}$, $\underline{Li_2Ge}$, $\underline{Li_3Ge}$, $\underline{Li_2Ge_2}$, $\underline{Li_4Ge}$	[184]/1981
Li + Pb	1124–1276	Li, Li_2, Pb, Pb_2, $\underline{LiPb}$	[185]/1980
Li + Sb	1180–1344	Li, Li_2, Sb, Sb_2, Sb_3, Sb_4, $\underline{LiSb}$, $\underline{LiSb_2}$, $\underline{Li_2Sb}$, $\underline{Li_2Sb_2}$, $\underline{LiSb_3}$	[186]/1982
Li + Bi, Li + Bi + Pb	1010–1239	Li, Li_2, Pb, Bi, Bi_2, Bi_3, $\underline{LiPb}$, $\underline{LiBi}$, $\underline{PbBi}$, $\underline{LiBi_2}$, $\underline{Li_2Bi}$, $\underline{LiBiPb}$, $\underline{Li_2Bi_2}$	[185]/1980, [187]/1982
	377–670		[147]/1983
Li + Sm	1089–1151	Li, Li_2, Sm, $\underline{LiSm}$	
Li + Eu	820–979	Li, Li_2, Eu, $\underline{LiEu}$	
Li + Tm	1109–1199	Li, Li_2, Tm, $\underline{LiTm}$	
Li + Yb	771–899	Li, Li_2, Yb, $\underline{LiYb}$	
Na	377–670	Na, Na_2, Na_3, $\underline{Na_4}$	[127]/1984, [225]
NaTl	948–976	Na, Na_2, Tl, $\underline{NaTl}$	[183]/1983
Na + Sb	915–1002	Na, Na_2, Sb, Sb_2, Sb_3, Sb_4, $\underline{NaSb}$, $\underline{NaSb_2}$, $\underline{Na_2Sb_4}$, $\underline{Na_6Sb_4}$	[189]/1985, [57]/1985
K + Sb	829–905	K, K_2, Sb, Sb_2, Sb_3, Sb_4, $\underline{K_2Sb_2}$, $\underline{K_2Sb_4}$, $\underline{K_4Sb_4}$	[189]/1985, [57]/1985, [171]

Rb + Au	1551–1713	Rb, Au, Au$_2$, RbAu	[190]/1981
Cs + Au	1551–1713	Cs, Au, Au$_2$, CsAu	[190]/1981, [191]/1979
Cs + Sb	990–1060	Cs, Cs$_2$, Sb, Sb$_2$, Sb$_4$, CsSb$_2$, Cs$_2$Sb, Cs$_2$Sb$_2$, Cs$_2$Sb$_2$, Cs$_2$Sb$_4$, Cs$_6$Sb$_4$	[192]/1982, [193]/1988
Cs + In + Sb	not specified	Cs, Cs$_2$, In, Sb, Sb$_2$, Sb$_4$, Cs$_2$InSb$_3$	[171], [193]/1988
Cs + Sn + Sb		Cs, Cs$_2$, Sn, Sb, Sb$_2$, Sb$_4$, CsSnSb$_3$, Cs$_2$Sn$_2$Sb$_2$	
Cs + Sb + Bi		Cs, Cs$_2$, Bi, Cs$_2$BiSb$_3$, Cs$_2$Bi$_3$Sb, Cs$_2$Bi$_2$Sb$_2$, Cs$_6$BiSb$_3$, Cs$_6$Bi$_2$Sb$_2$	[193]/1988
Cs + Sn + Sb + Bi	not specified	Cs, Cs$_2$, Sn, Sn$_2$, Sb, Sb$_2$, Sb$_4$, Bi, Bi$_2$, CsSnBiSb$_2$, CsSnBi$_2$Sb	[193]/1988
Mg	778–883	Mg, Mg$_2$	[182]/1983
Species with transition metals and no alkali or alkaline earth metals			
Sc + Y + Rh + C	1824–1939	Sc, Y, Rh, ScRh, YRh, and carbides	[194]/1980
Y + La + Ir + Pt + C	1955–2778	Y, La, Ir, Pt, LaPt, YIr, YPt	[195]/1980
Y + Ir + Au + C	1984–2787	Y, Ir, Au, Au$_2$, YIr, YAu, and carbides	[196]/1980
Y + Pd + C	1975–2286	Y, Pd, YPd, YC, YC$_2$, PdC	[197]/1984
Y + Pt + C	2259–2736	Y, Pt, Pt$_2$, YPt, and carbides	[198]/1981
Y + Si	2005–2224	Y, Si, Si$_2$, YSi, YSi$_2$	[199]/1986
Y + Ge	2031–2191	Y, Ge, YGe	
La + Ir + Au + C	2156–2828	La, Ir, LaIr, LaAu, and carbides	[200]/1979
La + Pt + C	1959–2466	La, Pt, LaPt	[201]/1981
Ti + Ir + Pt + C	2476–2817	Ti, Pt, TiPt	[202]/1979
V + Nb + Mo + Au	1824–2050	V, Au, Au$_2$, VAu	[203]/1979
Nb	2545–2677	Nb, Nb$_2$	[155]/1979
Cr	1542–1819	Cr, Cr$_2$	[126]/1987
Cr + Ge	1609–1808	Cr, Ge, Ge$_2$, CrGe, CrGe$_2$	[204]/1989

(Continued)

Table 3. (continued)

System in the cell	Temperature range in K	Gaseous species	Ref./Year
Cr + Sn	1409–1641	Cr, Sn, Sn$_2$, CrSn, CrSn$_2$	[204]/1989
Cr + Pb	1526–1745	Cr, Pb, Pb$_2$, CrPb	[205]/1985
Ni + Ge	1876	Ni, Ge, Ge$_2$, Ge$_3$, Ge$_4$, NiGe, NiGe$_2$	[206]/1984
Pd	1975–2030	Pd, Pd$_2$	[207]/1987
PdSi	2191–2368	Pd, Si, PdSi	[208]/1986
Pd + Ge	1595–1946	Pd, Ge, Ge$_2$, Ge$_3$, Ge$_4$, Ge$_5$, PdGe, PdGe$_2$, PdGe$_3$	
Cu	1238–1569	Cu, Cu$_2$	[209]/1979
Cu	1801–1897	Cu, Cu$_2$, Cu$_3$	[80]/1980
Cu + Si	1881–1980	Cu, Si, Si$_2$, CuSi	[146]/1981
Cu + Tb	1640–2041	Cu, Cu$_2$, Tb, CuTb	[209]/1979
Cu + Dy	1744–1956	Cu, Cu$_2$, Dy, CuDy	
Cu + Ho	1731–2020	Cu, Cu$_2$, Ho, CuHo	
Ag	1402–1599	Ag, Ag$_2$, Ag$_3$	[80]/1980
Ag + Si	1893	Ag, Si, Si$_2$, AgSi	[146]/1981
Au	1780–1985	Au, Au$_2$, Au$_3$	[80]/1980
Au + Si, Au + Si + Sc + Cu + Al impurity	1636–2096	Au, Au$_2$, Si, Al, AuSi, AuAl, Au$_2$Si, AuSi$_2$	[210]/1979
Au + Cu + Ge	1363–1921	Au, Au$_2$, Cu, Ge, Ge$_2$, Ge$_3$, Ge$_4$, AuCu, AuGe, Au$_2$Ge, AuGe$_2$, Au$_2$Ge$_2$, AuGe$_3$, AuGe$_4$	[211]/1979
Au + Sn		Au, Au$_2$, Sn, Sn$_2$, AuSn, AuSn$_2$, Au$_2$Sn, Au$_2$Sn$_2$, AuSn$_3$	[212]/1986
Hg		Hg, Hg$_2$	[213]/1982

Species made of group IIIb, IVb, Vb, VIb metals or metalloids, lanthanides and actinides

Species	Temp.	Vapor species	Ref./Year
In + Bi	1093–1260	In, Bi, Bi$_2$, Bi$_3$, Bi$_4$, InBi, InBi$_2$	[214]/1981
InTe	1106	Te$_2$, In$_2$Te, In$_2$Te$_2$	[215]/1984
In + AlAs	1483–1573	In, As$_2$, InAs	[216]/1982
Tl + Sb	1055–1127	Tl, Sb, Sb$_2$, Sb$_4$, TlSb, TlSb$_2$	[217]/1981
Tl + Sb	1127–1200	Tl, Tl$_2$, Sb, Sb$_2$, TlSb	[218]/1980
Sc + Ge, Pd + Ge, Pd + Ge	1427–2178	Ge, Ge$_2$	[219]/1986
Au + Sn	1371–1922	Sn, Sn$_2$, Sn$_3$	[179]/1986
Cu + Ag + Au + Sn	1400–1855	Sn, Sn$_2$, Sn$_3$	
Sn	1379–1651	Sn, Sn$_2$, Sn$_3$	
Pb + Sb	880–1030	Pb, Sb, Sb$_2$, PbSb	[220]/1981
Se	463–483	Se$_i$ (i = 2–8)	[221]/1982
Se	380–480	Se, Se$_i$ (i = 2–9)	[222]/1986
Te	590–680	Te$_i$ (i = 2–7)	

Table 4. Relative ion intensities $I(M_n^+)$ obtained with an ionizing electron energy E on vaporizing different metals M at the temperature T in a Knudsen cell

M	T/K	E/eV	$I(M^+)$	$I(M_2^+)$	$I(M_3^+)$	$I(M_4^+)$	Ref.
Li	1100	a	1	5.1×10^{-2}	5.5×10^{-5}	8.5×10^{-6}	[180]
Na	606	21	1	1.5×10^{-2}	3×10^{-7}		[127]
Nb	2578	23	1	1.5×10^{-4}			[155]
Cr	1666	60	1	5.1×10^{-6}			[126]
Cu	1413	21	1	3.6×10^{-4}			[209]

a 0.5 V above the appearance potential of Li^+, Li_2^+, Li_3^+; for Li_4^+ not specified

Extensive measurements were carried out by Hilpert and Ruthardt [126] to determine the enthalpy of dissociation of Cr_2 with low uncertainty according to the second-law method since the thermal population of low lying electronic states is assumed for this molecule and no experimental data regarding these states are available to date. The measurements carried out with two different Knudsen cell-mass spectrometer systems showed excellent agreement, and the value obtained according to the second-law method is given in Table 2. The existence of low-lying electronic states for Cr_2 is indirectly inferred from the enthalpy values obtained according to the second- and third-law methods [126]. The appearance potential of Cr_2^+ was determined for the first time to be 6.4 ± 0.2 V. The second-law determination of the enthalpy of dissociation under reliable conditions and the determination of the appearance potential of Cr_2^+ was made possible by the large sensitivity for the vapor pressure determination in the Knudsen cell.

A comparison between the two studies on Cr_2 carried out so far by the use of Knudsen effusion mass spectrometry, which demonstrates also the difference in sensitivity, is rendered possible by Fig. 6 showing the different temperature ranges of investigation [126].

The high sensitivity of the Knudsen cell-mass spectrometer system was, for example, also used to determine the enthalpy of dissociation of Cu_2 according to the second-law method with high accuracy and precision Hilpert [209]. The enthalpy of dissociation shown in Table 2 is obtained by taking into account the second- and third-law values agreeing well.

Comparatively large homonuclear gaseous species were determined by Grimley et al. [221] and Viswanathan et al. [222] on vaporizing selenium or tellurium under equilibrium conditions (see Table 3). Appearance potentials and ionization efficiency curves are given by Viswanathan et al. for the ions Se_i^+ ($i = 1$ to 9) and Te_i^+ ($i = 1$ to 7). The partial pressures or molar fractions of the different gaseous species listed in Table 3 were evaluated for one temperature each [221, 222]. Grimley et al. used the angular distribution technique (see Sect. 2.3) for the assignment of the ions to their neutral precursors. No evidence of the Se_3 molecule in the equilibrium vapor could be found by the use of this technique. Huang et al. [570] investigated the kinetics of vaporization of molten selenium. They found that the vaporization coefficient is about 0.1 for all species

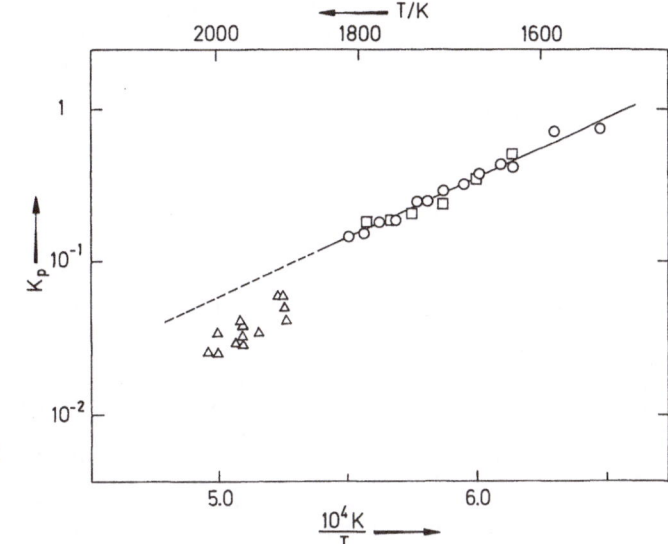

Fig. 6. Temperature dependence of the equilibrium constant for the reaction $2Cr(g) \rightleftarrows Cr_2(g)$ from Ref. 126, ○, □, and Ref. 228, △

$Se_i(g)$ (i = 1 to 8) and that the vapor compositions under equilibrium and non-equilibrium conditions are about the same.

3.3 Heteronuclear Gaseous Species

Tables 5 and 6 give an account of the enthalpies of dissociation of selected heteronuclear gaseous species observed during the investigations listed in Table 3.

Extensive investigations were carried out on the systems (alkali metal)–Sb. The system Li–Sb was investigated by Neubert et al. [186], the systems Na–Sb [57, 189], K–Sb [57, 171, 189], and Cs–Sb [192, 193] by Weil's group. Moreover, the ternary systems composed of Cs and Sb with In, Sn, or Bi [171, 193] and the quarternary system Cs–Sn–Sb–Bi [193] were studied by this group. A variety of different species was identified in the vapor of these systems by careful ionization efficiency measurements carried out by Weil's group. Appearance potentials and enthalpies of dissociation were determined for some of the vapor species. Comparatively small molecules of the type MSb, M_2Sb, MSb_2, M_2, M_2Sb_2; intermediate molecules M_2Sb_4; and clusters of the type M_6Sb_4 and M_4Sb_4 were observed for all or some of the binary systems investigated (cf. Table 3). The investigation of the ternary system Cs–Sb–Bi revealed that Sb can be replaced by Bi in the species $CsSb_4$ and Cs_6Sb_4 (cf. Table 3).

Particularly interesting are the intermetallic species such as Cs_2InSb_3 or $CsSnBiSb_2$ observed in the ternary systems Cs–Sb with In or Sn and the aforementioned quarternary system (cf. Table 3). They are first examples of gaseous complex Zintl compounds with alkali atoms as donor partners [193].

Table 5. Enthalpies of dissociation of heteronuclear diatomic molecules

Molecule	ΔH_o°	Ref.	Molecule	ΔH_o°	Ref.
	kJ mol^{-1}			kJ mol^{-1}	
LiMg	63.6 ± 6.3	[181]	LaIr	573 ± 12	[200]
LiCa	81.2 ± 8.4	[182]	LaPt	496 ± 21	[201]
LiTl	$76. \pm 4$	[183]	TiPt	394 ± 11	[202]
LiGe	202	[184]	VAu	238 ± 12	[203]
LiPb	74.9 ± 8.0	[185]	CrGe	150 ± 7	[204]
LiSb	169 ± 10	[186]	CrSn	139 ± 3	[204]
LiBi	135.7 ± 6.0	[185]	CrPb	102 ± 2	[204]
LiSm	45.3 ± 4.5	[147]	NiGe	278 ± 12	[205]
SiEu	63.2 ± 3.0	[147]	PdSi	257 ± 12	[207]
LiTm	65.1 ± 3.5	[147]	PdGe	252 ± 11	[208]
LiYb	33.4 ± 3.0	[147]	CuSi	217.7 ± 6.1	[146]
NaTl	68 ± 5	[183]	CuTb	187 ± 19	[209]
NaSb	138.4 ± 8	[57]	CuDy	140 ± 19	[209]
RbAu	239 ± 3	[190]	CuHo	139 ± 19	[209]
CsAu	249 ± 3.5	[190]	AgSi	174 ± 10	[146]
ScRh	440.3 ± 10.5	[194]	AuCu	224 ± 5	[211]
YRh	441.8 ± 10.5	[194]	AuSi	301 ± 6	[210]
YIr	453 ± 16	[196]	AuGe	270 ± 5	[211]
YPd	237 ± 15	[197]	AuSn	252.6 ± 7.2	[212]
YPt	470 ± 12	[198]	InAs	198 ± 10	[216]
YAu	304.1 ± 8.2	[196]	InBi	149.9 ± 1.7	[214]
YSi	254 ± 17	[199]	TlSb	123 ± 10	[217]
YGe	275 ± 11	[199]	PbSb	155 ± 10	[220]

Table 6. Enthalpies of dissociation of heteronuclear polyatomic molecules for the dissociation into atoms except for TlSb$_2$

Molecule	ΔH_o°	Ref.	Molecule	ΔH_o°	Ref.
	kJ mol^{-1}			kJ mol^{-1}	
Li$_2$Mg$_2$	381 ± 13	[181]	Au$_2$Si	583 ± 15	[210]
Li$_2$Ge	446.9	[184]	AuGe$_2$	532 ± 10	[211]
LiSb$_2$	472.2 ± 6.8	[186]	Au$_2$Ge	535 ± 12	[211]
LiSb$_3$	698 ± 30	[186]	Au$_2$Ge$_2$	927 ± 14	[211]
Li$_2$Sb	379 ± 10	[186]	AuGe$_3$	897 ± 20	[211]
Li$_2$Sb$_2$	689 ± 12	[186]	AuGe$_4$	1296 ± 30	[211]
LiBi$_2$	367.4 ± 7.0	[187]	AuSn$_2$	483 ± 12	[212]
Li$_2$Bi	326 ± 10	[187]	AuSn$_3$	788 ± 28	[212]
Li$_2$Bi$_2$	583 ± 15	[187]	Au$_2$Sn	541 ± 20	[212]
LiPbBi	341.2 ± 6.0	[187]	Au$_2$Sn$_2$	726 ± 35	[212]
YSi$_2$	800 ± 21	[199]	InBi$_2$	354.2 ± 3.6	[214]
CrGe$_2$	485 ± 11	[204]			
AuSi$_2$	602 ± 15	[210]	TlSb$_2$	390 ± 13[a]	[217]

[a] Reaction of dissociation: TlSb$_2 \rightleftarrows$ Tl + Sb$_2$

Different binding types (covalent, ionic, metallic) are realized in the different gaseous species of the binary, ternary, and quarternary systems mentioned above as discussed by Hartmann and Weil [171] considering also the possible structure of these species.

Systematic investigations of gaseous compounds containing lithium were carried out by Ihle's group [147, 181, 182, 184–187] (see Table 3). The investigations not listed in Table 3 published before 1980 are quoted in Ref. 185. The species LiTl was studied by Scheuring and Weil [183]. The resulting enthalpies of dissociation of the gaseous compounds were discussed on the basis of the Pauling model of a polar single bond. Summarizing considerations [185, 186] on the deviations between the experimental values and those predicted by this model show that the deviations from the experimental data range up to 76% but are in most cases below 20%. General trends are discussed [186], showing for example that the Pauling model fits all the better the lower the atomic number of the element M in the compound LiM. Large deviations of about 70% were observed for LiSm and LiYb in contrast to LiEu and LiTm [147].

The study of the molecules RbAu and CsAu (Busse and Weil [190, 191]) as well as LiTl and NaTl (Scheuring and Weil [183]) was carried out to obtain insights into solid state bond problems. The ionic contribution to the bond was estimated from the experimental enthalpy of dissociation and the covalent contribution according to Pauling [183]. It was concluded that the bonds in RbAu and CsAu are predominantly ionic in contrast to LiAu and NaAu for which the enthalpy of dissociation was obtained by Neubert and Zmbov [571]. This result reflects the change in bond character from metallic to ionic found for the solid and molten compounds MAu (M = Li, Na, Rb, Cs) [183]. By similar studies it could be shown that the ionicity of the NaTl molecule seems to be higher than that of LiTl(g) as indicated by the crystal structures of the corresponding solid compounds [183].

Systematic investigations were carried out by Gingerich's group to study the applicability of different empirical models for the prediction of bond energies and to reveal their limitations. The investigations, given in Table 3, of the diatomic heteronuclear multiply bonded molecules between the platinum-metals and the group III metals [194–198, 200, 201] as well as of the intermetallic molecule TiPt [202] were carried out to test the valence bond model and in some cases also the atomic cell model. In some cases the additional observation of gaseous metal carbides is reported (Table 3). Kingcade and Gingerich [199] as well as Zmbov et al. [220] compared the dissociation enthalpies determined for YSi and YGe [199] as well as PbSb [220] (see Table 5) with those estimated by the Pauling model, which cannot be used for multiply-bonded molecules. The enthalpy of dissociation obtained for LaIr (Table 5) was used to predict the hitherto unknown enthalpy of dissociation of the dimer W_2 as $\Delta_{dso}H_o^\circ = 535 \pm 40 \text{ kJ mol}^{-1}$ [200].

The interplay between theory and experiment is shown by the study of the molecules PdSi and PdGe (see Table 3). Shim et al. [207, 208] investigated these molecules by Knudsen effusion mass spectrometry and additionally determined six electronic states of the molecules by all electron ab initio Hartree-Fock and configuration interaction calculations. The chemical bond was interpreted in terms of donation and backdonation of charge.

Particularly large homonuclear gaseous species Si_n, Ge_n, Sn_n (n = 1 to 7), Pb_i (i = 1 to 4) were observed in the vapor of the corresponding IVb group elements under equilibrium conditions. It may be, therefore, possible that large mixed gaseous compounds can be studied by Knudsen effusion mass spectrometry if binary systems composed of a group IVb element and a further metal are studied. Reactions of the type $M + A_n \rightleftarrows MA_n$ describing the attachment of an atom M to a homonuclear group IVb polymer A_n are for example of interest. The investigation of the mentioned binary systems given in Table 3 was mainly carried out by the groups of Gingerich [199, 205, 207, 208, 210, 211, 212], Ihle [184, 185] and Hilpert [204]. Comparatively large heteronuclear species were observed for the systems Au–Ge [211], Au–Sn [212], Li–Ge [184], and Pd–Ge [208] (cf. Table 3). An account of the enthalpy changes referring to the aforementioned reaction, where M is a Ib group metal, is given in the recent paper by Kingcade and Gingerich [212]. The only heteronuclear species observed on the systematic investigation of the Cr–Ge, Cr–Sn, and Cr–Pb systems [204] are the mixed dimer and the trimer $CrGe_2$ and $CrSn_2$, although comparable Cr and Pb partial pressures were adjusted by the use of a two-compartment Knudsen cell to obtain optimal relative ion intensities of the heteronuclear species.

Extensive measurements were carried out to determine the enthalpy of dissociation of CrGe, $CrGe_2$, CrSn, and CrPb according to the second-law method with high accuracy and precision. The values are given in the Tables 5 and 6. Recently, all electron ab initio HF–CI calculations have been applied by Shim [538] to elucidate the electronic states and the nature of the chemical bond in CrGe. Gibbs energy functions of CrGe were computed from the predicted molecular parameters and a value of $\Delta H_{298}^\circ = 158.8 \pm 0.7 \text{ kJ mol}^{-1}$ resulted for the enthalpy of dissociation of CrGe by a third-law evaluation agreeing well with that by the second-law method $\Delta H_{298}^\circ = 154 \pm 7 \text{ kJ mol}^{-1}$ [204]. The compatibility of the theoretical [591] and experimental [204] studies is demonstrated by the agreement.

The same type of heteronuclear molecules AB and AB_2 were observed on the investigation of the two group IIIb/Vb systems In–Bi and Tl–Sb by Riekert et al. [214] and Balducci et al. [217]. Piacente and Gigli [216] studied the molecule InAs. Riekert et al. [214] compare the enthalpies of dissociation obtained for the molecule InSb with that of other diatomic heteronuclear molecules between the group IIIb and Vb elements. They find a decrease of the enthalpy of dissociation by replacing one of the elements forming these molecules by one of the same group with an increased ordinal. The value obtained by Balducci et al. for the molecule TlSb, not considered by Riekert et al. [214], is in agreement with this rule unlike the enthalpy of dissociation for InAs [216] which is slightly too high.

Srinivasa and Edwards [215] analyzed the gaseous phase over InTe(1) by the isothermal total exhaustion method. They determined the enthalpy change of the reaction $2In_2Te(g) + Te_2(g) \rightleftarrows 2In_2Te_2(g)$ to be $\Delta H^\circ_{298} = -310.9 \pm 0.5 \text{ kJ mol}^{-1}$.

3.4 Vaporization and Thermodynamic Properties of Alloys Obtained from Gas Phase Studies

Table 7 gives an account of the compounds and alloys investigated by Knudsen effusion mass spectrometry since about the year 1979. Alloys and compounds composed of up to four component elements are listed. The condensed phase in the Knudsen cell, its state, the solid compounds present in the given concentration range, the temperature range of the measurement, and the gaseous species observed are given. The investigations on technical multicomponent alloys are reported in Sect. 6. The investigations listed in Table 7 can be grouped under determination of thermodynamic properties of melts, of solid phases, of melts and solid phases, as well as under determination of phase boundaries. Some comments will be given on the investigations according to these items and in the given sequence.

3.4.1 Investigation of Alloys in the Liquid State

The thermodynamic activities of the alloy components are practically determined in all cases if only the liquid phase of the binary system was investigated (see Table 7). In addition, other mixing properties of these melts, such as excess Gibbs energies, heats of mixing or excess entropies were evaluated from the vaporization studies. All these properties were, for example, determined in the study of the Co–Ge melts by Tomiska et al. [246]. Heats of mixing were determined in addition to the thermodynamic activities or Gibbs energies of the melts for the Fe–Ni [243], Co–Cu [142], Ni–Pd [249], Ni–Cu [251], Cu–Ag [253], Cu–Si [254], Cu–Sn [136], Au–Al [256], and Al–Te [258] systems (see Table 7). Partial and integral quantities can be evaluated. The ion intensity ratio method (see Sect. 2.5) was used in the investigation of the melts except for the studies by Levin and Gel'd [245] and Said et al. [258]. Erdély et al. [256] investigated the Au–Al system by the use of two different mass spectrometer – Knudsen cell systems and obtained consistent results. Mass spectrometric Knudsen effusion and solid-state EMF measurements were carried out by Oishi et al. [249] in the study of Ni–Pd melts.

The mixing properties in binary melts depend on the different interactions A–A, B–B, and A–B between the alloy components A and B. Associates of short range order can be formed in the melt. Correlations between the thermodynamic functions and the existence of such associates are discussed for the Cu–Si [254,

Table 7. Mass spectrometric investigations for the determination of thermodynamic properties of alloys and their vaporization. (The solid compounds present in the given concentration range are underlined if the enthalpies of formation were determined)

Condensed phase in the cell	State of condensed phase	Solid compounds	Temperature range in K	Gaseous species	Ref./Year
$xLi + (1 - x)Pb$ $x = 0.05-0.95$	s, l	Li_8Pb_3, Li_3Pb, Li_7Pb_2, $Li_{22}Pb_5$	700–900	Li	[229]/1979
$xMg + (1 - x)Hg$ $x = 0.32-0.79$	s	$MgHg_2$, $MgHg$, Mg_5Hg_3, Mg_2Hg, Mg_5Hg_2, Mg_3Hg	294–711	Mg, Hg	[114]/1984, [132]/1980
Mg_3As_2	s	Mg_3As_2	980–1100	Mg, As_2, As_4	[230]/1985
Mg_3Bi_2	s	Mg_3Bi_2	750–950	Mg, Bi, Bi_2	[230]/1985
$xCa + (1 - x)Hg$ $x = 0.29-0.89$	s	$CaHg_3$, $CaHg_2$, $CaHg$, Ca_3Hg_2, Ca_5Hg_3, Ca_2Hg, Ca_3Hg	317–885	Ca, Hg	[113]/1984, [231]/1980
LaTe	s	LaTe	1670–2250	La, Te, LaTe	[232]/1982
$xTi + (1 - x)Ir$ $x = 0.93, 0.885, 0.796$	s	Ti_3Ir	1480–1770	Ti	[233]/1979
$xTa + (1 - x)Al$ $x = 0.3-0.9$	s	$TaAl_3$, Ta_2Al_3, Ta_2Al, Ta_3Al, Ta_4Al	1321–1713	Al	[234]/1986
$xV + (1 - x)Si$ $x = 0.63-0.89$	s	V_3Si, V_5Si_3, V_6Si_5	1696–2063	V, Si	[235]/1987
$xV + (1 - x)Si$ $x = 0.26-0.82$	s	V_3Si, V_5Si_3, V_6Si_5, VSi_2	1600–2000	V, Si	[236]/1985
$x_1Cr + x_2Ni + (1 - x_1 - x_2)Al$	s	Cr_5Al_8, NiAl, Ni_3Al	1423	Cr, Ni, Al	[237]/1985
$xCr + (1 - x)Cu$ $x = 0.03-0.94$	s, l		1675–1889	Cr, Cu	[238]/1982

xCr + (1 − x)Cu, x = 0.004–0.217	s, l		1523–1673	Cr, Cu	[136]/1984
xCr + (1 − x)Si, x = 0.16–0.66	s	CrSi₂, CrSi, Cr₅Si₃, Cr₃Si	1300–1600	Cr, Si	[239]/1986
WSe₂	s		1290–1720	Se, Se₂	[240]/1982
xFe + (1 − x)Co, x = 0.05–0.95	s, l		1373–1923	Fe, Co	[241]/1981
xFe + (1 − x)Co, x = 0.05–0.90	s		1650	Fe, Co	[242]/1986
xFe + (1 − x)Ni, x = 0.048–0.946	s, l		1373–1923	Fe, Ni	[241]/1981
xFe + (1 − x)Ni, x = 0.09–0.90	l		1873	Fe, Ni	[243]/1985
x_1Fe + x_2Co + (1 − x_1 − x_2)Ni	s, l		1473–1973	Fe, Co, Ni	[134]/1982
xFe + (1 − x)Cu, x = 0.26–0.85	l		1873	Fe, Cu	[142]/1981
xFe + (1 − x)Te, x = 0.58, 0.67	s	FeTe₀.₉	885–1048	Te, Te₂	[244]/1985
xRu + (1 − x)U, x = 0.78, 0.80	s	Ru₃U	1690–2100	U	[138]/1980
xCo + (1 − x)Cu, x = 0.06–0.95	l		1650–1850	Co, Cu	[142]/1981
xCo + (1 − x)Al, x = 0.5–0.9	l		1950	Co, Al	[245]/1979

(Continued)

Table 7. (continued)

Condensed phase in the cell	State of condensed phase	Solid compounds	Temperature range in K	Gaseous species	Ref./Year
$xCo + (1 - x)Ge$ $x = 0.10$–0.95	l		1650–1920	Ge, Ge_2, Co	[246]/1982
$xRh + (1 - x)Pu$ $x = 0.67$–0.75	s	Rh_3Pu, Rh_2Pu	1196–1560	Pu	[247]/1985
Ir_2Pu	s	Ir_2Pu	1673–2185	Pu	[248]/1985
$xNi + (1 - x)Pd$ $x = 0.1$–0.9	l		1750–1900	Ni, Pd	[249]/1986
$xNi + (1 - x)Cu$ $x = 0.14$–0.86	s		1350	Ni, Cu	[250]/1984
$xNi + (1 - x)Cu$ $x = 0.08$–0.86	l		1750	Ni, Cu	[251]/1983
$xNi + (1 - x)Al$ $x = 0.70$–0.92	s, l	$AlNi$, $AlNi_3$	1644–1734	Ni, Al	[120]/1987, [252]/1990
$xPt + (1 - x)Pu$ $x = 0.67$–0.90	s	Pt_5Pu, Pt_4Pu, Pt_3Pu, Pt_2Pu	1226–1462	Pu	[247]/1985
$xCu + (1 - x)Ag$ $x = 0.10$–0.90	l		1350–1550	Cu, Ag	[253]/1979
$xCu + (1 - x)Si$ $x = 0.11$–0.90	l		1500–1900	Cu, Si	[254]/1986
$xCu + (1 - x)Si$ $x = 0.10$–0.89	l		1250–1800	Cu, Si	[140]/1981

xCu + (1 − x)Sn x = 0.10–0.90	1		1500–1680	Cu, Sn	[136]/1984
xAg + (1 − x)Ge x = 0.10–0.93	1		1026–1425	Ag, Ge, Ag₂	[255]/1979
xAu + (1 − x)Al x = 0.15–0.95	1		1320–1740	Au, Al	[256]/1979
ZnSiAs₂, ZnGeAs₂, ZnSnAs₂	s		800–1000	Zn, As₄	[257]/1979
CdSiAs₂, CdGeAs₂, CdSnAs₂	s		595–820	Cd, As₄	[257]/1979
xAl + (1 − x)Te x = 0.37–0.97	1		1190	Al, Te, Te₂	[258]/1981
xIn + (1 − x)Bi x = 0.10–0.90	1		1030–1340	In, Bi₂	[259]/1982
xIn + (1 − x)Te x = 0.4–0.5	s	In₂Te₃	701–909	Te₂, In₂Te	[260]/1986
In₂Te₃	s	In₂Te₃	725–895	Te₂	[261]/1981
xSnTe + (1 − x)PbSe x = 0.25–0.72	s		850–970	SnTe, SnSe, PbTe, PbSe	[262]/1982
TeTm	s		1000–1800	Te, Te₂, Tm, TeTm	[263]/1981

140], In–Bi [259], Au–Al [256], and Al–Te [258] systems. Martin-Garin et al. [255] correlated the thermodynamic properties of the Ag–Ge melt obtained by them with the results on the structure of the melt evaluated from neutron diffraction measurements.

The Cu–Si melts were investigated by two groups (see Table 7). The thermo-dynamic activities obtained are very different. Those by Bergman et al. [254] are supported by literature data not determined by Knudsen effusion mass spectro-metry.

3.4.2 Investigation of Alloys in the Solid State

Equilibria between the gaseous phase and a solid compound, a solid solution or two coexisting solid phases were studied, if the state of the condensed phase investigated by Knudsen effusion mass spectrometry is solid (see Table 7).

Only solid solutions were present under the conditions of the investigation of the Fe–Co [242], Ni–Cu [250], and SnTe–PbSe [262] systems given in Table 7. The thermodynamic activities, excess Gibbs energies, mixing enthalpies and excess entropies were determined by the use of the ion intensity ratio method for the Fe–Co and Ni–Cu systems as described for the melts. The partial pressure of the molecules SnTe, SnSe, PbTe, and PbSe were obtained for different composi-tions of the quasi-binary system SnTe–PbSe using the isothermal evaporation technique.

The solid compounds present at the temperatures of the measurement in the concentration range investigated are given in Table 7. Thermodynamic proper-ties were determined for the equilibria between the gaseous and the condensed phases containing these compounds. Gibbs energies, enthalpies, and entropies of formation can be evaluated from these properties by the use of thermodynamic cycles and ancillary data. The enthalpies and Gibbs energies of formation obtained are listed in Table 8. Entropies of formation were additionally deter-mined for the given compounds of the Mg–Hg and Ca–Hg systems [113, 114], the compound Ti_3Ir [233], as well as the actinide/noble-metal compounds [138, 247, 248] also given in this table. The enthalpy of formation obtained by Hilpert [113, 114] for the compounds MgHg and CaHg is discussed and compared with those predicted by different models [264]. The difficulties of the investigation of gas/solid equilibria involving solid compounds and the procedure to obtain reliable results are for example described in Ref. 231.

The V–Si system was investigated by Myers and Kematick [235] as well as Storms and Myers [236]. The results obtained by the two groups agreed reasonably well. Those obtained for the enthalpies of formation of V_3Si and V_5Si_3 are given in Table 8. The comparatively rare property of two congruently vaporizing compositions is reported by Myers and Kematick for the V–Si system. One composition is that of V_5Si_3 at about 2050 K and above, the second is within the range of stoichiometry of the phase V_3Si at 1900 K and below. Between the two temperatures both phases will exhibit a convergent congru-ently vaporizing composition within some range of temperatures. Studies ana-

Table 8. Enthalpies of formation, $\Delta_f H^\circ_{298}/(m + n)$, and Gibbs energies of formation, $\Delta_f G^\circ_T/(m + n)$, at the temperature, T, of different solid compounds, $A_m E_n$, determined by the investigations listed in Table 7

Compound	$-\Delta_f H^\circ_{298}/(m + n)$	T^a/K	$-\Delta_f G^\circ_T/(m + n)$	Ref.
	kJ mol^{-1}		kJ mol^{-1}	
$MgHg_2$	20.2 ± 1.4	356	18.4 ± 0.3	[114]
$MgHg$	28.2 ± 0.8	550	25.3 ± 0.3	[114]
Mg_5Hg_3	27.1 ± 0.9	550	23.2 ± 0.3	[114]
Mg_2Hg	26.6 ± 0.9	550	22.3 ± 0.3	[114]
Mg_5Hg_2	23.7 ± 1.0	550	19.8 ± 0.3	[114]
Mg_3Hg	21.1 ± 1.0	550	17.5 ± 0.3	[114]
Mg_3As_2	87.8 ± 2.1			[230]
Mg_3Bi_2	26.3 ± 2.9			[230]
$CaHg_3$	45.0 ± 2.7	713	39.8 ± 0.4	[113]
$CaHg_2$	56.3 ± 3.3	713	47.4 ± 0.4	[113]
$CaHg$	56.8 ± 4.6	713	49.8 ± 0.4	[113]
Ca_3Hg_2	47.9 ± 5.5	713	42.3 ± 0.4	[113]
Ca_5Hg_3	45.1 ± 5.7	713	40.1 ± 0.4	[113]
Ca_2Hg	40.0 ± 6.1	713	36.2 ± 0.4	[113]
Ca_3Hg	29.6 ± 5.0	713	28.0 ± 0.4	[113]
$LaTe$	139.6 ± 5.0			[232]
Ti_3Ir	92 ± 15	1623	68.6 ± 8.4	[233]
$TaAl_3$	23.8 ± 2.5			[234]
Ta_2Al_3	18.9 ± 2.5			[234]
Ta_2Al	11.4 ± 2.5			[234]
Ta_3Al	9.3 ± 2.5			[234]
Ta_4Al	8.7 ± 2.5			[234]
V_3Si	44.6	1650	34.3	[235]
	44.8	1650	36.6	[236]
V_5Si_3	52.6	1650	45.6	[235]
	52.5	1650	45.5	[236]
V_6Si_5	49.3	1650	43.7	[236]
VSi_2	40.7	1650	35.3	[236]
Cr_3Si	28.2	1500	27.3	[239]
Cr_5Si_3	28.1	1500	31.0	[239]
$CrSi$	26.6	1500	30.1	[239]
$CrSi_2$	27.9	1500	25.5	[239]
$FeTe_{0.9}$	15.0	298	15.5	[244]
Ru_3U	12.3 ± 5.5	1896	3.19	[138]
Rh_3Pu	49.5 ± 1.9	1725	30.4	[247]
Rh_2Pu	59.8 ± 3.1	1519	44.6	[247]
Ir_2Pu	65.1	1929	64.9	[248]
Ni_3Al	38.3 ± 5.0	1600	35.1 ± 1.0	[252]
Pt_5Pu	44.1 ± 1.3	1614	26.9	[247]
Pt_4Pu	49.0 ± 1.8	1596	32.0	[247]
Pt_3Pu	54.8 ± 1.6	1579	37.7	[247]
Pt_2Pu	63.3 ± 2.5	1629	45.5	[247]

^{a} Mean temperature of the measurements except for $FeTe_{0.9}$

logous to the V–Si system were carried out by Myers et al. [239] for the Cr–Si system. One congruently vaporizing composition within the range of stoichiometry of the $CrSi_2$ phase was found for this system.

The vaporization studies of the actinide/noble-metal compounds Ru_3U, Ir_2Pu as well as those of the Rh–Pu and Pt–Pu systems (see Table 7) were carried out by an apparatus combining Knudsen effusion mass spectrometry and target collection. Target collection with radioanalysis of the exposed targets was generally used for the vapor pressure measurements of the alloys containing radioactive nuclides. Peterson and Starzynski [248] compared the Gibbs energies of formation of various actinide/noble-metal intermetallics obtained by different methods. The values obtained for the compound Rh_3U by the use of the aforementioned apparatus and by the EMF method are inconsistent.

The incongruent vaporization of the compounds of the Ta–Al system and of WSe_2 (see Table 7) was studied by an apparatus combining Knudsen effusion mass spectrometry and mass-loss measurements. Schiffman et al. [240] obtained an enthalpy change of $\Delta H_{298}^\circ = 311 \pm 3 \, kJ \, mol^{-1}$ for the reaction of evaporation $1/2 \, WSe_2(s) \rightleftarrows 1/2 \, W(s) + Se(g)$.

Two groups (Srinivasa and Edwards [260], Belousov et al. [261]) investigated the evaporation of the compound In_2Te_3 as indicated in Table 7. Srinivasa and Edwards used essentially the computer-automated simultaneous Knudsen-effusion and torsion-effusion method as well as Knudsen effusion mass spectrometry unlike Belousov et al. who only used the latter method. The vapor pressures by Srinivasa and Edwards are by a factor of 30 lower than those by Belousov et al. The incongruent vaporization process of In_2Te_3 determined by the two groups was also different. The vaporization process by Srinivasa and Edwards, according to which $Te_2(g)$ and a solid solution with $x_{In} = 0.42$ is formed, was determined by the additional use of X-ray analysis of the residue from the evaporation experiments. The solid solution also vaporizes incongruently producing $InTe(s)$ as well as $Te_2(g)$ and $In_2Te(g)$. The species $Te_2(g)$ and $In_2Te(g)$ are formed during the congruent vaporization of $InTe(s)$ [260].

The compounds $LaTe$, as well as Mg_3As_2 and Mg_3Bi_2, were investigated by Gordienko and Fenochka [232], as well as Alikhanyan et al. [230]. They found that the compounds $LaTe$ and Mg_3As_2 vaporize congruently whereas Mg_3Bi_2 dissociates forming $Bi(l)$. Frisch [263] observed that Tm vaporizes preferentially from a single crystal TeTm until a composition of $TeTm_{0.765}$ is reached at 1600 K. The congruency of vaporization was found to be a function of the temperature.

Oforka and Argent [237] determined the thermodynamic activities of the alloy components and the integral Gibbs energies of mixing for an isothermal section of the Ni–Cr–Al system at 1423 K. Some values were also reported for the three boundary systems Cr–Ni, Cr–Al, and Ni–Al.

3.4.3 Investigation of Alloys in the Liquid and the Solid State and Determination of Phase Boundaries

Investigations covering both the liquid and the solid phase are reported by Neubert [229] for the Li–Pb system, by Ono et al. [136] for the Cr–Cu system, by Rammensee and Fraser [241] for the Fe–Co and Fe–Ni systems, by Fraser and Rammensee [134] for the Fe–Co–Ni system, and by Hilpert et al. [252] for the Ni–Al system. Thermodynamic activities or activity coefficients of the alloy components were determined by all these investigations. Moreover, enthalpies of mixing are reported for the Li–Pb, Fe–Co, and Fe–Ni systems. The Gibbs energies and entropies were additionally obtained for the alloys of the Li–Pb system. Hilpert et al. [252] determined the partial and integral Gibbs energies, enthalpies, and entropies of formation for the Ni_3Al phase of stoichiometric composition by extensive measurements. A precise second-law evaluation was possible for the first time since the Al and Ni partial pressures could be determined over a temperature range of more than 200 K.

The activities for the Li–Pb alloys determined by Knudsen effusion mass spectrometry by Neubert agree well with those from EMF measurements [229]. The results obtained by Rammensee and Fraser [241] for solid Fe–Co alloys and the melt of the Fe–Ni system are not consistent with the investigations by Tomiska [242, 243] on these alloys also given in Table 7. The chemical activities determined by Hilpert et al. [252] for the nickel-rich part of the Ni–Al system (x_{Ni} = 0.7 to 0.92) are not very consistent with those obtained for only two compositions of this concentration range by Oforka and Argent's investigations [237] of the Cr–Ni–Al system.

Thermochemical properties of numerous different condensed phases were determined from the gas phase data as shown in the preceding section. In addition phase boundaries were also evaluated thereby improving our knowledge of phase diagrams. The liquidus line determined by Timberg and Toguri [238] as well as Ono et al. [136] for the copper-rich part of the Cr–Cu phase diagram deviates from that reported in the literature so far. Ono et al. additionally employed differential thermal analysis thereby supporting the liquidus line obtained by Knudsen effusion mass spectrometry. Differential thermal analysis and Knudsen effusion mass spectrometry were also used in the investigation of the Al–Te and Ni–Al phase diagrams by Said et al. [258] and Hilpert et al. [120], respectively, thereby illustrating the interplay between the two methods. The phase boundaries obtained by Knudsen effusion mass spectrometry in the aforementioned publications result from the use of the phase rule. The potential of this technique is particularly shown in Refs. 120, 258.

A different technique was used by Tomiska and Neckel [250] as well as Tomiska [242] to determine the Ni–Cu and Fe–Co phase diagrams, respectively, from the thermochemical data obtained by them for the solid alloys. They calculated the phase diagrams by the use of these data, of similar mass spectrometric data obtained by the same group for the corresponding liquid alloys, and of ancillary literature data.

4 Oxides

4.1 Introduction and General Aspects

Oxides are widely used and encountered in technical applications, such as high-temperature ceramics, nuclear fuels, glasses, and corrosion layers on superalloys.

In order to optimize the processes and materials used in these applications, the vaporization of condensed oxide phases, their thermochemistry as well as that of the gaseous oxide species has to be known. These demands initiated investigations of oxide vapors.

The investigations revealed the complex nature of such vapors. Numerous different molecules were identified such as polymers of the type $(MoO_3)_n$ ($n = 1$ to 5) (Table 12), ternary species such as Na_2CrO_3 (Table 9) or species with different relative oxygen content as Li_nO ($n = 1$ to 5) (Table 9) or

Table 9. Summary of mass spectrometric studies of oxides and hydroxides containing alkali and alkali earth metals except those listed in Table 18. (The gaseous species or solid compounds are underlined if their enthalpies of formation or of dissociation, only gaseous species, are given)

System in the cell	Temperature range in K	Gaseous species	Ref./Year
Oxides containing alkali metals			
$\underline{Li_2O}$	1225–1507	Li, O_2, $\underline{LiO}$, Li_2O, Li_3O, Li_2O_2	[278]/1980
Li_2O	1316–1663	Li, O_2, LiO, Li_2O, Li_3O, Li_2O_2	[279]/1979
Li_2O	1352–1663	Li, Li_2, $\underline{LiO}$, Li_2O, Li_3O, Li_2O_2	[280]/1978
Li_2O	1360–1670	Li, Li_2, LiO, Li_2O, $\underline{Li_3O}$, $\underline{Li_2O_2}$	[281]/1979
Li + O	1000–1300	Li, Li_2, Li_3, Li_3O, $\underline{Li_4O}$, $\underline{Li_5O}$	[282]/1987
$\underline{Li_2TiO_3}$	1180–1628	Li, O_2, $\underline{LiO}$, Li_2O, $\underline{Li_3O}$	[283]/1982
$\underline{LiCrO_2}$	1673–1873	Li, Cr, CrO, CrO_2, $LiCrO_2$	[284]/1979
$\underline{Li_2CrO_4}$	1140–1330	Li_2O, $\underline{Li_2CrO_4}$	[285]/1985
Li_2MoO_4	1160–1590	Li, Li_2O, MoO_3, $LiMoO_3$, Li_2MoO_4	[286]/1982
δ-Li_4MoO_5	1150–1445	Li, Li_2MoO_4	
Li_xWO_3	1073–1273	Li, LiO, WO_2, WO_3, Li_2WO_4	[287]/1982
Li_5FeO_4	1200–1500	Li, O_2, Li_2O	[288]/1978
$Li + B + O_2$	1758–2251	Li, B, BO, $\underline{LiBO}$, $LiBO_2$	[289]/1985
β-Li_5AlO_4	1262–1613	Li, O_2, LiO, Li_2O	[290]/1981
γ-$LiAlO_2$	1480–2020	Li, O_2, LiO, Li_2O	
$LiAl_5O_8$	1493–1920	Li, LiO, Li_2O	
β-Li_5AlO_4	1263–1625	Li, O_2, LiO, Li_2O	[291]/1980
γ-$LiAlO_2$	1300–1800	Li, O_2, Li_2O	
$LiAl_5O_8$	1300–1800	Li, O_2, Li_2O	
Li_2O + Si	1438–1583	Li, Li_2, SiO, SiO_2, Si_2O_2, $\underline{LiSiO}$	[292]/1980
Li_4SiO_4	1350–1629	Li, O_2, LiO, Li_2O, $\underline{Li_3O}$	[293]/1984
Li_2SiO_3	1166–1762	Li, $\underline{LiO}$, Li_2O, SiO, Li_2SiO_3	[294]/1981
Li_4SiO_4	1250–1670	Li, O_2, $\underline{LiO}$, Li_2O	[163]/1982
Li_2SiO_3	1300–1750	Li, O_2, $\underline{LiO}$, $\underline{Li_2O}$, Li_2SiO_3	
$LiSbO_3$	not specified	Sb_4O_6, $LiSbO_2$, $(LiSbO_2)_2$	[295]/1986
Na_2CrO_4	1070–1390	Na, $\underline{Na_2CrO_4}$	[285]/1985
Na_2MoO_4	1232–1334	Na, $\underline{Na_2MoO_4}$	[296]/1980
Na_2MoO_4	1050–1300	Na_2MoO_4	[297]/1976

Table 9. (continued)

System in the cell	Temperature range in K	Gaseous species	Ref./Year
Na_2WO_4	1401–1473	Na, Na_2WO_4	[296]/1980
$CaO + Fe_2O_3 + NiO +$ $Cr_2O_3 + Na_2SiO_3 +$ $U_3O_8 + Ir$	973–1923	Na, IrO_3, Na_2IrO_4	[298]/1977
$NaSbO_3$	not specified	Na, $NaSbO_2$, $(NaSbO_2)_2$	[295]/1986
$\underline{K_2O}$	580–825	K_2O, $\underline{(KO)_2}$	[299]/1979
$\underline{K_2CO_3}$	1087–1186 ⎫	K, O_2, $\underline{KO}$, $\underline{K_2O}$, CO_2, K_2CO_3	[300]/1977
K_2O	600–1000 ⎭		
KOH	703–1375 ⎫		
$K_2O + O_2(g)^a$	1173–1353 ⎬	K, H_2, O_2, $\underline{KH}$, $\underline{KO}$, OH, $\underline{K_2O}$, $\underline{KOH}$	[301]/1982
$K_2O + H_2(g)^a$	1325 and 1400 ⎭		
$CaO + K(g) + O_2(g)^a$	1880–1950	K, H_2, $\underline{KCaO}$, CaOH	[302]/1987
K_2CrO_4	1215–1286	K, $\underline{K_2CrO_4}$	[303]/1985
K_2CrO_4	1020–1370	K, $\underline{K_2CrO_4}$	[285]/1985
K_2MoO_4	1170–1310	$\underline{K_2MoO_4}$	[297]/1976
KBO_2	1053–1240	$\underline{KBO_2}$	[304]/1985
$K_2O + Al + Al_2O_3$	1330–1490	K, K_2, Al, Al_2O, $\underline{KAlO}$	[305]/1986
$SiO + Si + K(g)^a$	1600–1775	K, SiO, $\underline{KSiO}$	
$KSbO_3$	not specified	K, $KSbO_2$, $(KSbO_2)_2$	[295]/1986
Rb_2CrO_4	1000–1350	Rb, $\underline{Rb_2CrO_4}$	[285]/1985
$RbSbO_3$	not specified	Rb, $\underline{RbSbO_2}$, $(RbSbO_2)_2$	[295]/1986
Cs_2CrO_4	830–1380	Cs, $\underline{Cs_2CrO_4}$	[285]/1985
$CsAlSiO_4$	1242–1567	Cs	[306]/1980
$CsAlSi_5O_{12}$	1542–1803	Cs	
$CsAlSi_2O_6$	1324–1748	Cs	[307]/1980
Pollucite		Li, Na, K, Rb, Cs	
$CsSbO_3$	not specified	Cs, $\underline{CsSbO_2}$, $(CsSbO_2)_2$	[295]/1986

Oxides containing alkali earth and no alkali metals

System in the cell	Temperature range in K	Gaseous species	Ref./Year
$Be(ReO_4)_2$	704	Re_2O_7	[308]/1978
MgO	1820–1980	Mg, O_2, $\underline{MgO}$	[309]/1983
MgO	2020–2160	Mg, O_2, $\underline{MgO}$	[310]/1976
$Mg(ReO_4)_2$	998	Re_2O_7	[308]/1978
$MgO + MgIn_2O_4 + In$	1373	In, In_2O, $MgIn_2O_4$	[311]/1978
CaO	2080–2206	Ca, O_2, $\underline{CaO}$	[312]/1976
$CaO + H_2O(g)$ or $H_2(g)^a$	1720–1983	Ca, H_2, H_2O, $\underline{CaOH}$, $Ca(OH)_2$,	[302]/1987
$CaO + SrO + H_2O(g)^a$	1896–2001	H, H_2, Ca, Sr, $\underline{CaO}$, $\underline{SrO}$, $\underline{CaOH}$, $\underline{SrOH}$	[313]/1981
$Ca(ReO_4)_2$	1200–1250	Re_2O_7, $Ca(ReO_4)_2$	[308]/1978
$Ca_3Ga_2Ge_3O_{12}$	1240–1605	Ga, GeO, Ga_2O	[314]/1986
SrO	2010–2102	Sr, O_2, $\underline{SrO}$	[312]/1976

(Continued)

Table 9. (continued)

System in the cell	Temperature range in K	Gaseous species	Ref./Year
$Sr(ReO_4)_2$	1120–1270	Re_2O_7, $Sr(ReO_4)_2$	[308]/1978
$BaO + H_2O(g)^a$	1500–1702	H_2O, $Ba(OH)_2$	[315]/1981
$BaO + H_2(g)^a$	1303–1503	H_2, $\underline{BaOH}$	
$BaZrO_3$	1647–2118	Ba, BaO	[58]/1976
$BaCrO_4$	1200–1800	O_2, $BaCrO_2$, $BaCrO_3$, $BaCrO_4$	[316]/1982
$BaCrO_4$	1649–1762	Ba, BaO, $\underline{Ba_2O}$, $(BaO)_2$	[317]/1981
$BaCrO_4 + Mo$	2011 + 2037	BaO, $\underline{BaMoO_2}$, $\underline{BaMoO_3}$, $\underline{Ba_2MoO_4}$, $\underline{Ba_2MoO_5}$, $\underline{Ba_2Mo_2O_8}$	[318]/1982
Ba_2CaWO_6			[319]/1980
$Ba_3Y_2WO_9$ to			
$Ba_3Y_2(W_{0.2}Mo_{0.8})O_9$	1608–1922	Ba, Ca, BaO, $BaWO_3$, $BaWO_4$	
Ba_3WO_6			
$BaWO_4$			
$Ba(ReO_4)_2$	1000–1090	Re_2O_7, $\underline{Ba(ReO_4)_2}$	[308]/1978

^{a} Gaseous species leaked into the Knudsen cell

As_4O_n (n = 6 to 10) (Table 14). The vapor composition depends on the oxygen potential in the cell. This can, for example, be changed by the material of the inner wall of the cell. The intentional adjustment and control of different oxygen potentials in the cell and the direct mass spectrometric measurement of oxygen partial pressures with large sensitivity (cf. Sect. 2.6) can be useful for the study of oxide systems. This methodological aspect and the potential of Knudsen effusion mass spectrometry with multiple cells (cf. Sect. 2.6.2) for the determination of mixing properties of oxide melts is discussed by Allibert and Chatillon [265]. These authors give, moreover, an account of mass spectrometric Knudsen and Langmuir vaporization studies of pure binary oxides.

The recent reviews by Lamoreaux and Hildenbrand [97] and Lamoreaux et al. [98] describe the high-temperature vaporization behavior of binary alkali metal oxides and the binary oxides of the alkali earth metals as well as the group IIb, IIIb, and IVb elements except carbon. Tables with selected values for some thermochemical properties including enthalpies and entropies of formation of the condensed phases and the different vapor species are given. The data are used to generate plots of the equilibrium partial pressures of the vapor species over different condensed phases as a function of temperature. The plots are given for different oxygen potentials ranging from reducing to oxidizing. Maximum vaporization rates are reported. Pedly and Marshall [99] give an account of the thermochemical properties of gaseous diatomic monoxides. The same information on different gaseous species and condensed phases of binary oxides is listed in "The Oxide Handbook", edited by Samsonov [100]. Vapor pressures and vaporization rates are additionally tabulated.

Only binary oxides are also considered in the general review by Gingerich [21] and in the concise review on oxides by Kazenas et al. [266] published 6 years earlier than Ref. 21. The temperature range of the measurement and the vapor species identified are tabulated by Kazenas et al. [266] for the different condensed phases investigated. Results obtained by Knudsen effusion mass spectrometry were essentially considered by these authors. Rauh and Ackermann [267] discussed the determination of chemical and thermodynamic properties of refractory oxides above 1500 K by the study of their vapor and gas/condensed phase equilibria.

Data on ternary oxide species are given in the reviews by Büchler and Berkowitz-Mattuk [268], Choudary et al. [269] as well as Gorokhov and Semenov [28]. The review by Ngai and Stafford [270] on gaseous oxohalides, hydroxides and complex oxides of group III and transition elements is of interest.

Further compilations or reviews on oxide thermodynamics were for example carried out by Refs. 90–93, 95, 96, by the National Bureau of Standards, U.S. [271–276], by Brewer and Rosenblatt [101, 102], by Ackermann and Thorn [277], as well as Brewer [103].

The following sections give an account of the investigations carried out by Knudsen effusion mass spectrometry of binary and multicomponent oxides since about the year 1977 (Sects. 4.2 to 4.5) and 1979 (Sect. 4.6). The condensed phase in the Knudsen cell, the temperature range of the measurements and the vapor species identified are tabulated (Tables 9, 12, 14, 17, 18). The given investigations are commented and discussed. The enthalpies of dissociation determined for some of the vapor species are tabulated (Tables 10, 11, 13). A special section, Sect. 4.6, summarizes the investigations essentially carried out to determine the mixing properties of melts and solid solutions from gas phase studies.

The very few publications on gaseous hydroxides are additionally considered, since their investigation is closely related to that of the oxides.

4.2 Oxides Containing Alkali or Alkali Earth Metals

The breeding of tritium from lithium is one of the most important problems for deuterium-tritium nuclear fusion technology. This initiated investigations of the vapors and the gas phase chemistry over $Li_2O(s)$, $LiCrO_2(s)$, $Li_5FeO_4(s)$, as well as over the solid compounds of the systems $Li_2O-Al_2O_3$, Li_2O-SiO_2, and Li_2O-MoO_3 (cf. Table 9).

The vaporization of $Li_2O(s)$ has been investigated by four groups (Kimura et al. [278], Ikeda et al. [279], Kudo et al. [280], Wu et al. [281]) using Knudsen cells from Pt [278–281], graphite [280], Mo [279, 280], Ta [279, 280], Ni [279], and Nb [279]. Earlier studies are quoted in Ref. 278. Partial pressures of Li(g),

Table 10. Enthalpy change of the dissociation reactions for some molecules of binary alkali and alkali earth oxides and hydroxides

Molecule	Dissociation reaction	T/K	ΔH_T° kJ mol^{-1}	Ref.
Li_2O_2	$Li_2O_2 \rightleftarrows 2Li + 2O$	0	1087 ± 24	[278]
Li_3O	$Li_3O \rightleftarrows 3Li + O$	0	920 ± 49	[278]
Li_4O	$Li_4O \rightleftarrows 4Li + O$	0	1155 ± 25	[282]
Li_5O	$Li_5O \rightleftarrows 5Li + O$	0	1276 ± 27	[282]
KO	$KO \rightleftarrows K + O$	0	261.9 ± 4.0	[301]
K_2O	$K_2O \rightleftarrows 2K + O$	0	582 ± 12	[301]
KOH	$KOH \rightleftarrows K + OH$	0	352.3 ± 2.0	[301]
BeO	$BeO \rightleftarrows Be + O$	0	437 ± 13	[320]
MgO	$MgO \rightleftarrows Mg + O$	0	358 ± 13	[320]
CaO	$CaO \rightleftarrows Ca + O$	0	381 ± 8	[320]
CaOH	$CaOH \rightleftarrows Ca + OH$	0	404 ± 8	[302]
$Ca(OH)_2$	$Ca(OH)_2 \rightleftarrows Ca + 2OH$	0	853 ± 10	[302]
SrO	$SrO \rightleftarrows Sr + O$	0	412 ± 8	[320]
SrOH	$SrOH \rightleftarrows Sr + OH$	0	385 ± 17	[313]
BaO	$BaO \rightleftarrows Ba + O$	0	542 ± 4	[320]
Ba_2O	$Ba_2O \rightleftarrows Ba + BaO$	0	301 ± 40	[317]
$(BaO)_2$	$(BaO)_2 \rightleftarrows 2BaO$	0	360 ± 20	[317]
BaOH	$BaOH \rightleftarrows Ba + OH$	0	452 ± 4	[315]
$Ba(OH)_2$	$Ba(OH)_2 \rightleftarrows Ba + 2OH$	0	371 ± 2	[315]

Table 11. Enthalpy change of the dissociation reactions for some molecules of ternary oxides containing alkali and alkali earth metals

Molecule	Dissociation reaction	T/K	ΔH_T° kJ mol^{-1}	Ref.
LiBO	$LiBO \rightleftarrows Li + B + O$	0	1079 ± 25	[289]
LiSiO	$LiSiO \rightleftarrows Li + Si + O$	0	997 ± 21	[292]
Li_2CrO_4	$Li_2CrO_4 \rightleftarrows 2Li + Cr + 4O$	298	2743	[285]
Na_2CrO_4	$Na_2CrO_4 \rightleftarrows 2Na + Cr + 4O$	298	2613	[285]
KSiO	$KSiO \rightleftarrows K + SiO$	298	187 ± 42	[305]
K_2CrO_4	$K_2CrO_4 \rightleftarrows 2K + Cr + 4O$	298	2627	[285]
Rb_2CrO_4	$Rb_2CrO_4 \rightleftarrows 2Rb + Cr + 4O$	298	2634	[285]
Cs_2CrO_4	$Cs_2CrO_4 \rightleftarrows 2Cs + Cr + 4O$	298	2075	[285]
Ba_2MoO_5	$Ba_2MoO_5 \rightleftarrows BaO + BaMoO_4$	0	322 ± 60	[318]
$Ba_2Mo_2O_8$	$Ba_2Mo_2O_8 \rightleftarrows 2BaMoO_4$	0	351 ± 80	[318]

$Li_2O(g)$, $LiO(g)$, $Li_3O(g)$, and $Li_2O_2(g)$ measured over $Li_2O(s)$ in Pt cells are reported in Refs. 278–280. The pressures of $O_2(g)$ [278, 279] and $Li_2(g)$ [280] were additionally determined by the references given in brackets. The pressures obtained by the three groups with Pt cells are partly very different. They are discussed by Kimura et al. [278] who additionally evaluated from the pressures the enthalpy of formation of $Li_2O(s)$, $\Delta_f H_{298}^\circ = 592.6 \pm 8.6$ kJ mol^{-1}, agreeing very well with the values obtained by calorimetry. The measurements carried out with Knudsen cells made from Ni, Mo, Ta, Nb, and graphite show higher Li pressures than those with Pt cells thereby indicating reactions between the cell material and the Li_2O sample. The reactivity increases according to the following order: $Pt < Ni < Ta < Mo = Nb <$ graphite. The reduction of an oxide sample by the Knudsen cell material was also studied by Hilpert and Gerads [321] upon investigating the sublimation of BaO(s) with cells made of Mo, Pt, graphite, and Al_2O_3.

Phases of different quasi-binary systems $Li_2O-A_iO_k$ were investigated in addition to $Li_2O(s)$ (cf. Table 9). All these phases vaporize by decomposition. Obviously, also ternary gaseous oxide compounds such as $Li_2SiO_3(g)$, $Li_2MoO_4(g)$, and $LiCrO_2(g)$ were observed (cf. Table 9). In this case the phases additionally in part vaporize congruently. The partial pressures of the gaseous species over $Li_2SiO_3(s, l)$ (Table 9) were determined by two groups (Ikeda et al. [163], Nakagawa [294]) and compared in Ref. 163. The pressures of $Li_2SiO_3(g)$ and the most abundant species Li(g) in Ref. 294 exceed those in Ref. 163 by factors of about 3.5 and 1.5, respectively. The results by Brüning et al. [293] on the vaporization of $Li_4SiO_4(s, l)$ in part deviate from those by Ikeda et al. [163]. The partial pressures are compared in Ref. 293. The solid lithium aluminate compounds were studied twice by Ikeda et al. [290, 291]. The final results are given in Ref. 290.

The Li partial pressures over different lithium silicates, aluminates, and molybdates are summarized by Ikeda et al. [163]. Li(g) is the most abundant species in the equilibrium vapor over these phases.

Atomization energies of LiO(g), $Li_2O(g)$, $Li_3O(g)$, $Li_4O(g)$, $Li_5O(g)$, and $Li_2O_2(g)$ were determined by vaporization of Li_2O, Li with low oxygen content (< 500 ppm), Li_5AlO_4, $LiAlO_2$, Li_4SiO_4, Li_2SiO_3, and Li_2TiO_3 (cf. Table 9). The results obtained so far from 11 and 10 different investigations of LiO(g) and $Li_2O(g)$, respectively, are summarized by Ikeda et al. [163]. The given values for each of the two molecules generally agree well with each other and with those given by Gingerich [21] except for the enthalpy of atomization of LiO(g) determined by Kudo et al. [280] which exceeds the mean value by more than $30\ kJ\,mol^{-1}$. The mean values of the enthalpies of atomization for LiO(g) and $Li_2O(g)$ computed from all the given enthalpies deviate by less than 1.5% from those given by Gingerich [21].

The enthalpies of atomization for $Li_3O(g)$ determined so far are summarized by Nakagawa et al. [283] and those for $Li_2O_2(g)$ by Kimura et al. [278]. Those of $Li_2O_2(g)$ were evaluated by Wu et al. [281] and by Kimura et al. for different structures of $Li_2O_2(g)$ with and without O–O bond. The values of Wu et al. [281] are in each case larger by more than 10% than those of Kimura et al. [278]. The values with O–O bond exceed those without O–O bond by about 5%. The enthalpy of atomization determined by Kimura et al. [278] for $Li_2O_2(g)$ with O–O bond was selected in Table 10. This value exceeds the value given by Gingerich [21] by 5%. The enthalpies of atomization of $Li_3O(g)$ evaluated by Refs. 278, 280, 281, 283 agree within the given errors (cf. Ref. 283). The value obtained by Kimura et al. [278], obtained by a third-law evaluation, is given in

Table 10. Wu [282] identified the molecules Li_4O and Li_5O for the first time and determined their enthalpies of atomization (see Table 10) according to the third-law method.

Neubert [289] and Wu et al. [292] detected the molecules LiBO(g) and LiSiO(g), respectively (cf. Table 9). The atomization energies are given in Table 11. These authors additionally redetermined the atomization energies of $LiBO_2$(g) and Si_2O_2(g). The values agreed within the given error limits with those in the JANAF tables [90] for $LiBO_2$(g) and listed by Gingerich [21] for Si_2O_2(g).

Ermilova et al. [297] studied the vaporization of Na_2MoO_4(s, l) as well as K_2MoO_4(s, l) with Pt Knudsen cells, Spoliti et al. [296] that of Na_2MoO_4(l) as well as $NaWO_4$(l) using Ta cells lined with Al_2O_3, and Kuligina and Semenov [285] that of A_2CrO_4(s) (A = Li, Na, K, Rb, Cs) using Pt cells. These phases vaporize according to the reaction

$$A_2XO_4(s, l) \rightleftarrows A_2XO_4(g) \ (A = \text{alkali metals}; X = Mo, W, Cr). \quad (19)$$

Partial pressures and enthalpies of vaporization were determined. Spoliti et al. showed by ionization efficiency measurements that Na_2MoO_4(l) and Na_2WO_4(l) in addition to reaction Eq. (18) in part also vaporize by decomposition and/or reduction to Na(g). The experimental conditions, such as sample purity and crucible material, are extremely important for the decomposition. Also M_2CrO_4(s) (M = Na, K, Rb, Cs) vaporizes in part to M(g) [285]. Li_2O(g) is formed in addition to Li_2CrO_4(g) on vaporizing Li_2CrO_4(s) [285]. The results obtained for the vaporization of K_2CrO_4(s) in Ref. 285 agree well with those by Rudny et al. [303]. The Na_2MoO_4 pressure at 1250 K of Ermilova et al. [297] exceeds that of Spoliti et al. [296] by a factor of 2. Spoliti et al. combined the mass spectrometric and infrared analysis of Na_2MoO_4(g) and Na_2WO_4(g) to determine their enthalpies of vaporization on the basis of a consistent set of partial pressures and molecular parameters. The values obtained for the first time by using both the second- and the third-law methods agree well. Ermilova et al. [297] applied only the second-law method.

Semenov and Smirnova [295] showed by the use of Pt Knudsen cells that the alkali metal antimonates, with the exception of $LiSbO_3$(s), vaporize congruently according to the reaction

$$MSbO_3(s) \rightleftarrows MSbO_2(g) + 1/2O_2(g) \ (M = Na, K, Rb, Cs) \quad (20)$$

thereby forming gaseous antimonites.

Farber et al. [304] determined the enthalpy of formation at 298 K of KBO_2(g) according to the second- and third-law methods as $-672.8 \pm 10 \, \text{kJ mol}^{-1}$ and $-672.4 \pm 10 \, \text{kJ mol}^{-1}$, respectively, thereby improving the value given in the JANAF tables [90].

Cs(g) is the most abundant species observed by Odoj and Hilpert [306, 307] upon vaporizing the synthetic compounds $CsAlSiO_4$(s), $CsAlSi_2O_6$(s), and $CsAlSi_5O_{12}$(s) from Mo Knudsen cells. In addition three pollucite minerals of different origin were studied. The Cs pressures determined over the synthetic $CsAlSi_2O_6$(s) and the minerals agreed reasonably well [307].

Gas inlet Knudsen cells were used by Farber et al. [301, 302], Murad [313], as well as Farber and Srivastava [315] to introduce $O_2(g)$, $H_2(g)$, or $H_2O(g)$ into the cell for the determination of bond energies of $KH(g)$ as well as oxides and/or hydroxides of potassium, calcium, strontium, and barium. Improved bond energies for $KO(g)$, $K_2O(g)$, $KOH(g)$, $CaOH(g)$, $Ca(OH)_2(g)$ (see Table 10) as well as the first thermochemical bond energy for $KH(g)$ were obtained by Farber et al. These values are essentially in agreement with those by Gingerich [21] ($KO(g)$, $KH(g)$) and JANAF [90] ($KO(g)$, $KH(g)$, $KOH(g)$, $CaOH$, $Ca(OH)_2$). Hastie et al. [322] studied the vapor species KOH and $(KOH)_2$ by the use of transpiration mass spectrometry. Thermodynamic data of these species were redetermined. Butman et al. [323] investigated ion-molecule equilibria in the vapor over potassium hydroxide. The same method was used by Kudin et al. [324] to elucidate the vapor composition over the alkali metal hydroxides MOH (M = Na, K, Sr, Cs). Partial pressures of ionic and neutral species were determined. The species $KCaO(g)$, $KAlO(g)$, and $KSiO(g)$ can be obtained by replacing hydrogen by potassium in the corresponding hydroxides thereby producing an ionic bonding (Farber et al. [302, 305]).

The dissociation energies of the alkaline earth monoxides have been reviewed by Srivastava [320] and recommended values are given (Table 10). The recommended values agree within the given errors with those selected by Gingerich [21] and JANAF [90], with that determined by Kazenas et al. [309] for $MgO(g)$ as well as with those by Murad [313] for $CaO(g)$ and $SrO(g)$. The dissociation energies for $CaOH(g)$, $SrOH(g)$, $BaOH(g)$, and $Ba(OH)_2$ in Table 10 in general agree well with those selected by JANAF which were only derived from flame spectral data for $CaOH(g)$ and $SrOH(g)$.

Semenov et al. [308] showed that upon vaporizing the alkali earth metal perrhenates in a Pt cell only those of Ca, Sr, and Ba exist as a stable form in the gas phase (cf. Table 9).

Kudin et al. [316, 318] as well as Kudin [317] evaporated $BaCrO_4$ from Knudsen cells made from Mo [318] and from Ta lined with ZrO_2 [316, 317]. At first $O_2(g)$ was observed using the lined Ta cell. Thereafter and at increased temperatures also Cr, Ba, Ba_2O and $(BaO)_2$ were detected. The molecules $BaCrO_2(g)$, $BaCrO_3(g)$, and $BaCrO_4(g)$ were identified after fast heating up of the sample.

Odoj and Hilpert [58] showed the incongruent vaporization of $BaZrO_3(s)$ to $BaO(g)$. They determined the enthalpy of formation of $BaZrO_3(s)$ from the component oxides as $\Delta_f H_o^\circ = -106 \pm 21$ kJ mol^{-1} by evaluating the enthalpies of vaporization according to the second- and third-law methods.

4.3 Oxides of Transition Metals

The titanium-oxygen system is particularly complex among the binary metal-oxygen systems. It exhibits many stoichiometric and widely non-stoichiometric solid phases as well as two gaseous compounds, $TiO(g)$ and $TiO_2(g)$. The complexity explains the large number of investigations carried out

Table 12. Summary of mass spectrometric studies of oxides and hydroxides of transition metals except those listed in Table 18. (The gaseous species or solid compounds are underlined if their enthalpies of formation or of dissociation, only gaseous species, are given)

System in the cell	Temperature range in K	Gaseous species	Ref./Year
$Sc(ReO_4)_3$, $Y(ReO_4)_3$, $La(ReO_4)_3$	≤ 1473	ReO_3, Re_2O_7	[325]/1983
TiO_x $(O \leq x < 2)$	1964 and 2290	Ti, O, TiO, TiO_2	[326]/1982
Ti_2O to TiO_2	1700–2200	Ti, TiO, TiO_2	[148]/1982
TiO_x $(x = 0.065, 0.080, 0.092,$ and $0.851)$	1693–1920	Ti, TiO	[327]/1981
TiO_x $(x = 0.86 - 1.25)$	1684–1948	Ti, TiO	[328]/1980
TiO	1694–1919	Ti, TiO	[329]/1977
$Al_2O_3 + BaTiO_3 +$ $Ti + TiO_2$	1874–2222	Ba, Ti, Al, BaO, $\underline{TiO}$, TiO_2, AlO,	[330]/1976
$CoTiO_3$ or $CoO + TiO_2$	2200–2400	TiO, TiO_2, $\underline{Ti_2O_3}$, $(TiO_2)_2$	[331]/1985
ZrO_x $(x = 1.98, 1.96, 1.84, 1.74)$	1850–2860	Zr, O, ZrO, $\underline{ZrO_2}$	[332]/1981
ZrO_{2-x}	2365–2950	O, ZrO, ZrO_2	[333]/1979
$Zr + ZrO_2$	1900–2500	Zr, O, ZrO, ZrO_2	[334]/1979
$V_2O_3 + VO$, V_2O_3	1860–2182	V, $\underline{VO}$, VO_2	[335]/1985
$Eu_2O_3 + V_2O_5 + V$	1913–2282	V, Eu, $\underline{VO}$, VO_2, $(VO_2)_2$, EuO	[336]/1983
Li depleted $LiNbO_3$	1877–2288	NbO, NbO_2, $\underline{Nb_2O_4}$, $\underline{Nb_2O_5}$, $\underline{Nb_4O_9}$, $\underline{Nb_4O_{10}}$	[337]/1986
Nb_2O_5	1726–1988	NbO_2	[338]/1983
$NbO_{2+x} +$ liquid	1925–2271	NbO, NbO_2	
NbO	1948–2085	$\underline{NbO, NbO_2}$	[339]/1982
NbO + Nb	1971–2175		
NbO_2	1953–2323	O, $\underline{NbO}$, NbO_2	[340]/1981
$NbO_{2\pm x}$	1958–2326	NbO, NbO_2	[341]/1981
$NbO + NbO_2$	2011–2083	NbO, NbO_2	
$Eu + Cr + P + O + W$	1685–2340	Cr, P_2, Eu, $\underline{CrO}$, WO_2, WO_3, PO, PO_2, EuO, $CrPO_2$, $EuWO_4$	[342]/1981
MoO_2	1540–1780	MoO_2, $\underline{MoO_3}$, $\underline{(MoO_3)_2}$, $\underline{(MoO_3)_3}$	[343]/1983
Mo_4O_{11}	840–940	$(MoO_3)_3$, $(MoO_3)_4$, $(MoO_3)_5$	
MoO_3	850–940	$(MoO_3)_3$, $(MoO_3)_4$, $(MoO_3)_5$	
MoO_2	1500–1730	$\underline{MoO_2}$, $\underline{MoO_3}$, $\underline{(MoO_3)_2}$, $\underline{(MoO_3)_3}$	[344]/1976
WO_3	1310–1530	$(WO_3)_2$, $\underline{W_3O_8}$, $(WO_3)_3$, $(WO_3)_4$,	[345]/1985
WO_3	1303–1376	$(WO_3)_3$, $(WO_3)_4$, $(WO_3)_5$	[346]/1980
WO_2	1400–1600	$(WO_3)_2$, $(WO_3)_3$, $(WO_3)_4$	[347]/1976
$Fe_2O_3 + MnSe + EuS$	1644–2061	Mn, Fe, Se Se_2, $\underline{MnO}$, $\underline{FeO}$, SO, SeO, SeS	[348]/1984

Table 12. (continued)

System in the cell	Temperature range in K	Gaseous species	Ref./Year
$Fe_2O_3 + H_2O(g)^a$	1743–1808	H, H_2, Fe, H_2O, FeO, FeOH	[349]/1980
$Fe_2O_3 + H_2(g)^a$	1723–1815	H, H_2, Fe, H_2O, FeO, FeOH	
RuO_2	300–1400	O_2	[350]/1983
$CoTiO_3$, $CoTi_2O_5$	1300–1700	Co, O_2, CoO	[351]/1984
$CuReO_4 + CuI$	750	Re_2O_7, Cu_3X_3, ReO_3X, $Cu_2(ReO_4)_2$,	[352]/1982
$CuReO_4 + CuBr$	not specified	Cu_2XReO_4, $Cu_3X_2ReO_4$, (X = Br, I)	
ZnO	1203–1546	Zn, O_2, ZnO	[353]/1984
ZnO	1446	Zn, O_2, ZnO	[354]/1982
CdO	989–1337	Cd, O_2, CdO	[353]/1984
CdO	886–1090	Cd, O_2	[355]/1981
CdO	1250	Cd, O_2, CdO	[354]/1982
CdO	1130–1320	Cd, O_2, CdO	[356]/1978
HgO	657	Hg, O_2, HgO	[354]/1982

^{a} Gaseous species leaked into the Knudsen cell

on this system (cf. Table 12). Banon et al. [326] elucidated the fragmentation patterns of TiO(g) and TiO_2(g) on electron impact ionization by vaporizing various samples of different composition by the use of a single and a multiple Knudsen effusion cell. Their results are of basic interest for all investigations of the condensed Ti–O system by Knudsen effusion mass spectrometry. Partial pressures and partial molal enthalpies of vaporization of Ti(g) and TiO(g) were determined for stoichiometric TiO(s) (Sheldon and Gilles [329]), for nearly the entire homogeneity range of this compound (Heidemann et al. [328]), as well as for one sample of this homogeneity range and three metallic samples (Granier and Gilles [327]) (cf. Table 12). Integral free energies as well as enthalpies and entropies of formation were calculated for the monoxide homogeneity range [328]. The Ti and TiO pressures above stoichiometric TiO(s) measured by Refs. 328 and 329 agree well. Approximate pressures were estimated in Ref. 327. Free energies of formation at 1900 K of TiO(s), Ti_3O_5(s), Ti_4O_7(s), Ti_5O_9(s), and $Ti_{10}O_{19}$(s) were evaluated from measured chemical activities using multiple Knudsen cells by Banon et al. [148]. The same authors also investigated the liquid Ti–O system in the range x_{Ti} = 0.33 to 0.44. Chemical activities of Ti, TiO (x_{Ti} = 0.5), Ti_2O_3 (x_{Ti} = 0.40) and TiO_2 (x_{TiO} = 0.33) were evaluated for the given composition range. Thermodynamic data for assumed species in the melt were evaluated.

Hildenbrand [330] redetermined by isomolecular exchange reactions the atomization energy of TiO_2(g) (Table 13) and the enthalpy of dissociation of TiO(g), D_0 = 662.7 ± 6.3 kJ mol^{-1}, which he combined with other recent work to establish a final value given in Table 13. The results of a second- and third-law evaluation agreed well. Balducci et al. [331] identified for the first time the

Table 13. Enthalpy change of the dissociation reactions for some molecules of binary transition metal oxides and hydroxides

Molecule	Dissociation reaction	T/K	ΔH_T° kJ mol^{-1}	Ref.
TiO	TiO $\rightleftarrows$ Ti + O	0	661 ± 8	[330]
TiO$_2$	TiO$_2$ $\rightleftarrows$ Ti + 2O	0	1272 ± 12	[330]
Ti$_2$O$_3$	Ti$_2$O$_3$ $\rightleftarrows$ 2Ti + 3O	0	2312 ± 56	[331]
(TiO$_2$)$_2$	(TiO$_2$)$_2$ $\rightleftarrows$ 2Ti + 4O	0	3917 ± 47	[331]
VO	VO $\rightleftarrows$ V + O	0	625.5 ± 8.5	[336]
VO$_2$	VO$_2$ $\rightleftarrows$ V + 2O	0	1177 ± 18	[336]
(VO$_2$)$_2$	(VO$_2$)$_2$ $\rightleftarrows$ 2V + 4O	0	2880 ± 23	[336]
(NbO$_2$)$_2$	(NbO$_2$)$_2$ $\rightleftarrows$ 2Nb + 4O	0	3322 ± 45	[337]
Nb$_2$O$_5$	Nb$_2$O$_5$ $\rightleftarrows$ 2Nb + 5O	0	3910 ± 59	[337]
Nb$_4$O$_9$	Nb$_4$O$_9$ $\rightleftarrows$ 4Nb + 9O	0	7958 ± 82	[337]
Nb$_4$O$_{10}$	Nb$_4$O$_{10}$ $\rightleftarrows$ 4Nb + 10O	0	8595 ± 71	[337]
CrO	CrO $\rightleftarrows$ Cr + O	0	436.2 ± 8.9	[342]
MnO	MnO $\rightleftarrows$ Mn + O	0	369 ± 8	[348]
FeO	FeO $\rightleftarrows$ Fe + O	0	401 ± 8	[349]
FeOH	FeOH $\rightleftarrows$ Fe + OH	0	322 ± 17	[349]
ZnO	ZnO $\rightleftarrows$ Zn + O	0	282.7 ± 4.2	[353]
HgO	HgO $\rightleftarrows$ Hg + O	0	221 ± 33	[354]

molecules Ti$_2$O$_3$(g) and (TiO$_2$)$_2$(g) and determined their enthalpies of atomization.

Stoichiometric ZrO$_2$(s) can vaporize incongruently until a congruently vaporizing composition ZrO$_{2-x}$(s) (x = 0 to 0.4) is reached. Rauh and Garg [332] determined the congruently vaporizing composition of zirconium dioxide in W and Re Knudsen cells. In addition, they obtained the partial pressures of Zr(g), ZrO(g), ZrO$_2$(g), and O(g) as a function of the temperature of four different compositions of the cubic phase of substoichiometric ZrO$_{2-x}$ by using Knudsen effusion mass spectrometry and the mass-loss effusion method. These investigations complement analogous studies of the univariant system Zr(s, l) + ZrO$_2$(s) by Ackermann et al. [334]. Belov and Semenov [333] also studied the vaporization of ZrO$_2$(s), thereby evaluating equations for the ZrO$_2$, ZrO, and O pressures over congruently evaporating ZrO$_{2-x}$(s).

The enthalpies of atomization of VO(g) and VO$_2$(g) by Balducci et al. [336] were selected in Table 13 since these authors studied all gas reactions in contrast to Banchorndhevakul et al. [335]. The value for VO$_2$(g) deviates from that adopted by JANAF unlike that for VO(g).

The vaporization of phases of the system Nb–O was conducted by Kamegashira et al. [340] as well as Matsui and Naito [338, 339, 341]. By analyzing the solid residue after vaporization Kamegashira et al. showed that NbO$_2$(s) vaporizes congruently, and NbO(s), Nb$_{12}$O$_{29}$(s) as well as Nb$_2$O$_5$(s) by decomposition. Partial pressures of the species shown in Table 12 were determined over the two-phase mixture Nb(s) + NbO(s) and NbO(s) [339], over the two-phase mixture NbO(s) + NbO$_2$(s) [341] and NbO$_{2+x}$ ($-0.028 \leq x$

≤ 0.037) [340, 341], as well as over the two-phase mixture NbO_{2+x} + liquid ($0.063 \leq x \leq 0.386$) and Nb_2O_5(s, l) [338]. The pressures evaluated (total pressure, partial pressures of NbO_2(g), NbO(g), O(g), O_2(g)) for the Nb–O system ($0 < 0/Nb < 2.05$) are summarized in [338]. The pressures showed that NbO_2(s) is the only phase vaporizing without decomposition and that stoichiometric $NbO_{2.000}$(s) is the congruently vaporizing composition. Thermodynamic properties were evaluated from the partial pressures, such as partial molar enthalpies and entropies of oxygen as well as enthalpies of formation of $NbO_{2\pm x}$(s) [341], the enthalpy and entropy of fusion of Nb_2O_5 [338], phase boundaries [338], as well as enthalpies of formation and dissociation energies of NbO(g) and NbO_2(g) [339, 340]. The data on NbO(g) and NbO_2(g) are reviewed and partly reevaluated in Ref. 339. Balducci et al. [337] determined the dissociation energies of complex gaseous niobium molecules (Table 13) by studying the equilibrium over a condensed phase resulting from extensive vaporization of $LiNbO_3$.

Mixtures of different compounds were heated by Balducci et al. [342] in different Knudsen cells to show the existence of $CrPO_2$(g) and to study gas phase equilibria containing this molecule and CrO(g). The atomization energies of $CrPO_2$(g) and CrO(g) (Table 13) were evaluated from second- and third-law values agreeing very well.

Chizhikov et al. [344, 347] reinvestigated the disproportionation of MoO_2(s) [344] as well as WO_2 [347]; Marushkin et al. [345] as well as Aleshko-Ozhevskaya et al. [346] the congruent sublimation of WO_3(s). Fragmentation patterns were elucidated in the recent study of Marushkin et al. [345] yielding in the main improved data for the composition of the vapor and the congruently subliming phase.

Ikeda et al. [343] made extensive measurements on the sublimation of different molybdenum oxides and determined the partial pressures of the species listed in Table 12. The dissociation enthalpies obtained are based on a second- and third-law evaluation and agree fairly well with those given in the review by Gingerich [21].

The dissociation enthalpies of FeO(g) and MnO(g) (Table 13) were redetermined by Smoes and Drowart [348] from gas phase metathesis reactions, and relevant literature data are also critically reviewed. There is good agreement between the data of the second- and third-law evaluation obtained by Ref. 348. The dissociation energy of FeO(g) obtained by Murad [349] with a gas inlet cell agrees well with that by Drowart and Smoes. Murad additionally determined the dissociation energy of FeOH (Table 13).

Frisch and Dai [350] reinvestigated the vaporization of RuO_2(s) and showed that the O_2-decomposition pressures over a single crystal is smaller by about a factor of three than that of a powder sample.

Group IIb oxides were studied by three groups (E.K. Kazenas et al. [353], Grade and Hirschwald [354], Behrens and Mason [355], as well as Grade et al. [356]). These oxides vaporize predominantly by dissociation into gaseous metal atoms and O_2, and a much smaller amount of MO molecules (M = Zn, Cd, Hg).

The dissociation energy of ZnO(g), $D_0^\circ = 280 \pm 14\,\mathrm{kJ\,mol^{-1}}$, in Ref. 354 agrees well with the value by Ref. 353 given in Table 13. The dissociation energies of CdO(g) by Ref. 354, 230.3 kJ mol^{-1}, by Ref. 356, 373.2 kJ mol^{-1}, and by Ref. 353, 273.2 $\pm$ 4.2 kJ mol^{-1}, deviate from each other. The values by Refs. 353 and 354 were obtained by the third-law method. A second-law evaluation, as used by Ref. 356, is difficult due to the very small relative abundancy of CdO in the vapor.

Ovchinnikov et al. [325] observed no gaseous ternary oxides upon vaporizing $M(ReO_4)_3$(s) (M = Sc, Y, La) in Pt-Knudsen cells (Table 12) thereby demonstrating that these compounds vaporize by decomposition. This was also observed by Balducci et al. [351] upon vaporizing $CoTiO_3$(s) and $CoTi_2O_5$(s) in a molybdenum Knudsen cell lined with platinum except for the lid. The enthalpy of formation of $CoTi_2O_5$(s) was evaluated from the partial pressures and ancillary data as $\Delta_f H_{298}^\circ = 2162.7 \pm 34.3\,\mathrm{kJ\,mol^{-1}}$.

4.4 Oxides of Group IIIb, IVb, Vb Elements and Their Quasi-binaries with Transition Metals

The equilibrium vapor over Al_2O_3 and its free vaporization was studied by Chervonnyi et al. [358] as well as Paule [359], respectively. Chervonnyi et al. also detected ions in the vapor (Table 14). The existence of the molecule AlO_2(g) upon vaporizing Al_2O_3 is discussed by Kashireninov et al. [393] as well as by Farber and Srivastava [394].

Gomez et al. [145] reinvestigated the vaporization of In_2O_3(s) (Table 14) under controlled oxygen pressure and reevaluated the relevant literature data. The dissociation energy of InO(g) and the atomization energy of In_2O(g) obtained by them are not given in Table 15, since there are still inconsistencies in the thermodynamics of the indium-oxygen system (cf. Ref. 145). This is the case though the enthalpy of formation of In_2O_3(s) determined in Ref. 145, $\Delta_f H_{298}^\circ = -919 \pm 27\,\mathrm{kJ\,mol^{-1}}$, agrees well with the literature values obtained by different methods. Kaposi et al. [362] used data on the stability of InO(g) for the estimation of the enthalpies of formation and dissociation of In_2MoO_4(g). Kligina et al. [363] concluded that Tl_2CrO_4 may vaporize congruently and evaluated the enthalpy of atomization of Tl_2CrO_4(g) (Table 16).

The dissociation energy of SiO(g) determined by Kvande and Wahlbeck [364] according to the second-law method, $D_0^\circ = 805.8 \pm 10.9\,\mathrm{kJ\,mol^{-1}}$, is in agreement with the values adopted in the tables of Gingerich [21] and JANAF [90]. There is no basis for a revision of these values as follows from a critical assessment of the literature data (cf. Ref. 364). Plies [367] showed the existence of the molecules $GeWO_4$ and GeW_2O_7 and determined enthalpies of dissociation by the second-law method (Table 16). The dissociation energies of $(PbO)_n$(g) (n = 2, 3, 4 ... 6) were determined for the first time by Drowart et al. [395]. They were redetermined by Semenikhin et al. [369] (Table 14) by the use of the second- and third-law methods. The data obtained by the two groups agree

Table 14. Summary of mass spectrometric studies of oxides of group IIIb, IVb, Vb elements and their quasi-binaries with transition metals except those listed in Table 18. (The gaseous species or solid compounds are underlined if their enthalpies of formation or of dissociation, only gaseous species, are given)

System in the cell	Temperature range in K	Gaseous species	Ref./Year
B_2O_3	1550	BO, $\underline{BO_2}$, B_2O_2, B_2O_3	[357]/1980
Al_2O_3	2300–2600	Al, O, $\underline{AlO}$, Al_2O, $\underline{Al_2O_2}$, Al^+, Al_2O^+	[358]/1977
Al_2O_3	~2450	Al, O, AlO, $\underline{Al_2O}$, $\underline{Al_2O_2}$	[359]/1976
$Al_2O_3 + H_2(g)^a$	2030–2422	H, Al, O, AlO, Al_2O, $\underline{AlOH}$	[360]/1986
$Al(ReO_4)_3$, $Ga(ReO_4)_3$, $In(ReO_4)_3$	<1000	Re_2O_7	[361]/1981
$\underline{In_2O_3}$	1322–1520	In, O_2, $\underline{InO}$, In_2O,	[145]/1982
$In_2O_3 + MoO_2 + MoO_3$	1283–1478	In, In_2O, $(\underline{MoO_3})_2$, $(MoO_3)_3$, $\underline{In_2MoO_4}$	[362]/1985
Tl_2CrO_4	800–1000	Tl_2CrO_4	[363]/1986
$Si + SiO_2$	1243–1489	$\underline{SiO}$	[364]/1976
GeO_2	1453–1643	GeO, GeO_2, Ge_2O_2, Ge_3O_3	[365]/1981
GeO_2	1653	GeO, GeO_2, Ge_3O_3	[366]/1979
$GeO_2 + WO_2$	1258–1383	GeO, WO_3, $(WO_3)_2$, W_3O_8, $(WO_3)_3$, $\underline{GeWO_4}$, $\underline{GeW_2O_7}$	[367]/1982
SnO_2	1296–1419	O_2, SnO, Sn_2O_2	[368]/1976
PbO, $PbO - B_2O_3$	950–1150	Pb, O_2, PbO, $(PbO)_i$ (i = 2, 3, 4, 5, 6)	[369]/1983
$Pb + PbO + TeO_2$	900<	Pb, PbO, $\underline{PbO_2}$	[370]/1978
$PbMoO_4$, $PbWO_4$	<1200	Pb, $(MO_3)_3$, PbO, $(PbO)_2$, $PbMO_4$, Pb_2MO_5, PbM_2O_7, (M = Mo, W)	[371]/1984
As_2O_3	600	As_4O_6	[372]/1984

(Continued)

Table 14. (continued)

System in the cell	Temperature range in K	Gaseous species	Ref./Year
As_2O_4, As_2O_5, $UO_2(AsO_3)_2$, $(UO_2)_2As_2O_7$, $(UO_2)_3(AsO_4)_2$, $(UO_2)_3(As_2O_4)_2 + Fe_2O_3 + Bi_2O_3$, $(UO_2)_3(As_2O_4)_2 + Fe_2O_3 + PbO$	600–1245	Pb, Bi, O_2, PbO, Bi_2O_3, AsO, AsO_2, As_2O_3, As_4O_6, As_4O_{7+n} (n = 0 to 3), $PbAs_2O_4$, $Pb_2As_2O_5$, $Pb_3As_2O_6$	[372]/1984
$(Sb_{2-x}As_x)O_3$ (x = 1.9, 1.5, 1.0)	650, 780, 800	As_4O_6, $Sb_{4-r}As_rO_6$ (r = 0 to 4)	
As_2O_5	862–939	O_2, As_4O_6, As_4O_{7+n} (n = 0 to 3)	[373]/1983
As_2O_3, $As_2O_3 + CaO$	317–372	AsO, AsO_2, As_2O_3, As_4O_i (i = 3 to 6)	[44]/1982
Sb_2O_3	580–730	Sb_4O_6	[374]/1983
BiO_3	1003–1193	Bi, Bi_2, O_2, BiO, Bi_2O, Bi_2O_2, Bi_2O_3, Bi_3O_4, Bi_4O_6	[375]/1980
$MnSe + EuS + Fe_2O_3$	1304–1807	Mn, Fe, Se, Se_2, MnO, FeO, SO, SeO, SeS	[376]/1984
not specified	370–411	Se, SeO, SeO_2	[377]/1978

ª Gaseous species leaked into the Knudsen cell

Table 15. Enthalpy change of the dissociation reactions for selected molecules of binary group IVb, Vb, VIb, and lanthanide oxides

Molecule	Dissociation reaction	T/K	ΔH_T° kJ mol^{-1}	Ref.
PbO_2	$PbO_2 \rightleftarrows Pb + 2O$	0	772 ± 21	[370]
AsO_2	$AsO_2 \rightleftarrows As + 2O$	0	828 ± 7	[372]
As_2O_3	$As_2O_3 \rightleftarrows 2As + 3O$	0	1676 ± 8	[372]
As_4O_7	$As_4O_7 \rightleftarrows As_4O_6 + O$	0	342 ± 5	[372]
As_4O_8	$As_4O_8 \rightleftarrows As_4O_7 + O$	0	328 ± 6	[372]
As_4O_9	$As_4O_9 \rightleftarrows As_4O_8 + O$	0	313 ± 6	[372]
As_4O_{10}	$As_4O_{10} \rightleftarrows As_4O_9 + O$	0	296 ± 6	[372]
Sb_4O_6	$Sb_4O_6 \rightleftarrows 4Sb + 6O$	298	3762 ± 33	[374]
SeO	$SeO \rightleftarrows Se + O$	0	426 ± 6	[376]
PrO	$PrO \rightleftarrows Pr + O$	0	736 ± 13	[382]
NdO	$NdO \rightleftarrows Nd + O$	0	697 ± 13	[380]
SmO	$SmO \rightleftarrows Sm + O$	0	569 ± 8	[381]
EuO	$EuO \rightleftarrows Eu + O$	0	467 ± 7	[387]
GdO	$GdO \rightleftarrows Gd + O$	0	709 ± 13	[378]
HoO	$HoO \rightleftarrows Ho + O$	0	603 ± 13	[378]
ErO	$ErO \rightleftarrows Er + O$	0	602 ± 13	[378]
TmO	$TmO \rightleftarrows Tm + O$	0	510 ± 13	[378]
LuO	$LuO \rightleftarrows Lu + O$	0	667 ± 8	[378]

Table 16. Enthalpy change of the dissociation reactions for selected ternary oxide molecules of group IIIb, IVb elements, lanthanides and transition metals

Molecule	Dissociation reaction	T/K	ΔH_T° kJ mol^{-1}	Ref.
Tl_2CrO_4	$Tl_2CrO_4 \rightleftarrows 2Tl + Cr + 4O$	0	2445 ± 20	[363]
$GeWO_4$	$GeWO_4 \rightleftarrows GeO + 1/2W_2O_6$	1323	141 ± 4	[367]
GeW_2O_7	$GeW_2O_7 \rightleftarrows GeO + W_2O_6$	1338	323 ± 17	[367]
$EuTiO_3$	$EuTiO_3 \rightleftarrows Eu + Ti + 3O$	0	2278 ± 28	[383]
$EuVO_2$	$EuVO_2 \rightleftarrows Eu + V + 2O$	0	1504 ± 11	[384]
$EuVO_3$	$EuVO_3 \rightleftarrows Eu + V + 3O$	0	2193 ± 16	[384]
$EuTaO_3$	$EuTaO_3 \rightleftarrows Eu + Ta + 3O$	0	2463 ± 34	[385]
$EuCrO$	$EuCrO \rightleftarrows Eu + Cr + O$	0	928 ± 45	[385]
$EuCrO_2$	$EuCrO_2 \rightleftarrows Eu + Cr + 2O$	0	1427 ± 45	[385]
$EuWO_3$	$EuWO_3 \rightleftarrows Eu + W + 3O$	0	2383 ± 29	[387]
$EuWO_4$	$EuWO_4 \rightleftarrows Eu + W + 4O$	0	2990 ± 27	[387]
Eu_2WO_5	$Eu_2WO_5 \rightleftarrows 2Eu + W + 5O$	0	4017 ± 46	[387]
EuW_2O_7	$EuW_2O_7 \rightleftarrows Eu + 2W + 7O$	0	5340 ± 38	[387]

within the given error limits. Zmbov and Miletic [370] identified mass spectrometrically the molecule PbO_2 and determined its enthalpy of atomization (Table 15).

Three groups (Drowart [372], Plies and Jansen [373], Brittain et al. [44]) investigated the complex As–O system. Vaporization processes far from equilibrium as well as the thermochemistry of gas phase oxygen metathesis reactions and other equilibria were elucidated in Ref. 372. The dissociation energies obtained are given in Table 15. Saturated and unsaturated superheated vapors were used in Ref. 44 to identify thermodynamically stable As/O molecular species. The study of Sidorov et al. [375] shows that the Bi–O system exhibits complexity similar to the As–O system (cf. Table 14). Heats of formation of gaseous bismuth oxides were estimated.

Smoes and Drowart [376] determined the dissociation enthalpy of SeO by the study of an isomolecular exchange reaction. There is excellent agreement between the second- and third-law evaluation. The dissociation enthalpy obtained agrees excellently with the spectroscopic value and is given in Table 15. By comparison with spectroscopic data an enthalpy of dissociation, $\Delta H_0^\circ(SeO)$ = $424.7^{+0.6}_{-6.0}$ kJ mol^{-1}, is selected as best estimate [376].

4.5 Oxides of Lanthanides or Actinides and Their Quasi-binaries with Transition Metals

The dissociation enthalpies of some lanthanide monoxides were redetermined by Murad and Hildenbrand [378], Murad [380], Hildenbrand [381], as well as Balducci et al. [387] by the study of isomolecular exchange reactions. The enthalpies of dissociation obtained are listed in Table 15. The errors of these values are considerably less than those given in the review by Gingerich [21] for these dissociation enthalpies. Various aspects of the bonding of the lanthanide monoxide series are discussed by Murad and Hildenbrand [378].

The dissociation enthalpies of the lanthanide dioxide molecules (cf. Table 17) obtained by Kordis and Gingerich [382] are given in the review by Gingerich [21].

Balducci et al. [383–387] identified numerous gaseous ternary euro-pium-containing high-temperature species of the systems Eu–X–O (cf. Table 17). The thermochemical properties of these species and of further molecules of the composition $EuXO_n(g)$ (n = 1 to 4; X = IVa, Va, VIa, group metals) are summarized and discussed by Balducci et al. [385]. Some are given in Table 16.

Storms [389] redetermined the uranium activity for different compositions between the phases $U(1) + UO_{2-x}$ and $UO_{2.0}$. Partial pressures of U, UO, UO_2, UO_3, O, and O_2 are given. The measured dependence of the enthalpy of vaporization on the composition does not agree with that of previous measurements. Matsui and Naito [391] studied the vaporization of $(U_{1-y}Nb_y)O_{2+x}$ and of $UO_{2.000}$ due to its importance for fuel-cladding interactions in nuclear reactors. The UO_3 and UO_2 partial pressures determined by these authors over $UO_{2.000}$ agree reasonably well with those by Storms [389]. Belov et al. [390] showed that heating U_3O_8 in vacuum was accompanied by partial distillation of oxygen with the formation of a congruently vaporizing suboxide phase of composition $UO_{1.97}$ at 2400 K. UO_2 partial pressures were determined for this composition and temperature.

Table 17. Summary of mass spectrometric studies of oxides of lanthanides or actinides and their quasi-binaries with transition metals except those listed in Table 18. (The gaseous species are underlined if their enthalpies of formation or of dissociation are given)

System in the cell	Temperature range in K	Gaseous species	Ref./Year
$Gd_2O_3 + Y + Y_2O_3$	1908–1977	Y, Gd, YO, GdO	[378]/1980
$Gd + Gd_2O_3 + TiO_2$	1957–2024	Ti, Gd, TiO, GdO	
$Ho_2O_3 + Er_2O_3 + Ti + TiO_2$	1855–2178	Ti, Ho, Er, TiO, HoO, ErO	
$Tm_2O_3 + Al_2O_3$	2249–2364	Al, Tm, AlO, TmO	
$Lu + Lu_2O_3 + Y + Y_2O_3$	1849–1944	Y, La, LuO, YO	
Pr_2O_3, $Pr_2O_{3.67}$	2270 and 2250	PrO, PrO_2	[379]/1979
$Pr_6O_{11} + Ti$	1750–2000	Ti, Pr, TiO, PrO	[380]/1978
$Nd_2O_3 + Sc_2O_3 + Ti$	1850–2000	Sc, Ti, Nd, ScO, TiO, NdO	
$Sm_2O_3 + Al_2O_3$	2087–2298	Al, Sm, AlO, SmO	[381]/1977
$Eu_2O_3 + Sm_2O_3 + Ti + TiO_2$	2120–2242	Ti, Sm, Eu, TiO, SmO, EuO	[382]/1977
HoN, GdN, EuN, Tb, Lu, Lu + Ag + Au (all with O-traces)	1957–2701	M, Lu, MO, M_2O, M_2O_2, GdO_2, HoO_2, LuO, Lu_2O, (M = Eu, Gd, Tb, Ho)	
$Eu_2O_3 + TiO_2 + Mo$	2220–2490	Eu, TiO, TiO_2, EuO, EuTiO_3	[383]/1985
$Eu_2O_3 + V_2O_5 + V + Mo$	1915–2282	Eu, VO, VO_2, EuO, EuVO_2, EuVO_3	[384]/1983
$Eu_2O_3 + Ta$	2000–2400	TaO_2, EuO, EuTaO_3	[385]/1986
Eu_2O_3 and/or $EuPO_4 + (Cr + Cr_2O_3) + W$	2000–2400	Cr, CrO, CrO_2, WO_2, WO_3, EuO, EuCrO, EuCrO_2, EuWO_4	
$Eu_2O_3 + EuPO_4 + Mo$, $Eu_2O_3 + Mo$, $EuPO_4 + Mo$	1750–2380	Eu, MoO_2, MoO_3, PO, PO_2, EuO, EuMoO_3, EuMoO_4	[386]/1977
$Eu_2O_3 + EuPO_4 + W$, $Eu_2O_3 + W$, $EuPO_4 + W$	1945–2255	Eu, WO_2, WO_3, EuO, EuWO_3, EuWO_4, Eu_2WO_5, EuW_2O_7	[387]/1977

(Continued)

Table 17. (Continued)

System in the cell	Temperature range in K	Gaseous species	Ref./Year
ThO_2	2420–2950	ThO, ThO_2	[333]/1979
$^{231}Pa + {}^{231}PaO_2$	2000–2400	PaO, PaO_2	[388]/1981
$U(l) + UO_{2-x}$, UO_{2-y} ($-0.004 \leq y \leq 0.058$)	1667–2175	U, UO, UO_2, UO_3	[389]/1985
U_3O_8	1650–2450	UO, UO_2, UO_3	[390]/1984
UO_2	2182–2343	UO_2, UO_3	[391]/1985
$(U_{1-y}Nb_y)O_{2+x}$ ($y = 0.01, 0.05$, $x = 0.000–0.022$)	2025–2343	NbO, NbO_2, UO_2, UO_3	[391]/1985
U–O, Ce–O, U–Ce–O, U–La–O, Ce–La–O, Ce–La–Y–O	1500–2300	Y, La, Ce, U, YO, LaO, CeO, CeO_2, UO, UO_2, UO_3	[392]/1981

Younes et al. [392] studied different systems (Table 17) to understand the high-temperature behavior of nuclear fuels. The enthalpy changes of different isomolecular gas-phase oxygen-exchange reactions between the molecules listed in Table 17 were determined according to the second-law method.

4.6 Investigation of Melts, Glasses, and Solid Solutions of Quasi-binary, -ternary, and -quarternary Systems

Table 18 gives an account of the investigations since about 1979. They are grouped under glass-forming and non-glass-forming systems. The term glass-forming-system means that the system is composed of a network-building (e.g. SiO_2, B_2O_3) and a network-modifying oxide (e.g. Na_2O, K_2O) or of a network-building oxide only. The determination of properties of condensed phases was the aim of these studies by investigating their equilibrium vapor. The behavior of evaporation was elucidated and mixing properties of the condensed phase were generally evaluated. Knudsen effusion mass spectrometry is an important method for determining such properties. However, the method was not yet so widely used to study oxide systems as for metal alloys.

4.6.1 Glass-forming Systems

The knowledge of the thermodynamic properties of glass components at high temperatures in both the condensed and gaseous state are important for technical applications. The study of melts renders possible predictions concerning the solid state. The knowledge of the high temperature vaporization is, moreover, useful since glasses are generally produced by high-temperature processes.

Shul'ts et al. [396, 397, 399, 400] as well as Stolyarova and Semenov [407] studied the melts of the quasi-ternary Na_2O–B_2O_3–GeO_2 system as well as its constituent two component systems Na_2O–B_2O_3, Na_2O–GeO_2, and B_2O_3–GeO_2 (see Table 18). Partial pressures [399, 407] as well as the excess integral thermodynamic functions of the components of the melts [396] were determined. These functions are negative for the Na_2O–B_2O_3 and Na_2O–GeO_2 systems, but positive for the B_2O_3–GeO_2 system.

Altemose and Tong [398] measured Na and O_2 partial pressures over different Na_2O/SiO_2 samples, Plante [135] those of K over K_2O/SiO_2 samples. The concentration ranges are given in Table 18. Pressures are reported in detail by Ref. 135 unlike Ref. 398. Enthalpy changes of reactions of vaporization are given in Ref. 398 for different compositions.

Kowalska et al. [409] combined the results of their measurements on the K_2O–Cu_2O–SiO_2 system with those by Plante [135] and other workers to determine the isoactivity curves of SiO_2 in the melt of this quasi-ternary system.

Kambayashi and Kato [401, 402] determined the chemical activities of MgO and SiO_2 at temperatures of 1873 and 1973 K. Data were obtained for the

Table 18. Mass spectrometric investigations of melts, glasses, and solid solutions of quasi-binary, ternary, and quarternary systems

System in the cell	Temperature range in K	Gaseous species	Ref./Year
Glass forming systems			
$xNa_2O + (1-x)B_2O_3$ ($x = 0-0.5$)	1240–1380	B_2O_3, $NaBO_2$, $(NaBO_2)_2$	[396]/1980, [397]/1979
$xNa_2O + (1-x)SiO_2$ ($x = 0.015-0.26$)	1170–1620	Na, O_2	[398]/1980
$xNa_2O + (1-x)GeO_2$ ($x = 0-0.5$)	1490–1550	Na, O_2, GeO	[396]/1980, [399]/1979, [400]/1979
$xK_2O + (1-x)SiO_2$ ($x = 0.15-0.33$)	1300–1800	K	[135]/1979
$xMgO + (1-x)SiO_2$ ($x = 0.20-0.78$)	1873, 1973	Mg, SiO	[401]/1984, [402]/1983
$xB_2O_3 + (1-x)SiO_2$ ($x = 0-1$)	1350–1590	B_2O_3	[403]/1987, [404]
$xB_2O_3 + (1-x)GeO_2$ ($x = 0.1-0.9$)	1390–1480	O_2, B_2O_3, GeO	[396]/1980, [399]/1979
$xB_2O_3 + (1-x)Bi_2O_3$ ($x = 0-0.9$)	1120–1310	Bi, Bi_2, O_2, B_2O_3, BiO, Bi_2O, Bi_2O_2, Bi_2O_3, Bi_3O_4, Bi_4O_6	[405]/1981
$xSiO_2 + (1-x)GeO_2$ ($x = 0-0.95$)	1176–2000	O_2, SiO, SiO_2, GeO	[398]/1980
$xGeO_2 + (1-x)Sb_2O_3$ ($x = 0.25, 0.5, 0.75$)	580–1450	O_2, GeO, Sb_2O_2, Sb_2O_4, Sb_3O_3, Sb_4O_6	[406]/1981
$x_1Na_2O + x_2B_2O_3 + (1-x_1-x_2)Ge_2O_3$	1270–1370	Na, O_2, B_2O_3, GeO, $NaBO_2$, $(NaBO_2)_2$	[396]/1980, [399]/1979, [400]/1979, [407]/1979
$0.2Na_2O + 0.2B_2O_3 + 0.6SiO_2$	915–1172	$NaBO_2$, $(NaBO_2)_2$	[408]/1986
$0.18Na_2O + 0.27B_2O_3 + 0.55SiO_2$	936–1158	$NaBO_2$, $(NaBO_2)_2$	
$0.27Na_2O + 0.18B_2O_3 + 0.55SiO_2$	915–1144	Na, $NaBO_2$, $\underline{(NaBO_2)_2}$	
$x_1K_2O + x_2Cu_2O + (1-x_1-x_2)SiO_2$ ($x_2 < 0.10, 1-x_1-x_2 = 0.60-0.854$)	1340–1390	K, Cu	[409]/1986

System	Temperature	Species	Reference
$xNaAlSi_3O_8 + (1-x)KAlSi_3O_8$ (x = 0–1)	1150–1880	Na, K	[410]/1982, [411]/1983
$xNaAlSi_4O_{10} + (1-x)KAlSi_4O_{10}$ (x = 0–1)	1450–1850	Na, K	[412]/1985
$xNaAlSi_5O_{12} + (1-x)KAlSi_5O_{12}$ (x = 0–1)	1520–1770	Na, K	
Non-glass-forming systems			
$xMgO + (1-x)Al_2O_3$ (x = 0.5, 0.33, 0.26)	1850–2300	Mg, Al, AlO	[413]/1981
$xCaO + (1-x)Al_2O_3$ (x = 0.35–0.65)	2060	Ca, Al, O, AlO, Al$_2$O	[414]/1981
$xZrO_2 + (1-x)M_2O_3$ (M = Gd, Dy, Ho, Er, Y, Lu)	2700–2900	Zr, ZrO$_2$, MO except for M = Yb, additionally in some cases: M, O	[415]/1984, [416]/1981, [417]/1980
$xZrO_2 + (1-x)Pr_2O_{3.64}$ (x = 0.35, 0.45)	2370–2850	ZrO, ZrO$_2$, PrO$_2$, PrO$_2$	[379]/1979
$xU_3O_8 + (1-x)ZrO_2$	1900–2400	UO, UO$_2$, UO$_3$	[390]/1984
$xU_3O_8 + (1-x)Y_2O_3$	1900–2400	UO, UO$_2$, UO$_3$	
$xU_3O_8 + (1-x)(ZrO_2 + Y_2O_3)$	199–2400	UO, UO$_2$, UO$_3$	
$xLiReO_4 + (1-x)CsReO_4$ (x = 0–1)	720–840	(LiReO$_4$)$_n$, (CsReO$_4$)$_n$, n = 1, 2; LiCs(ReO$_4$)$_2$	[156]/1979
$xKReO_4 + (1-x)RbReO_4$ (x = 0.094–0.857)	775–875	(KReO$_4$)$_n$, (RbReO$_4$)$_n$, n = 1, 2; KRb(ReO$_4$)$_2$	[418]/1980
$xKReO_4 + (1-x)CsReO_4$ (x = 0.1–0.9)	765–840	(KReO$_4$)$_n$, (CsReO$_4$)$_n$, n = 1, 2; KCs(ReO$_4$)$_2$	[419]/1980
$x_1CaF_2 + x_2CaO + (1-x_1-x_2)Al_2O_3$	1600, 1700	CaF, CaF$_2$, AlF, AlOF	[420]/1979

melt and the two-phase regions $MgO + Mg_2SiO_4$, Mg_2SiO_4 + liquid, and SiO_2 + liquid being present at these temperatures for different compositions. The phase boundaries between the melt and the two-phase regions Mg_2SiO_4 + liquid and SiO_2 + liquid were redetermined at the measurement temperatures. Gibbs energies of formation resulted for $Mg_2SiO_4(s)$. Thermodynamic data for assumed species in the melt were evaluated.

A negative deviation from ideal behavior results from the data by Shul'ts et al. [403] for the chemical activity of B_2O_3 in the melt of the B_2O_3–SiO_2 system whereas Boike et al. [404] obtained a positive deviation (cf. Fig. 7). The result by Boike et al. [404] is in agreement with the phase diagram of the B_2O_3–SiO_2 system by Rockett and Foster [421] indicating a miscibility gap. Moreover, it is supported by calorimetric measurements by Hervig and Navrotski [422] showing positive enthalpies of mixing. The result by Shul'ts et al. [403] might be caused by the procedure of the measurement used, which is not based on several successive comparisons between the pressures over pure B_2O_3 and the B_2O_3/SiO_2 melt for each composition as carried out by Boike et al. [404].

Minaeva et al. [405] measured the chemical activity of Bi_2O_3 for the B_2O_3–Bi_2O_3 system. The results support the existence of a miscibility gap between 10 and 30 mol% Bi_2O_3 at 1230 K.

Asano and Yasue [408] determined the vapor species and their partial pressures over sodium borosilicate glasses with different Na_2O/B_2O_3 ratios which are of interest as stable host matrices for the storage and disposal of high-level radioactive wastes (cf. Table 18). The total vapor pressure is the highest over the glass with the Na_2O_3/B_2O_3 ratio larger than unity.

Extensive measurements were carried out to determine the mixing properties of different aluminosilicate melts. Two groups, Rammensee and Fraser [410] as well as Rogez et al. [411], studied the pseudobinary join $NaAlSi_3O_8$–$KAlSi_3O_8$

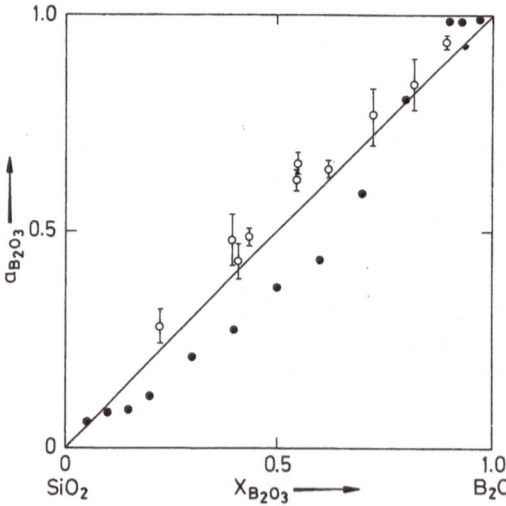

Fig. 7. Chemical activity of B_2O_3 at different compositions of the melt (Ref. 404 at 1475 K, ○; evaluated from Ref. 403 at 1588 K, ●)

with Mo and Pt Knudsen cells, respectively. Fraser and Bottinga [423] reevaluated the chemical activities of $NaAlSi_3O_8$ and $KAlSi_3O_8$ obtained by Rogez et al. and showed that they agree very well with those by Rammensee and Fraser. It is also shown that the Knudsen cell measurements and the calorimetric data of the same system are in good agreement. The different aluminosilicate melts of this system mix close to ideally over most of the composition range. Near ideal mixing behavior was also observed by Fraser et al. [412] across the two joins $NaAlSi_4O_{10}$–$KAlSi_4O_{10}$ and $NaAlSi_5O_{12}$–$KAlSi_5O_{12}$. Fraser and Bottinga [423] also discuss the pros and cons of Knudsen effusion mass spectrometry and solution calorimetry for the aforementioned investigations.

4.6.2 Non-glass-forming Systems

Sasamoto et al. [413] determined the Mg partial pressures over the MgO/Al_2O_3 spinel solid solution system (see Table 18) by the use of W Knudsen cells with a ThO_2 liner. Chemical activities of MgO were evaluated by comparing the Mg pressures over the solid solution with those obtained on vaporizing MgO(s) from the same cell.

Allibert et al. determined the chemical activities of CaO and Al_2O_3 at 2060 K for the melt of the CaO–Al_2O_3 system [414] as well as Allibert and Chatillon those of CaO and CaF_2 at 1600 and 1700 K for the liquid range of the CaF_2–CaO–Al_2O_3 system [420]. Heats of mixing for the melt as well as Gibbs energies of formation for the solid interoxide compounds were evaluated for the CaO–Al_2O_3 system by combining the measured activities with appropriate literature data [414]. Isoactivity curves are given in the CaF_2–CaO–Al_2O_3 system [420].

The use of ZrO_2 as a high-refractory structural material is only possible after its stabilization as a cubic solid solution formed by the addition of certain oxides such as those of the lanthanides and Y_2O_3. The thermal stability of these solid solutions is thus an important practical problem and initiated the following studies.

Belov et al. [379, 415–417] investigated the evaporation of solid solutions between ZrO_2 and different lanthanide oxides as well as Y_2O_3 (see Table 18) by measuring the ion intensities of the vapor species effusing from the cell at different constant temperatures for several hours. Partial pressures of ZrO, YO, and lanthanide monoxide vapor species were determined except for the ZrO_2–Yb_2O_3 solid solution. Two extreme cases can be distinguished for the evaporation process: (a) complete evaporation of the lanthanide oxide from the solid solution by decomposition forming ZrO_2(s) in practical terms and (b) formation of an azeotropic composition and total evaporation of the sample at invariable vapor composition. An example of case (a) is the vaporization of the ZrO_2–$Pr_2O_{3.6}$ solid solution [379]. The ZrO_2–Lu_2O_3 solid solution vaporizes according to case (b) with the azeotropic composition $x_{Lu_2O_3} = 0.5$ at 2700 K. Extensive measurements were carried out for the ZrO_2–Lu_2O_3 solid solution by vaporizing samples of the complete composition range of the system [415]. These measurements showed an initial distillation of oxygen from the samples leading to an oxygen depletion in the solid phase as observed for the pure component oxides. Samples with compositions of the ZrO_2 rich part of the phase diagram up to $X_{M_2O_3} = 0.2$ were vaporized to study the ZrO_2–M_2O_3 solid solutions (M = Y, Gd, Dy, Ho, Er, Yb, Lu).

The U_3O_8-containing solid solutions given in Table 18 evaporate at 2300 K by practically total distillation of U_3O_8 from these solutions forming finally condensed phases made of the remaining oxides (Belov et al. [390]). The vaporization of U_3O_8 proceeds in the same way as for pure $U_3O_8(s)$ (see Sect. 4.5). The UO_2 partial pressures determined over the different solutions at 1400 K are directly proportional to the U_3O_8 concentration in the condensed phase.

Lukas and Kowalska [418] as well as Lukas et al. [156, 419] investigated the vaporization of the perrhenate systems $MReO_4$–$M'ReO_4$ with different alkali metals M, M'. Melts as well as solid solutions or both are present at temperatures of 806 and 855 K ($KReO_4$–$RbReO_4$, Ref. 418), 833 K ($LiReO_4$–$CsReO_4$, Ref. 156), as well as at 806 K ($KReO_4$–$CsReO_4$, Ref. 419) for which the mixing properties were evaluated. Positive ($KReO_4$–$RbReO_4$, $KReO_4$–$CsReO_4$) and negative ($LiReO_4$–$CsReO_4$) deviations from ideal behavior are obtained for the systems given in parentheses.

5 Halides

5.1 General Review and General Aspects

Metal halides in general vaporize forming monomer molecules of the type MX_i or $M'X_j$ and gaseous homo-complexes such as $(MX_i)_n$ and $(M'X_j)_n$, $n = 2, 3, 4$. In addition, gaseous hetero-complexes, such as $MM'X_{i+j}$, can be present in the vapor. The metallic ions are bound by halogen bridges in these complexes. The formation of gaseous complexes from the monomeric metal halide species has often been observed and is a very interesting phenomena in inorganic gas phase chemistry.

The different homo- and hetero-complexes observed and their thermochemical data are summarized in various review articles by McPhail et al. [424], Schäfer [425, 426], Øye and Gruen [427], Hastie [428, 429], Novikov and Gavryuchenkov [430], Büchler and Berkowitz-Mattuk [268], as well as Bauer and Porter [59]. Gaseous metal halide species is the only subject of these articles. Rules were developed to predict many properties of complexes (see Refs. 426, 428, 430). Brooker and Papatheodorou [431] as well as Papatheodorou [432] give an account of the vibrational properties and spectroscopic studies of the complexes. Theoretical and experimental investigations of alkali halide clusters are reviewed by Martin [433].

The experimental and theoretical studies on alkali halides are discussed and summarized in a book edited by Davidovits and McFadden [434]. Molecular parameters and thermodynamic properties including partial pressures are given for alkali halides and metal dihalides by Brewer and Brackett [435] as well as Brewer et al. [436], respectively.

The study of gaseous metal halide complexes is of fundamental interest for the chemistry of coordination compounds. In addition, gaseous metal halide species are of practical importance for material science and technical applications. These practical aspects are reported briefly in the recent review by Schäfer [425] and in detail in the book by Hastie [428]. The potential of enhanced vapor phase material transport is for example an important practical aspect and has been described in the monograph by Schäfer [437]. It is of interest for metal halide lamps, chemical vapor deposition, and metallurgical processes. The practical significance of complexation for metal halide lamps and the experimental methods used in addition to Knudsen effusion mass spectrometry for the study of metal halide vapors are reported in the recent review article by Hilpert [438].

The aforementioned review papers [59, 268, 424–430] reveal that a large number of metal fluoride and metal chloride complexes have been identified and investigated so far whereas our knowledge of the complexation of metal bromides and iodides is comparatively scanty.

An account of the investigations carried out by Knudsen effusion mass spectrometry since 1980 is given by Tables 19 to 21. The condensed phases vaporized in the Knudsen cell, the temperature range of the measurements, and the gaseous species determined in the equilibrium vapor are listed. Some of the measurements in Tables 19 to 21 are discussed in Sects. 5.2, 5.3, and 5.4. Sections 5.5 and 5.6 summarize some of the results obtained.

5.2 Vapors of Alkali Metal Halides

The evaporation of lithium fluoride (cf. Table 19) was reinvestigated by Yamawaki et al. [439] due to its importance for molten salt reactors and for blanket materials of a fusion reactor. They summarize the results obtained so far by other workers. Their results agree reasonably well with those given in the JANAF tables [90]. Particular emphasis was laid on the elucidation of the fragmentation observed by electron impact ionization. Mohazzabi and Searcy [480] used the flux gradient of a porous barrier made from nickel powder for such investigations whereas Grimley et al. [481] employed the angular distribution technique (cf. Sect. 2.3) which has proven particularly valuable in the determination of fragmentation patterns.

The species $(NaBr)_i(g)$ (i = 1, 2, 3) were identified by Hilpert and Miller [463] in the vapor over sodium bromide. A consistent set of thermodynamic data based on a second- and third-law evaluation was obtained unlike Refs. 448, 464. The mole fraction of the dimer at the mean temperature of 750 K of Hilpert and Miller's measurements is 52 percent smaller than that given by the JANAF tables [90].

Table 19. Summary of mass spectrometric Knudsen effusion studies of binary halides since about the year 1980. (The gaseous species or solid compounds are underlined if enthalpies of dissociation or formation are given)

System in the cell	Temperature range in K	Gaseous species	Ref./Year
Fluorides			
LiF	1006–1200	LiF, $(LiF)_2$, $(LiF)_3$	[439]/1982
ScF_3	1159–1411	ScF_3	[440]/1980
MoF_4	470–970	MoF_5	[441]/1985
WF_4	325–430	WF_6	
$Fe + MgF_2$	1338–1552	MgF, $\underline{FeF}$, $\underline{FeF_2}$	[442]/1985, [443]/1986
$Ni + NiF_2$	1278–1482	$\underline{NiF}$	[443]/1986
$Pt + F(g)^a$ and $Pt + MnF_3$	550–1120	MnF_3, MnF_4, PtF_2, PtF_4	[444]/1983
$Pt + TbF_4$	650–720	F_2, PtF_4, PtF_6	[445]/1986
$Au + XeF_2 \cdot 2MnF_4$	494–976	AuF_3, $(AuF_3)_2$, $\underline{(AuF_3)_3}$	[446]/1986
$BF_3(g)^a$	1780–1900	BF, $\underline{BF_2}$, $\underline{BF_3}$	[447]/1984
Chlorides			
NaCl	873–1163	$NaCl$, $\underline{(NaCl)_2}$, $\underline{(NaCl)_3}$	[448]/1981
KCl	818–1305	KCl, $\underline{(KCl)_2}$, $\underline{(KCl)_3}$	
RbCl	788–1113	$RbCl$, $\underline{(RbCl)_2}$, $\underline{(RbCl)_3}$	
CsCl	708–1043	$CsCl$, $\underline{(CsCl)_2}$, $\underline{(CsCl)_3}$	
$CaCl_2$	1173	$CaCl_2$, $\underline{(CaCl_2)_2}$	[449]/1982
MCl_2 ($M = Sr$, Ba)	1400	MCl_2, $\underline{(MCl_2)_2}$	
ZrCl	887–976	$ZrCl_4$	[450]/1985
$CrCl_3$	not specified	$CrCl_3$, $CrCl_4$	[451]/1981
MCl_2 ($M = Ti$, V, Cr, Mn, Fe, Co, Ni, Zn), VCl_3, $ScCl_3$	510–1040	$TiCl_3$, $TiCl_4$, VCl_3, VCl_4, MCl_2, $(MCl_2)_2$ ($M = Cr$, Mn, Fe, Co, Zn), $(CrCl_2)_3$, $NiCl_2$, $(ScCl_3)_i$ ($i = 1, 2, 3$)	[452]/1980

Pd_6Cl_{12}, Pt_6Cl_{12}	~423	Pd_6Cl_{12}, Pt_6Cl_{12}	[453]/1984
$ZnCl_2$	455–570	$ZnCl_2$	[454]/1987
$CdCl_2$	560–700	$CdCl_2$	[455]/1987
$AlCl_3$	73, 146, 383	$AlCl_3$, $(AlCl_3)_2$, $(AlCl_3)_3$	[456]/1986, [457]/1980
$Ge + GaCl_3(g)$[a]	673–873	$GaCl$, $GaCl_3$, Ga_2Cl_4, Ga_2Cl_6, $GaCl_2$, $GaGeCl_5$	[494]/1980
$TlCl$	467–607	$TlCl$, $(TlCl)_2$	[458]/1984
$SnCl_2$	576–694	$SnCl_2$, $SnCl_4$, $(SnCl_2)_2$	[459]/1987
$BiCl_3$	390–475	$BiCl_3$	[460]/1982
$EuCl_2$	not specified	$EuCl_2$	[461]/1982
UC or (UC + Cu) + $Cl_2(g)$[a] or $CaCl_2(g)$[a]	610–2440	Ca, Cu, Cl_2, U, $CaCl$, $CuCl$, UCl, UCl_2, UCl_3, UCl_4, UCl_5	[462]/1982
Bromides			
$NaBr$	673–1103	$NaBr$, $(NaBr)_2$, $(NaBr)_3$	[463], [464]/1986, [448]/1981
KBr	783–1103	KBr, $(KBr)_2$, $(KBr)_3$	[448]/1981
$RbBr$	758–1088	$RbBr$, $(RbBr)_2$, $(RbBr)_3$	
$CsBr$	678–993	$CsBr$, $(CsBr)_2$, $(CsBr)_3$	
$CaBr_2$	1023–1243	$CaBr_2$, $(CaBr_2)_2$	[449]/1982
$SrBr_2$	1133–1283	$SrBr_2$, $(SrBr_2)_2$	
$BaBr_2$	1203–1233	$BaBr_2$, $(BaBr_2)_2$	
$CdBr_2$	560–670	$CdBr_2$	[455]/1987
$TlBr$	510–622	$TlBr$, $(TlBr)_2$	[458]/1984
$Si + SiBr_4(g)$[a]	1054–1603	$SiBr$, $SiBr_2$, $SiBr_3$, $SiBr_4$	[465]/1980
$SnBr_2$	373–573	$SnBr_2$, $(SnBr_2)_2$	[466]/1990, [464]/1986
$BiBr_3$	400–478	$BiBr_3$	[460]/1982
$UC + Br_2(g)$[a] or $UBr_4 + Br_2(g)$[a]	632–2405	Br, Br_2, U, UBr, UBr_2, UBr_3, UBr_4, UBr_5	[467]/1987

(Continued)

Table 19. (continued)

System in the cell	Temperature range in K	Gaseous species	Ref./Year
Iodides			
NaI	495–1048	NaI, $(NaI)_2$, $(NaI)_3$, $(NaI)_4$	[468]/1986, [51]/1984, [469]/1983, [470]/1982
KI	748–1073	KI, $(KI)_2$, $(KI)_3$, $(KI)_4$	[468]/1986, [470]/1982
RbI	733–978	RbI, $(RbI)_2$, $(RbI)_3$	[470]/1982
CsI	592–958	CsI, $(CsI)_2$, $(CsI)_3$, $(CsI)_4$	[469]/1983, [470]/1982, [109]/1985, [50]/1984, [471]/1982
CaI_2	869–1108	CaI, $(CaI_2)_2$	[449]/1982, [472]/1986
SrI_2	943–1053	SrI_2, $(SrI_2)_2$	[449]/1982
BaI_2	1063–1213	BaI_2, $(BaI_2)_2$	
ScI_3	734–897	ScI_3, $(ScI_3)_2$	[52]/1985, [473]/1985
FeI_2	563–725	I, I_2, FeI_2, FeI_3, $(FeI_2)_2$, $(FeI_2)_3$	[60]/1985, [474]/1984
AgI	750–860	I, I_2, AgI, $(AgI)_3$, $(AgI)_4$	[65]/1980
ZnI_2	474–554	ZnI_2, $(ZnI_2)_2$	[53]/1985
CdI_2	460–590	CdI_2, $(CdI_2)_2$	[455]/1987
HgI_2	340–400	HgI_2	[475]/1986
Hg_2I_2	340–410	Hg, HgI_2	
TlI	493–615	TlI, $(TlI)_2$	[458]/1984
SnI_2	474–582	SnI_2, $(SnI_2)_2$	[52]/1985, [476]/1981
PbI_2	539–670	PbI_2, $(PbI_2)_2$	[52]/1985
AsI_3	300–900	I, I_2, As, As_4, AsI, AsI_2, AsI_3, $(AsI_3)_2$	[501]/1980
DyI_3	830–1150	DyI_3, $(DyI_3)_2$, $(DyI_3)_3$	[477]/1983, [538]
HoI_3	935–1055	HoI_3, $(HoI_3)_2$	[478]/1986
ThI_4	617–760	I_2, ThI_4	[479]/1986

ᵃ Gaseous species leaked into the Knudsen cell

Table 20. Summary of mass spectrometric Knudsen effusion studies since about the year 1980 of quasi-binary systems of metal halides. (The gaseous species or solid compounds are underlined if enthalpies of dissociation or formation are given)

System in the cell	Temperature range in K	Gaseous species	Ref./Year
Systems composed of fluorides			
$MF + MnF_2$	1145, 1170, 1182	MF, $(MF)_2$, MnF_2, $MMnF_3$, $KMnF_4$, $MMnF_4$, M_2MnF_4	[502]/1982
$MF + MnF_2 +$ traces of AlF_3 ($M = $ Li, Na, K)	1000, 1044, 1125	MnF_3^-, $Mn_2F_5^-$, AlF_4^-, $Al_2F_7^-$, $MAl_2F_8^-$, $LiMnAlF_7^-$, $Li_2MnAlF_8^-$	
$MF + AlF_3$ ($M = $ Li, Na, K, Rb, Cs)	800–890	$MAlF_4$	[503]/1980
$xNaF + (1-x)ThF_4$	1163, 1204, 1241	NaF_c, $(NaF)_2$, ThF_4, $NaThF_5$	[504]/1983
$NaF + ThF_4 +$ traces of ZrF_4	1005–1123	ThF_5^-, ZrF_5^-, $Zr_2F_9^-$	
$xKF + (1-x)ZrF_4$	863–1165	KF, $(KF)_2$, ZrF_4, $KZrF_5$, K_2ZrF_6, KZr_2F_9	[505]/1981
$CsF + ZrF_4$	810, 827	CsF, ZrF_4, $CsZrF_5$, Cs_2ZrF_6, $CsZr_2F_9$	[506]/1980
K_2PtF_6	980–1120	F, KF	
$KF + GaF_3 +$ traces of AlF_3	819–913	AlF_4^-, $Al_2F_7^-$, GaF_4^-, $Ga_2F_7^-$, $GaAlF_7^-$, KGa_2F_8, KAl_2F_8	[507]/1985
$xRbF + (1-x)ZrF_4$	839–1122	RbF_c, ZrF_4, $RbZrF_5$, $RbZr_2F_9$	[508]/1983
$xRbF + (1-x)ZrF_4 +$ traces of AlF_3	947–979	ZrF_5^-, $Zr_2F_9^-$, AlF_4^-, $RbAl_2F_8^-$	
Systems composed of chlorides			
$xLiCl + (1-x)MgCl_2$	1023	$LiCl$, $(LiCl)_2$, $(LiCl)_3$, $MgCl_2$, $(MgCl_2)_2$, $LiMgCl_3$, Li_2MgCl_4	[509]/1983
$MCl + MgCl_2$ ($M = $ Li, K)	not specified	MCl, $(MCl)_2$, $MgCl_2$, $MMgCl_3$, $MMgCl_2$, M_2MgCl_4	

(Continued)

Table 20. (continued)

System in the cell	Temperature range in K	Gaseous species	Ref./Year
$LiCl + CuCl$	1027–1211	$LiCl, (LiCl)_2, (CuCl)_3, (CuCl)_4, LiCu_2Cl_3, Li_2CuCl_3$	[510]/1981
$MCl + CuCl$ (M = Li, Na, K, Rb, Cs)	not specified	$MCl, (MCl)_2, CuCl, (CuCl)_3, (CuCl)_4, MCu_2Cl_3$	[511]/1982
$MCl + GaCl_3$ (M = Li, K, Cs)	536–944	$MCl, GaCl_3, (GaCl_3)_2, MGaCl_4, M_2GaCl_5, M_2Ga_2Cl_8$	[512]/1981
$MCl + PbCl_2$ (M = Li, Na, K)	not specified	$MCl, (MCl)_2, PbCl_2, MPbCl_3$	[513]/1984
$xNaCl + (1-x)KCl$	793–1083	$NaCl, (NaCl)_2, KCl, (KCl)_2, NaKCl_2$	[514]/1982
$MAlCl_4 + MCl$ (M = Li, Na, K)	200–700	$AlCl_3, (AlCl_3)_2, MAlCl_4, M_2Al_2Cl_8$	[515]/1984
$NaAlCl_4 + NaCl$	450–550	$AlCl_3, (AlCl_3)_2, NaAlCl_4, Na_2Al_2Cl_8$	[516]/1987
$NaAlCl_4$	not specified	$AlCl_3, NaAlCl_4, Na_2AlCl_5$	[517]/1983
$NaCl + SnCl_2$	713–963	$NaCl, (NaCl)_2, SnCl_2, (SnCl_2)_2, NaSnCl_3$	[518]
$NaCl + DyCl_3$	not specified	$DyCl_3, NaDyCl_4$	[513]/1984
$KCl + MCl_2$ (M = Sr, Ba)	1073	$KCl, (KCl)_2, SrCl_2, BaCl_2, KMCl_3, K_2SrCl_4$	[509]/1983
$xKCl + (1-x)MnCl_2$	900–1100	$KCl, (KCl)_2$	[519]/1983
$KCl + CuCl$	973–861	$KCl, (KCl)_2, (CuCl)_3, (CuCl)_4, KCu_2Cl_3$	[510]/1981
$KAlCl_4$	not specified	$KAlCl_4, K_2AlCl_5$	[517]/1983
$KCl + DyCl_3$	not specified	$KDyCl_4$	[513]/1984
$MCl + CaCl_2$ (M = Rb, Cs)	not specified	$MCl, (MCl)_2, CaCl_2, MCaCl_3, M_2CaCl_4$	[509]/1983
$MCl_3 + (AlCl_3)_2(g)$ (M = Ti, V, Sc, Nd)	568–1123	$(AlCl_3)_2, MAlCl_6$	[520]/1981

System	Temperature	Species	Reference/Year
$MoCl_3 + (AlCl_3)_2(g)$	410–870	no hetero-complex detected	
$ZrCl_4 + (AlCl_3)_2(g)$	378–441	$ZrCl_4$, $(AlCl_3)_2$, $ZrAlCl_7$, $ZrAl_2Cl_{10}$	
$TaCl_5 + (AlCl_3)_2(g)$	362–375	$TaCl_5$, $(AlCl_3)_2$, $TaAlCl_8$	
$WCl_6 + (AlCl_3)_2(g)$	413–500	no hetero-complex detected	
$MCl_2 + (AlCl_3)_2(g)$	377–684	$(AlCl_3)_2$, MAl_2Cl_8, $CdAlCl_5$	[521]/1981
(M = Fe, Zn, Cd, Pt)			
$Pd_6Cl_{12} + Pt_6Cl_{12}$		$Pd_nPt_{6-n}Cl_{12}$ (n = 0, 1, 2 ... 6)	[453]/1984
$HgCl_2 + (AlCl_3)_2(g)$	363, 381	$HgCl_2$, $(AlCl_3)_2$, $HgAlCl_5$	[522]/1981
Systems composed of bromides			
$LiBr + CuBr$	not specified	$LiBr$, $(LiBr)_2$, $CuBr$, $(CuBr)_3$, Li_2CuBr_3, $LiCu_2Br_3$	[510]/1981
$NaBr + CaBr_2$	not specified	$NaBr$, $(NaBr)_2$, $CaBr_2$, $NaCaBr_3$, Na_2CaBr_4, $NaCa_2Br_5$	[509]/1983
$xNaBr + (1-x)SrBr_2$	1073	$NaBr$, $SrBr_2$, $NaSrBr_3$, Na_2SrBr_4	[523]/1984
$NaBr + SnBr_2$	653–893	$NaBr$, $(NaBr)_2$, $SnBr_2$, $(SnBr_2)_2$, $NaSnBr_3$	[518]
$KBr + CaBr_2$	not specified	KBr, $(KBr)_2$, $CaBr_2$, $KCaBr_3$, K_2CaBr_4, KCa_2Br_5	[509]/1983
$xKBr + (1-x)MBr_2$	1073	KBr, MBr_2, $KMBr_3$, K_2MBr_4	[523]/1984
(M = Sr, Ba)			
$xRbBr + (1-x)SrBr_2$	1073	$RbBr$, $(RbBr)_2$, $SrBr_2$, $RbSrBr_3$, Rb_2SrBr_4	[509]/1983
$MBr + M'Br_2$	not specified	MBr, $(MBr)_2$, $M'Br_2$, $MM'Br_3$, $M_2M'Br_4$	
(M = Rb, Cs; M' = Ca, Sr, Ba)			
Systems composed of iodides			
$NaI + CsI$	730–900	NaI, $(NaI)_2$, $(NaI)_3$, CsI, $(CsI)_2$, $(CsI)_3$, $NaCsI$, $NaCs_2I_3$, Na_2CsI_3, $Na_2Cs_2I_4$	[524]/1983
$xNaI + (1-x)CsI$	930–1050	NaI, $(NaI)_2$, $(NaI)_3$, CsI, $(CsI)_2$, $(CsI)_3$, $NaCsI_2$, $NaCs_2I_3$, Na_2CsI_3	[525]/1986
$xNaI + (1-x)CaI_2$	not specified	NaI, $(NaI)_2$, CaI, $NaCaI_3$, Na_2CaI_4, $NaCa_2I_5$	[509]/1983

(Continued)

Table 20. (continued)

System in the cell	Temperature range in K	Gaseous species	Ref./Year
$xNaI + (1 - x)SrI_2$	1073	NaI, SrI_2, $NaSrI_3$, Na_2SrI_4	[523]/1984
$NaI + ScI_3$	613–848	NaI, $(NaI)_2$, ScI_3, $(ScI_3)_2$, $NaScI_4$, Na_2ScI_5	[64]/1990
$NaI + FeI_2$	574–683	NaI, $(NaI)_2$, FeI_2, $(FeI_2)_2$, $NaFeI_3$, Na_2FeI_4	[61]/1987
$NaI + CuI$	840–950	NaI, $(NaI)_2$, CuI, $(CuI)_3$, $NaCuI_2$, Na_2CuI_3, $NaCu_2I_3$	[510]/1981
$NaI + SnI_2$	450–840	NaI, $(NaI)_2$, SnI_2, $(SnI_2)_2$, $NaSnI_3$	[62]/1987. [526]/1986, [48]/1983, [527]/1983
$NaI + PbI_2$	562–669	NaI, $(NaI)_2$, PbI_2, $(PbI_2)_2$, $NaPbI_3$	[61]/1987
$NaI + DyI_3$	743–933	NaI, $(NaI)_2$, $(NaI)_3$, $(NaI)_4$, DyI_3, $NaDyI_4$, Na_2DyI_5	[63]/1988, [528]/1987, [529]/1983
$xNaI + (1 - x)DyI_3$	980–1020	NaI, $(NaI)_2$, DyI_3, $NaDyI_4$, Na_2DyI_5	[530]/1990
$xKI + (1 - x)CaI_2$	1073	KI, $(KI)_2$, CaI_2, $KCaI_3$, K_2CaI_4, KCa_2I_5	[509]/1983
$xRbI + (1 - x)CaI_2$	1073	RbI, $(RbI)_2$, CaI_2, $RbCaI_3$, Rb_2CaI_4	
$CsI + CaI_2$	973	CsI, $(CsI)_2$, CaI_2, $CsCaI_3$, Cs_2CaI_4, $CsCa_2I_5$	
$CsI + DyI_3$	880–1000	CsI, $(CsI)_2$, $(CsI)_3$, DyI_3, $(DyI_3)_2$, $CsDyI_4$	[529]/1983
$CsI + HoI_3$	770–870	CsI, $(CsI)_2$, HoI_3, $CsHoI_4$, Cs_2HoI_5	[528]/1987
$ScI_3 + SnI_2$	726–873	ScI_3, $(ScI_3)_2$, SnI_2, $ScSnI_5$	[62]/1987

Table 21. Summary of studies of quasi-binary systems of metal halides and metal cyanides with different anions since about the year 1980. (The gaseous species are underlined if enthalpies of dissociation are given)

Condensed phase	Temperature range in K	Gaseous species identified	Ref./Year
$xNaCl + (1 - x)NaBr$	773–879	$NaCl$, $(NaCl)_2$, $NaBr$, $(NaBr)_2$, Na_2ClBr	[531]/1987, [539]/1986
$NaBr + NaI$	693–933	$NaBr$, $(NaBr)_2$, $(NaBr)_3$, NaI, $(NaI)_2$, $(NaI)_3$, Na_2BrI, Na_3Br_2I, Na_3BrI_2	[532]
$NaI + SnBr_2$	723–805	$NaBr$, $(NaBr)_2$, NaI, $(NaI)_2$, $SnBr_2$, SnI_2, $SnBrI$, Na_2BrI, $NaSnBr_3$, $NaSnI_3$, $NaSnBrI_2$, $NaSnBr_2I$	[533]
$xKCl + (1 - x)KBr$	745–862	KCl, $(KCl)_2$, KBr, $(KBr)_2$, K_2ClBr	[534]/1985, [539]/1986
$xKCl + (1 - x)KCN$	872	KCl, KCN, K_2ClCN	[535]/1984
$xKCN + (1 - x)KCNO$	777	KCN, $KCNO$, $K_2(CN)(CNO)$	
$xTlCl + (1 - x)TlBr$	540–600	$TlCl$, $(TlCl)_2$, $TlBr$, $(TlBr)_2$, Tl_2ClBr	[540]/1981
$xTlCl + (1 - x)TlI$	520–580	$TlCl$, $(TlCl)_2$, TlI, $(TlI)_2$, Tl_2ClI	[536]/1987
$xTlBr + (1 - x)TlI$	491–594	$TlBr$, $(TlBr)_2$, TlI, $(TlI)_2$, Tl_2BrI	[537]/1983
$SnBr_2 + SnI_2$	413–583	$SnBr_2$, $(SnBr_2)_2$, SnI_2, $(SnI_2)_2$, $SnBrI$, Sn_2BrI_3, $Sn_2Br_2I_2$, Sn_2Br_3I	[466]

The investigations by Emons et al. on alkali metal chlorides, bromides [448], and iodides [470] give a coarse survey of the relative abundances of dimers and trimers in the equilibrium vapor. The abundances obtained for these polymers of CsI(g) and NaI(g) in general agree reasonably well with those determined by Hilpert [51], Viswanathan and Hilpert [50] as well as Lelik et al. [469]. However, the enthalpies of vaporization obtained by Emons et al. [448, 470] according to the second-law method for the monomer and dimer as well as the dimerization enthalpies do not lead to new values. This is also the case for the enthalpy data evaluated according to the second-law method by Lelik et al. [469]. The enthalpy data of both groups show in part large deviations from well established values.

The aforementioned preliminary mass spectrometric Knudsen effusion study by Emons et al. [448], which includes also the vapor over KCl(s, l), is complemented by Hastie's work on KCl(l) [322] using transpiration mass spectrometry. The equilibrium partial pressures of KCl(g) and $(KCl)_2(g)$ obtained by Hastie agree reasonably well with those given in the JANAF tables [90] largely based on extrapolated data of KCl(s). Equilibrium pressures of $(KCl)_3(g)$ are given for the first time [322].

Hilpert [51], Viswanathan and Hilpert [50] as well as Hilpert and Bencivenni [109] presented a consistent data set (see Table 19) based on a second- and third-law evaluation agreeing well for the monomer, dimer, and trimer gaseous species of sodium and cesium iodide. The results obtained so far on the composition and thermochemistry of the equilibrium vapor of these iodides are reported in Refs. 51 and 50. In addition, the experimental dissociation enthalpies for the dimer and trimer are compared with the energies obtained by theoretical computations using different models such as the Pauling model for the alkali halides [482] and the simple shell model [483]. The Pauling model for alkali halides was applied to predict dissociation energies for $(NaI)_3(g)$ [51] and other trimer alkali halides [86]. Figure 8 shows as an example the contour plot of surface energy obtained for this molecule. The $(NaI)_3$ molecule has a ring structure as $(CsI)_3(g)$ (cf. Fig. 9). The Na^+ and I^- ions are arranged on cycles with different radii; r in Fig. 8 is the radius belonging to Na^+; δ is the difference between the radii for Na^+ and I^-.

In addition, the tetramers $(NaI)_4(g)$ and $(CsI)_4(g)$ were identified by Hilpert and coworkers over sodium and cesium iodide. The relative abundances of the ions $Na_4I_3^+$ and $Cs_4I_3^+$ representing the tetramer agree well with those reported by Lelik et al. [469]. The mole fraction of the tetramer $(NaI)_4(g)$, 3.7×10^{-5}, and the trimer $(NaI)_3(g)$, 1.5×10^{-3}, and of the dimer $(NaI)_2(g)$, 2.0×10^{-1}, at 830 K evaluated from the ion intensities shows good agreement with that obtained by Demercurio and Grimley [468] by the use of the angular distribution technique (Sect. 2.3). This technique was also used to study the species $(KI)_i(g)$ (i = 1, 2, 3, 4) over potassium iodide [468].

Figure 9 shows the different structures of cesium iodide polymers which are typical of alkali halides. Obviously, the tetramer shows a cubic or a ring structure with T_d of D_{4h} symmetry, respectively. The determination of the

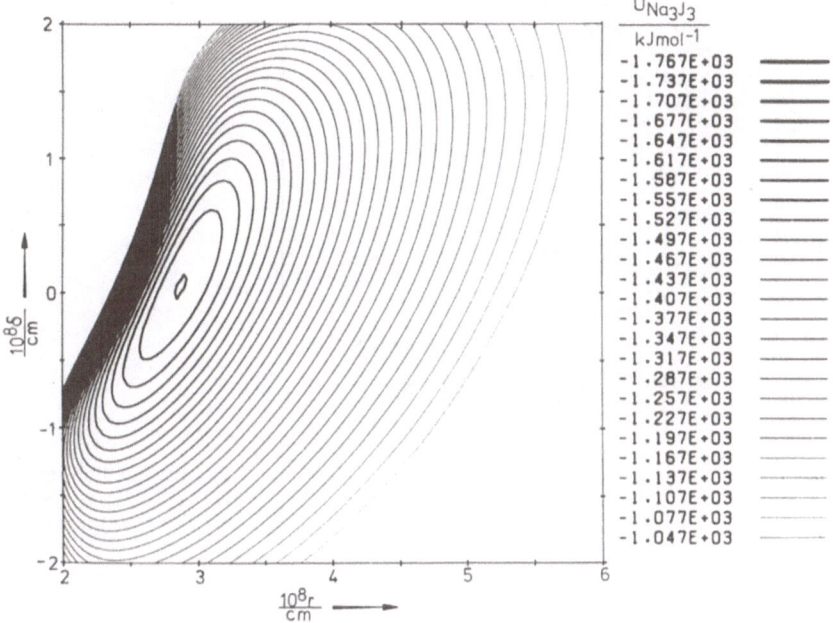

Fig. 8. Contour plot of surface energy of the molecule $(NaI)_3(g)$ relative to the completely separated ions Na^+ and I^- [51]

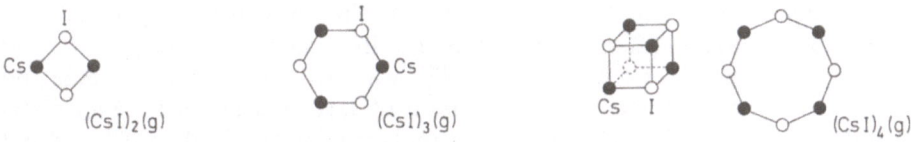

Fig. 9. Structures of cesium iodide homo-complexes

enthalpy of sublimation for the tetramer according to the third-law method by using these two structures and comparison with the corresponding second-law sublimation enthalpy supports the cubic structure for $(CsI)_4(g)$ at the temperatures of the measurement 748 to 813 K [109].

Cordfunke [484] recently investigated the vaporization of cesium iodide in the vicinity of the melting point by the transpiration method.

Figure 10 shows a comparison of the equilibrium constants for the dissociation reaction $(NaI)_2(g) \rightleftarrows 2NaI(g)$ obtained by different methods. The equilibrium constant by Hilpert and Seehawer (Fig. 10) was obtained from the data reported by these authors in Ref. 485. It agrees excellently with the results of Miller and Kusch [486] determined by the velocity distribution method as well as those of Demercurio and Grimley [468] obtained by the use of the angular distribution technique. The equilibrium constants of Datz et al. [487] as well as

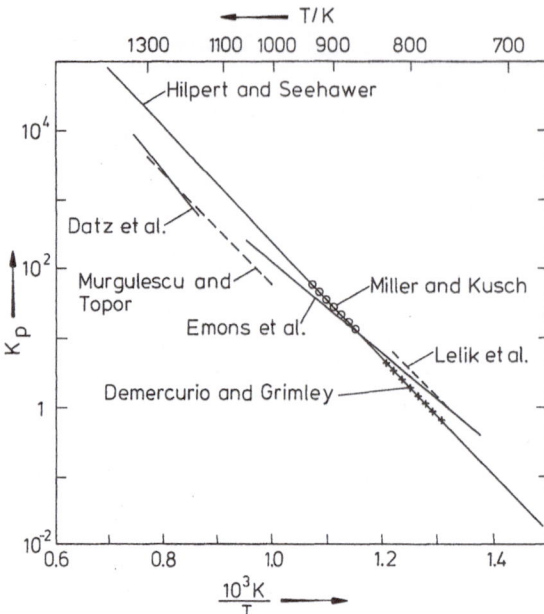

Fig. 10. Equilibrium constant $K_p = p^2 [\text{NaI}]/p\,[(\text{NaI})_2]$ for the reaction $(\text{NaI})_2 \rightleftarrows 2\text{NaI}$

Murgulescu and Topor [488] obtained in an indirect way are comparatively small. The results of this work and those of Ref. 468 are based on a direct measurement of the NaI and $(\text{NaI})_2$ partial pressures. Work [489] as well as Mucklejohn and O'Brien [490] used the equilibrium constant by Datz et al. [487] for the computation of partial pressures from the total vapor pressure and the vapor density over sodium iodide. The vaporization of sodium iodide and the thermochemistry of the vapor species is of particular importance for metal halide lamps [438]. The equilibrium constants obtained by Lelik et al. [469] and Emons et al. [470] result from investigations by Knudsen effusion mass spectrometry which in part deviate from well established data.

Kappes et al. [491] studied dialkali monohalides such as NaKCl(g) and K_2Cl(g) generated in supersonic nozzle beams by photoelectron spectroscopy and determined their ionization potentials. The first direct detection of a dialkali monohalide was carried out by Peterson et al. [492].

On summarizing, one can state that in particular our knowledge of the thermochemistry of trimer and tetramer alkali halide species is increased by the recent investigations (cf. Table 19). Further investigations, carried out earlier than those in Table 19 and yielding thermodynamic data, have been reported for the species $(\text{NaCl})_i$ (i = 3, 4) [493], $(\text{LiF})_3$, and $(\text{LiCl})_3$ [90].

5.3 Vapors of Halides Except Those of the Alkali Metals

Reinhart and Behrens [440] reinvestigated the vaporization of ScF_3. The ScF_3 partial pressure was determined and the enthalpy of vaporization to ScF_3(g) was evaluated according to the second- and third-law methods. Malkerova et al. [441] elucidated the thermochemistry of the incongruent vaporization of $MoF_4(s)$ and $WF_4(s)$.

Two compartment Knudsen cells were used by Gesenhues and Wendt [456] as well as Schäfer and Flörke [457] to study the vapor over $AlCl_3(s)$. The electron impact ionization mass spectra of $AlCl_3(g)$ and $(AlCl_3)_2(g)$ as well as the trimerization enthalpy of $AlCl_3(g)$ resulted.

Pittermann and Weil [65] showed that $I(g)$ and $(AgI)_3(g)$ are the most abundant species in the equilibrium vapor over silver iodide at a temperature of 750 K. The fragmentation of the different vapor species by electron impact ionization was studied and the enthalpy of trimerization of $AgI(g)$ evaluated according to the second- and third-law methods. The value obtained agrees within the given error limits with the enthalpies of trimerization for $AgCl(g)$ and $AgBr(g)$ [65] determined by the same group in the course of systematic investigations on the vaporization of silver halides [495–498].

Hilpert and coworkers studied the vaporization of $SnBr_2(s, l)$ [466], $CaI_2(s)$ [472], $ScI_3(s)$ [52], $ZnI_2(s)$ [53], $SnI_2(s)$ [52], and $PbI_2(s)$ [52]. A consistent data set is given for the evaporation and the thermochemistry of the monomer and dimer species of these iodides. The set is composed of molecular parameters, thermodynamic properties, and partial pressures. There is good agreement between the enthalpy changes evaluated according to the second- and third-law methods.

Vibrational frequencies were obtained by the F,G matrix method from estimated force fields. Different structures were assumed on the basis of electron diffraction studies of related molecules. Tables with the thermodynamic functions, such as enthalpy and entropy increments, computed by the rigid rotator harmonic oscillator approximation, are given in Refs. 52, 485 ($(NaI)_2(g)$, $(NaI)_3(g)$, $SnI_2(g)$, $(SnI_2)_2(g)$, $PbI_2(g)$, $(PbI_2)_2(g)$, $ScI_3(g)$, $(ScI_3)_2(g)$), Ref. 472 ($CaI_2(g)$, $(CaI_2)_2(g)$), and Ref. 466 ($SnBr_2(g)$, $(SnBr_2)_2(g)$) for the species given in parentheses. The enthalpies of dissociation are reported in Sect. 5.6. In contrast to Hilpert et al. [52] Hirayama and Kleinosky [476] detected no ions originating from $(SnI_2)_2(g)$ upon vaporizing SnI_2. The appearance potentials determined for SnI_2^+, SnI^+, and Sn^+ in Ref. 476 are about 3V larger than those of Hilpert et al. [52] which are in agreement with the results of investigations by photoelectron spectroscopy. Ricart et al. [499] report results on the structure, the vibrational frequencies, and the ionization potentials of SnI_2 obtained by an ab initio SCF and CI study.

The vaporization of FeI_2 was studied by Hilpert et al. [60] as well as Grade and Rosinger [474]. The most complete study carried out is that described in Ref. 60. Six gaseous species were identified in the equilibrium vapor and their partial pressures were determined. The gaseous species $FeI_3(g)$ and $(FeI_2)_3(g)$ were identified for the first time. The existence of the molecule $FeI_3(g)$ has so far been predicted by Schäfer and Hönes [500] to explain the results of their transpiration experiments. The enthalpy changes evaluated for the reaction

$$FeI_2(s) + 1/2 I_2(g) \rightleftarrows FeI_3(g) \qquad (21)$$

by using the transpiration method and Knudsen effusion mass spectrometry agree well [60]. Grade and Rosinger [474] attributed the ion $Fe_2I_2^+$ to the species $Fe_2I_2(g)$ and determined its partial pressure and enthalpy of formation. According to Ref. 60 this ion is generated by fragmentation from the dimer $(FeI_2)_2(g)$, due to its comparatively high appearance potential and the temperature dependence of its intensity agreeing within the given error limits with that of the ions $Fe_2I_4^+$ and $Fe_2I_3^+$ (cf. Table 1).

Two-compartment Knudsen cells and Knudsen cells with a gas inlet system were used to study the thermal dissociation of $PtF_4(g)$ (Korobov et al. [445]), $BF_3(g)$ (Farber and Srivastava [447]), $(AlCl_3)_3(g)$ (Schäfer and Flörke [457]), $UCl_5(g)$ (Hildenbrand and Lau [462]), $SiBr_4(g)$ (Farber and Srivastava [465]), and $AsI_3(g)$ (Alikhanyan et al. [230]) to their gaseous component metals and gaseous component halides. Different techniques were used. The subhalides $AsI_2(g)$ and $AsI(g)$ of $AsI_3(g)$ as well as the components $As_4(g)$, $As_2(g)$, $I_2(g)$, and $I(g)$ became for example detectable by superheating $AsI_3(g)$, at 500 to 890 K generated in the comparatively cold cell of a two-compartment Knudsen cell at about 340 K (Alikhanyan et al. [501]). The study of the equilibria between $SiBr_4(g)$, $SiBr_3(g)$, $SiBr_2(g)$, $SiBr(g)$, and $Si(c)$ was rendered possible at cell temperatures between 1054 to 1603 K by using a Knudsen cell with a gas inlet system and metering $SiBr_4(g)$ into the effusion cell containing silicon (Farber and Srivastava [465]).

Sapegin et al. [592] evaluated the atomization energies of all lanthanide chlorides of the types $LnCl_3(g)$, $LnCl_2(g)$, and $LnCl(g)$ from the appearance potentials determined by them for the reaction

$$LnCl_n(g) + e^- \longrightarrow Ln^+ + nCl + 2e^-. \tag{22}$$

Kleinschmidt and Ward [593] obtained the enthalpy changes of the reactions $PuF_n \rightleftarrows PuF_{n-1} + F$ (n = 0 to 6) in a similar way.

5.4 Vapors of Quasi-binary Metal Halide Systems

The notation, A + B, in the Tables 20 and 21 for the condensed phase means that the two materials A and B were inserted into the Knudsen cell. The study of gaseous species is generally the aim of such investigations. In contrast to this, thermochemical data are evaluated from the partial pressures for the condensed phases of the system A–B over the complete composition range, if the condensed phase is given as, $xA + (1 - x)B$.

Hetero-complexes such as $MM'X_{i+j}$ or MM'_2X_{i+2j} were detected in the equilibrium vapor over the quasi-binary MX_i–$M'X_j$ systems (cf. Table 20). Hetero-complexes with mixed anions are given in Table 21. In addition, the monomers and homo-complexes of the type $(MX_i)_n$ or $(M'X_j)_n$ (n = 1, 2, 3) were observed as they are present in the vapor over the pure component metal halides.

Only quasi-binary metal fluoride systems containing one alkali fluoride component have been investigated since about the year 1980 (see Table 19). Some of the investigations were carried out by the study of ion molecule equilibria (see Sect. 5.5).

Knowledge is most extensive on the complexation of metal chlorides. Different aspects of metal chloride hetero-complexes are particularly described in the reviews by Schäfer [425, 426] as well as Øye and Gruen [427]. The studies of the hetero-complex formation with $GaCl_3$ or $AlCl_3$ [cf. Table 20] are important for chemical vapor transport. Gesenhues and Wendt [515] investigated the vaporization of $MCl/AlCl_3$ melts (M = Li, Na, K) by the use of the transpiration method, Knudsen effusion vapor deposition, and Knudsen effusion mass spectrometry. Itoh et al. [541] determined the chemical activities in solid and liquid solutions over the complete composition range of the NaCl–KCl system. Similar studies were carried out for melts of the $KCl–MnCl_2$ system (cf. Table 20).

The most complete investigations on the complexation of metal bromides and iodides were carried out after the year 1980 and are thus contained in Tables 20 and 21. In the years 1974 to 1979 the following bromine and iodine hetero-complexes were detected by Knudsen effusion mass spectrometry: Ag_3Cl_2Br, Ag_3ClBr_2 [498]; $RbCaI_3$, Rb_2CaI_4 [542]; $NaScI_4$, Na_2ScI_5, $CsCeI_4$, $CsLaI_4$ [543], and $LiScI_4$, $CsScI_4$ [544]. About the same number is mentioned in Hastie's book [428] and review [429] published in the years 1975 and 1971, respectively.

Many of the bromine and iodine systems listed in the Tables 20 and 21 were studied due to their practical use for metal halide lamps [438]. Particularly important for such lamps are the systems $NaI–ScI_3$, $NaI–SnI_2$, and $NaI–DyI_3$ on account of the radiation emitted by the elements Na, Sc, and Dy.

The complexation in the $NaI–ScI_3$ system (cf. Table 20) has been investigated by us for the first time by the use of two-compartment Knudsen cells. By this means a substantially improved assignment of the various ions in the mass spectrum to their neutral precursors is rendered possible, which is a pre-requisite for the determination of reliable thermodynamic data. Two-compartment Knudsen cells were also used for the study of the systems $NaX–SnX_2$ (X = Cl, Br, I) (Table 20) as well as the systems $NaBr–SnI_2$ and $SnBr_2–SnI_2$ (Table 21) due to the different volatilities of the sodium and tin halides and in order to elucidate fragmentation patterns especially for systems with mixed anions.

Mucklejohn and O'Brien [464] detected the complex ions $NaSnCl_2^+$, $NaSnClBr^+$ ($NaBr–SnCl_2$) and $NaSnCl_2^+$ ($NaI–SnCl_2$) over phases of the systems given in parentheses. The enhancement of the sodium concentration in the gas phase by complex formation compared to NaI(s) was estimated.

The study of the mixing properties of the solid systems $xNaCl + (1 - x)NaBr$ and $xKCl + (1 - x)KBr$ by Miller and Skudlarski (Table 21) is complemented by the evaluation of the thermodynamic properties for the gaseous species Na_2ClBr and K_2ClBr by the same authors [539].

The thermodynamic data of the hetero-complexes of the $NaI–DyI_3$ system have been determined by Hilpert and Miller [63], Kaposi et al. [529], and

Gavrilin et al. [528]. The dissociation enthalpies obtained by the three groups agree fairly well. The measurements by Hilpert and Miller were carried out over a larger temperature range (743 to 933 K) than those of Gavrilin et al. (770 to 870 K) and Kaposi et al. (860 to 920 K). Moreover, Hilpert and Miller determined the dissociation entropy and evaluated quantitative fragmentation patterns.

Hilpert and coworkers investigated the NaI–SnI_2 system extensively by the use of one- [48, 526] and two-compartment [62] Knudsen cells.

5.5 Studies of Equilibria Involving Negative Ions

The potential of Knudsen effusion mass spectrometry for the study of ion-molecule and ion-ion reactions in the Knudsen cell is discussed and the instrumentation described in the recent review article by Sidorov et al. [572]. In this section the results determined for halides by studying equilibria involving negative ions are reviewed as an example for this interesting variant of the method.

Partial pressures of neutral molecules and negative ions present in the Knudsen cell were measured mass spectrometrically and enthalpy changes were evaluated according to the second- and third-law methods for the ion-molecule equilibria observed. In general, the equilibrium of isomolecular ion exchange reactions was investigated (e.g. Refs. 504, 508, 545). Such ion exchange reactions often involve the ion AlF_4^- as standard since its enthalpy of formation is known well. Enthalpies of formation of the negative ions identified, as well as affinities of electrons and of F^- anions for molecules were obtained. These affinities correspond to the enthalpy changes of the reactions

$$A^- \longrightarrow A + e^- \tag{23}$$

and

$$AF^- \longrightarrow A + F^-, \tag{24}$$

respectively, where A is a molecule.

The enthalpies of formation and electron affinities obtained since the review by Sidorov et al. [546] are summarized in Table 22. The electron affinity of a molecule AF, EA(AF), was evaluated by computing the difference between the enthalpies of formation of the negative ion $\Delta_f H_0^\circ(AF^-)$ and the neutral molecule $\Delta_f H_0^\circ(AF)$ (cf. Ref. 554) or by the use of the electron affinity of F, EA(F), as well as the enthalpy changes $\Delta H_0^\circ(A-F)$ and $\Delta H_0^\circ(A-F^-)$ necessary to split off one F or F^- species (cf. Ref. 550). A remarkably high value of EA = 7.9 + 0.8 eV was estimated by Pyatenko and Gorokhov [554] for the extremely unstable molecule U_2F_{12} indicating a comparatively high stability of the ion $U_2F_{12}^-$.

The enthalpies of formation of ZrF_5^- determined by Pyatenko et al. [548] and Skokan et al. [549] agreed well (cf. Table 22). The value for FeF_4^- has been determined by Chilingarov et al. [551], Sorokin et al. [552], and Sidorov et al.

Table 22. Enthalpies of formation $\Delta_f H^\circ_{298}(AF^-)$ and electron affinities EA(AF) for the fluoride species AF^- and AF, respectively

Ion AF^-	$-\Delta_f H^\circ_{298}(AF^-)$	EA(AF)	Ref.
	kJ mol^{-1}	eV	
ScF_4^-	2013 ± 13		[545]
YF_4^-	2000 ± 21		[547]
$Y_2F_7^-$	3548 ± 29		[547]
LaF_4^-	1938 ± 17		[548]
ZrF_5^{2-}	2343 ± 17		[548]
ZrF_5^-	2340 ± 14		[549]
$Zr_2F_9^-$	4228 ± 15		[549]
MoF_5^-		3.6 ± 0.2	[550]
MoF_6^-		3.6 ± 0.2	[550]
$MoOF_4^-$		4.0 ± 0.4	[550]
MnF_3^-	1217		[502]
MnF_4^{2-}	1463 ± 60		[551]
FeF_4^-	1475 ± 15		[551]
FeF_3^-	1138 ± 14	4.3 ± 0.2	[552]
FeF_4^-	1412 ± 14	5.45 ± 0.20	[552]
$Fe_2F_5^-$	1769 ± 17ᵃ		[552]
$Fe_2F_7^-$	2276 ± 18ᵃ		[552]
FeF_3^-	1069 ± 21ᵃ	3.62 ± 0.13	[553]
FeF_4^{2-}	1423 ± 18ᵃ		[553]
$Fe_2F_5^-$	1740 ± 43ᵃ	3.8 ± 0.4	[553]
$Fe_2F_6^-$	2071 ± 38ᵃ	4.45 ± 0.24	[553]
$Fe_2F_7^-$	2379 ± 37ᵃ	5.0 ± 0.2	[553]
RhF_4^-	482 ± 19		[551]
GaF_4^-	1638 ± 32		[507]
$Ga_2F_7^-$	2777 ± 48		[507]
$GaAlF_7^-$	3092 ± 37		[507]
ThF_5^-	2432 ± 13		[504]
$U_2F_{10}^-$	4490 ± 30ᵃ	4.5 + 0.4	[554]
$U_2F_{11}^-$	4850 ± 40ᵃ	6.1 + 0.7	[554]
UOF_5^-	2530 ± 40ᵃ		[555]
$UO_2F_3^-$	2090 ± 40ᵃ		[555]

ᵃ $\Delta_f H^\circ_0(AF^-)$

[553]. Chilingarov et al. [446] obtained the enthalpy change for the reaction $AuF_3 + F^- \rightleftarrows AuF_4^-$ as $-426 + 25$ kJ mol^{-1}.

The enthalpies of formation as well as the enthalpies of dissociation for the hetero-complexes $MGaF_4(g)$ (Zhuravleva et al., Ref. 507), $MThF_5(g)$ (Sidorov et al., Ref. 504), $MZrF_5(g)$ (Skokan et al., Ref. 549), as well as $MFeF_3(g)$ and $MFeF_4(g)$ (Sorokin et al., Ref. 552) (M = Li, Na, K, Rb, Cs) are reported by the references quoted. The enthalpy changes for the dissociation to two gaseous metal fluorides and for the dissociation to the alkali metal cation and the metal fluoride anion are given. The enthalpies of formation and dissociation were obtained by using estimated molecular parameters of the hetero-complexes which are additionally reported. The main subject of the aforementioned articles was, however, the determination of the enthalpies of formation of negative ions (cf. Table 22). By employing the determined enthalpies of formation for GaF_4^-, ThF_5^-, ZrF_5^-, FeF_3^-, and FeF_4^- the computation of the enthalpy changes for the dissociation of the hetero-complexes to ions was rendered possible. An analogous study by Sidorov and Gubarevich [502] yielded the enthalpy of dissociation of the hetero-complexes $MMnF_3(g)$ and $MMnF_4(g)$ (M = Li, Na, K, Rb, Cs) to the M^+ cations and the MnF_3^- or MnF_4^- anions, respectively. In addition, the enthalpies of formation of $MMnF_3(g)$, $MMnF_4(g)$, and MnF_3^- as well as the enthalpy of dissociation of the hetero-complexes to the fluorides is given.

Ion molecule equilibria were used by Skokan et al. [508] to determine the activity of ZrF_4 in the $RbF-ZrF_4$ system. They measured for example the ion intensity ratio $I(Zr_2F_9^-)/I(ZrF_5^-)$ and computed the ZrF_4 partial pressure by employing the known equilibrium constant of the reaction

$$Zr_2F_9^-(g) \rightleftarrows ZrF_5^-(g) + ZrF_4(g). \tag{25}$$

The potential of the new method and its pros and cons are discussed in Ref. 508.

Examples showing the variety of negative ionic species present in the equilibrium vapor of a Knudsen cell over different condensed phases are given in Table 20. Traces of ZrF_4 (see Ref. 504 in Table 20) or AlF_3 (see Refs. 502, 507, 508 in Table 20) were added to enable the study of ion exchange reactions involving ZrF_5^- or AlF_4^- as a standard without changing the activities of the main system components significantly.

Lelik et al. [556] and Kaposi et al. [557] studied ion molecule equilibria in the gaseous phase over DyI_3 and HoI_3 as well as over the $CsI-DyI_3$, $CsI-HoI_3$, and $CsI-NaI-DyI_3$ systems. The ions NaI_2^-, CsI_2^-, I^-, MI_4^-, $M_2I_7^-$, $CsM_2I_8^-$ (M = Dy, Ho), and $NaDy_2I_8^-$ were identified in the equilibrium vapor. The enthalpies of formation are reported for the ions underlined.

5.6 Thermochemical Properties

Tables 23 and 24 show some selected enthalpies and entropies of dissociation determined by Knudsen effusion mass spectrometry. In addition, partial pressures and equilibrium constants were evaluated. An account of partial pressures and equilibrium constants is given in a recent review by Hilpert [438].

The enthalpies of dissociation of the homo-complexes (Table 23) were evaluated according to the second- and third-law methods if they are tabulated at 298 K. Otherwise, only the second-law method was used. The values obtained by the use of the second- and the third-law methods in general agreed excellently (see references quoted in Table 23).

The data for the hetero-complexes in Table 24 were obtained by a second-law evaluation. The accuracy of these data is checked by the computation of second-law dissociation enthalpies of dimer homo-complexes additionally present in the vapor of the quasi-binary systems. The dissociation enthalpies obtained in this way in general agree very well with the data in Table 23 resulting by vaporizing pure metal halides (cf. e.g. Ref. 61).

Table 23. Enthalpy and entropy changes at the temperature T for the dissociation of different iodine and bromine homo-complexes $(MX_i)_n$ according to the reaction $(MX_i)_n(g) = nMX_i(g)$. (The given errors are probable uncertainties)

Complex	T/K	$\Delta_d H_T^\circ$	$\Delta_d S_T^\circ$	Ref.
		kJ mol^{-1}	J mol^{-1} K^{-1}	
Iodides				
$(NaI)_2$	298	172.2 ± 3.8	125.3 ± 6.9	[51]
$(NaI)_3$	298	307.2 ± 5.2	241.0 ± 9.1	[51]
$(CsI)_2$	298	149.2 ± 4.4	119.5 ± 7.9	[50]
$(CsI)_3$	770	289.9 ± 8.6	228.3 ± 10.9	[109]
$(CaI_2)_2$	298	192 ± 14	134 ± 16	[472]
$(ScI_3)_2$	298	189.8 ± 14.7	165.3 ± 19.3	[52]
$(FeI_2)_2$	298	157.1 ± 6.4	144.0 ± 10.9	[60]
$(FeI_2)_3$	298	296.5 ± 10.3	263.2 ± 16.8	[60]
$(AgI)_3$	770	416		[65]
$(ZnI_2)_2$	298	93.2 ± 4.2	141.0 ± 9.5	[53]
$(SnI_2)_2$	298	101.0 ± 4.4	158.3 ± 10.2	[52]
$(PbI_2)_2$	298	115.9 ± 4.7	164.6 ± 9.0	[52]
$(DyI_3)_2$	298	178 ± 7		[538]
$(HoI_3)_2$	1000	203 ± 12		[478]
Bromides				
$(NaBr)_2$	298	189.1 ± 4.9	130.0 ± 7.6	[463]
$(NaBr)_3$	298	332.1 ± 8.4	237.3 ± 14.1	[463]
$(SnBr_2)_2$	298	106.8 ± 3.6	153.7 ± 8.8	[466]

Table 24. Enthalpy and entropy changes of gas phase reactions for the dissociation of different iodine, bromine, and chlorine hetero-complexes. (The given errors are probable uncertainties. δH see text)

Reaction	$\Delta_d H_{298}^\circ$	$\Delta_d S_{298}^\circ$	δH	Ref.
	kJ mol^{-1}	J mol^{-1} K^{-1}	kJ mol^{-1}	
$NaScI_4 = NaI + ScI_3$	228.2 ± 4.8	165.2 ± 7.9	47 ± 9	[64]
$Na_2ScI_5 = 2NaI + ScI_3$	388.1 ± 8.7	310.1 ± 12.8		[64]
$NaDyI_4 = NaI + DyI_3$	217.1 ± 5.9	130.5 ± 9.9	42 ± 7	[63]
$Na_2DyI_5 = 2NaI + DyI_3$	372.4 ± 10.5	269.2 ± 17.9		[63]
$NaFeI_3 = NaI + FeI_2$	183.6 ± 5.0	142.9 ± 7.8	19 ± 7	[61]
$Na_2FeI_4 = 2NaI + FeI_2$	333.0 ± 8.7	274.4 ± 13.7		[61]
$NaSnI_3 = NaI + SnI_2$	166.3 ± 4.5	144.2 ± 6.5	30 ± 3	[62]
$NaPbI_3 = NaI + PbI_2$	168.3 ± 5.1	151.2 ± 8.7	24 ± 6	[61]
$ScSnI_5 = ScI_3 + SnI_2$	172.3 ± 7.7	181.1 ± 10.2	27 ± 9	[62]
$NaSnBr_3 = NaBr + SnBr_2$	189.8 ± 5.5	165.4 ± 7.4		[518]
$NaSnCl_3 = NaCl + SnCl_2$	196.6 ± 5.7	153.1 ± 6.7		[518]

The dissociation enthalpies of the 1,1 hetero-complexes can be predicted by the semiempirical rule [425, 430],

$$\Delta_d H_{298}^\circ [M'MI_{j+k}] = 1/2\, \Delta_d H_{298}^\circ [(M'I_j)_2]$$

$$+ 1/2\, \Delta_d H_{298}^\circ [(MI_k)_2] + \delta H, \qquad (26)$$

where $M'I_j$ and MI_k are different metal iodides; $\Delta_d H^\circ_{298}$ [$M'MI_{j+k}$], $\Delta_d H^\circ_{298}$ [$(M'I_j)_2$], and $\Delta_d H^\circ_{298}$ [$(MI_k)_2$] are dissociation enthalpies of hetero-complexes and homo-complexes. The term δH is the enthalpy change of the reaction

$$M'MI_{j+k} = \tfrac{1}{2}(M'I_j)_2 + \tfrac{1}{2}(MI_k)_2 \tag{27}$$

and its physicochemical meaning can be explained by the structure of the complexes involved [61, 425, 430]. The δH values should be positive, since there is an additional Coulomb attraction if the two cations M and M' are different; the values of δH should be small if the coordination numbers of the cations are the same in the reacting hetero- and homo-complexes (cf. Eq. 27). The δH values in Table 24 in general support these rules. The comparatively high δH values for the dissociation of $NaScI_4(g)$ and $NaDyI_4(g)$ might indicate a change of the coordination numbers [64]. The values of δH in Table 24 can be valuable for their prediction for other iodine hetero-complexes thereby increasing the accuracy of the estimation by Eq. (26).

In addition to the dissociation of the homo- and hetero-complexes containing the same anions the following isomolecular reaction for the formation of a bromoiodide has been investigated [466]:

$$SnI_2(g) + SnBr_2(g) \rightleftarrows 2SnBrI(g). \tag{28}$$

The enthalpy and entropy changes of this reaction result at 500 K as $\Delta H^\circ_{500} = 0.3 \pm 2.2 \text{ kJ mol}^{-1}$ and $11.6 \pm 4.1 \text{ J mol}^{-1} \text{ K}^{-1}$. It follows that the change of the Gibbs energy of the reaction depends practically only on the entropy changes. This is typical for such reactions giving rise to the formation of mixed halides with a decrease in symmetry caused by the exchange of halogen atoms. This was observed first by Bernauer et al. [498] on studying the reaction

$$(AgCl)_3(g) + (AgBr)_3 \rightleftarrows Ag_3Cl_2Br + Ag_3ClBr_2 \tag{29}$$

Similar results were obtained by Miller and Skudlarski [539] for the reaction

$$(MCl)_2 + (MBr)_2 \rightleftarrows 2M_2ClBr \ (M = Na, K). \tag{30}$$

Thermodynamic properties of condensed phases can be evaluated from gas phase data. They show for example a small tendency to immiscibility for the solid solution of the NaBr–NaI system (Miller and Hilpert [532]) as observed for the KCl–KBr system (Miller and Skudlarski [534]). In contrast to this, the Gibbs energy of the melt of the $NaI–DyI_3$ system for NaI > 0.53 at 1000 K indicates a tendency to compound formation (Hilpert et al. [530]).

6 Technical Materials

The investigations reported in Sects. 3, 4, and 5 were mainly carried out to determine fundamental thermodynamic properties of gaseous species and condensed phases as for example enthalpies of dissociation or mixing properties. In contrast to these fundamental studies some work was carried out on real technical systems.

Hastie and Plante. [558] studied different real MHD slags in addition to synthetic oxide mixtures to elucidate the thermochemistry of MHD slags. The rate of alkali vaporization from complex alkali silicate systems, such as coal slags and glass, can be greatly enhanced by the presence of chloride-containing species as for example $HCl(g)$ (Plante et al. [559]). A gas inlet Knudsen effusion system was used to study the influence of $HCl(g)$ on the alkali vaporization. The chemical reaction taking place is $M_2O(slag) + 2HCl(g) \rightleftarrows 2MCl(g) + H_2O(g)$ (M = alkali metal).

Thermal decomposition limits the use of materials at high temperatures. The sialons have a potential as high-strength and high-temperature structural ceramics. Moon et al. [560] determined the vaporization of two sialons. The enthalpies of formation of $SiAl_2O_2N_2(s)$ and $SiAl_4O_2N_4(s)$ were evaluated from the partial pressures obtained. Vitreous sodium polyphosphate is used as a chemical binder in the manufacture of magnesian unroasted refractories, concretes, and tamping as well as guniting materials because it ensures the desired strength of the artifacts at 1073 to 1473 K. Semenov et al. [561] therefore analyzed the vapor over phosphate-bonded periclase. The vapor over different synthetic micas was studied by Anikin et al. [562].

The investigations by Asano and Yasue [408] on glass host matrices for the storage and disposal of high-level radioactive wastes (cf. Sect. 4.6.1) are complemented by vaporization studies of Na_2O–B_2O_3–Al_2O_3–SiO_2 matrices doped with up to 26 to 36 different modifiers. The investigations by Itoh and Nakamura [563] are useful for the safety requirements for the design of off-gas systems in the vitrification processes and storage facilities. Moreover, the pressure is one parameter for the stability of the inclusion of fission products in the matrix.

The sorption of the fission products Cs and Sr by the graphitic materials, from which the core and the fuel elements of High-Temperature Gas-Cooled Reactors (HTGRs) are made, is important for the prediction of fission product release in the case of an accident. Hilpert et al. [564, 565] determined, therefore, Cs and Sr partial pressures over such graphitic materials with different Cs and Sr concentrations. The vaporization enthalpies obtained showed a strong chemisorption of Cs and Sr by these materials. The vaporization enthalpy of Sr exceeds that of the pure Sr metal by about $210 \, kJ \, mol^{-1}$ at 1500 K, if fuel element matrix graphite with a Sr concentration of $4.0 \, mmol \, kg^{-1}$ is considered [564]. This value for Cs amounts to about $230 \, kJ \, mol^{-1}$ at 1250 K for a similar concentration of $4.2 \, mmol \, kg^{-1}$ [565]. In addition, sorption isotherms were evaluated.

The influence of $H_2O(g)$ or $I(g)$ on the evaporation of Cs, Sr, Ag, Rb, and Ba from matrix graphite was studied by Hilpert [133] by using a Knudsen cell with a gas inlet system.

The partial pressures and the chemical activities of the component elements of technical alloys are important parameters for the use of these materials.

Chemical activities are necessary to predict or to resolve compatibility problems. Hilpert and Ali-Khan [566] as well as Hilpert et al. [567] determined the partial pressures of the component elements given in parentheses over the solid cobalt-containing nickel base alloys inconel 643 (Ni, Cr, Fe, Co), inconel 617 (Ni, Cr, Co), and nimonic PE 13 or hastelloy X (Ni, Cr, Fe, Co). The chemical activities obtained are discussed by Hilpert [568] with respect to the microstructure of the alloys investigated. Hirai et al. [569] determined the partial pressures of the components Mn, Fe, and Cr over the austenitic steel SUS 304 L. In addition to the chemical activities, partial free energies and excess partial free energies of mixing of the alloy components were evaluated [566, 567, 569]. The mentioned technical alloys were investigated due to their possible use in fusion or fission reactors. Figure 11 shows as an example the results obtained for alloy inconel-643. The partial pressures obtained with different diameters of the effusion orifice in the Knudsen cell agree well. This shows the existence of thermodynamic equilibrium in the cell in practical terms.

A clear decision whether investigations are to be mentioned in this section or in one of the others is sometimes not possible. Also numerous investigations reported in Sects. 3, 4, and 5 were initiated by their technological significance.

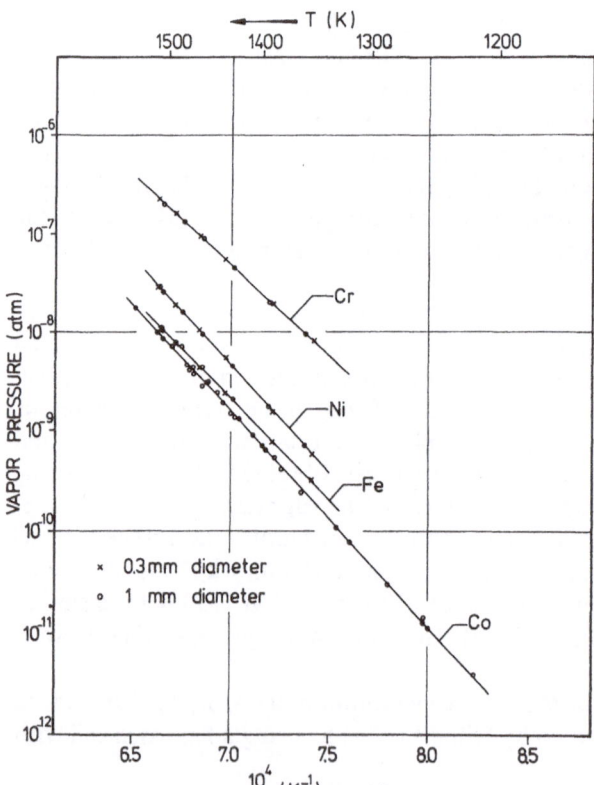

Fig. 11. Partial pressures of Ni, Cr, Fe, and Co for the nickel base alloy inconel 643 obtained by the use of Knudsen cells with diameters of the effusion orifice of 1 and 0.3 mm [567]

The vaporization of barium tungstates (Sect. 4.2) is of interest for thermionic emitters used in certain types of arc lamps, that of ZrO_2-based solid solutions for high-refractory ceramics (Sect. 4.3), and that of glasses for their production (Sect. 4.6.1). The vaporization of some metal halides (Sect. 5) and amalgams (Sect. 3.4.2) is important for the development of lamps. Finally, the investigations (Sect. 4) initiated by the development of fusion and fission reactors should be mentioned as further examples.

7 Summary and Conclusions

An account of the gaseous species observed by Knudsen effusion mass spectrometry in the eqilibrium vapor of metals, alloys, oxides, halides, and technical systems is given. The fundamentals and recent developments of this method are briefly reported. Dissociation and atomization enthalpies of selected gaseous species are tabulated. Accounts of the equilibrium studies by Knudsen effusion mass spectrometry in order to obtain thermodynamic properties for condensed phases from gas phase data are additionally given for the aforementioned materials. Table 8 shows as an example the enthalpies and Gibbs energies of formation for different solid intermetallic compounds. A special section (Sect. 5.5) summarizes the results obtained for halides by studying equilibria between gaseous species involving negative ions by Knudsen effusion mass spectrometry.

Many new results have been obtained during the period considered, beginning from the years 1977 to 1980. Some of them are given in the following as examples:

- In the field of metals or alloys gaseous complex Zintl compounds were identified for the first time. The combination of the results of all electron ab initio computations and Knudsen effusion mass spectrometry have made the determination of improved values for bond energies possible. Moreover, bond energies for transition metal trimers were obtained for the first time.
- In the field of oxides the extensive investigations of the Ti–O and the As–O system should be mentioned as well as the numerous investigations of Li_2O and the quasi-binary oxide systems containing Li_2O as one component.
- In the field of metal halides there has been a particularly large increase of our knowledge of gaseous metal iodide homo- and hetero-complexes. Thermodynamic data of gaseous species and condensed phases of quasi binary systems with mixed anions have been determined.

Most of the vaporization studies by Knudsen effusion mass spectrometry, carried out for the determination of thermodynamic data of condensed phases, refer to alloys and to a smaller extent to oxide systems as shown by this article. In comparison to this only a few metal halide systems have been investigated. In addition to the study of elements and binary oxides or halides by far most of the investigations were carried out on binary metal as well as quasi-binary oxide and metal halide systems. There have been only a very few investigations of systems with an increased number of components compared to those mentioned.

Moreover, this article shows that numerous gaseous species have been identified and dissociation or atomization enthalpies evaluated by investigating metals as well as alloys, oxides or metal halide systems. Among the new species identified for the first time and not belonging to the aforementioned groups considered are the molecules OPF [573], OPBr [574], SPCl [575], OAsCl, and OSbCl [576], which are similar to nitrosyl chloride. In addition, a new class of gaseous rare earth carbides such as $(RE)C_n$ or $(RE)_2C_n$ with $n = 1$ to 8 or $M(RE)C_n$ with $n = 1$ to 4 were observed by the extensive investigations of Gingerich's group. These results and the knowledge of the stability of gaseous rare earth molecules were reviewed by Gingerich [577]. The study of the first gaseous alkaline earth carbide, BaC_2 [578], and of several gaseous transition metal carbides [579–582] should also be mentioned in this context.

The emphasis of this article has been chosen on the basis of the author's research work using Knudsen effusion mass spectrometry. A disadvantage of this method is its limitation to comparatively low pressures of about up to 10 Pa (cf. Sect. 2.4). However, direct mass spectrometric sampling of high-pressure systems (e.g. 10 to 10^6 Pa) is possible by the use of high-pressure mass spectrometry described in the reviews by Milne and Green [583, 584]. Bonnell and Hastie [585, 586] developed transpiration mass spectrometry by coupling the transpiration method and high-pressure mass spectrometry. Applications of this method for the study of KCl and KOH are given in Ref. 322.

Limitations to high-temperature materials chemistry research due to the non-availability of suitable container materials have been overcome by laser induced vaporization mass spectrometry (see e.g. Ref. 587). This technique couples laser heating of refractory materials under vacuum with the mass spectrometric analysis of the vapor plume. Hastie et al. [588] have recently investigated the vaporization of graphite by this technique. The investigations by Ohse's group on the laser induced vaporization of fast breeder oxide and carbide fuels should also be mentioned in this context (see Refs. 589, 590 and references quoted therein).

More than 30 years have passed since the pioneering mass spectrometric vaporization study by Chupka and Inghram [6] in the year 1953. Our knowledge of the chemistry of inorganic vapors has increased substantially during this period. There is, however, a need for a continuation of the investigations. The investigations could, for example, have the following aims:

1) Study of complex vapor species thereby contributing to the active field of cluster research to bridge the gap between the gas and the condensed phase,
2) study of compounds or systems for applied research, and
3) improvement of the existing data on gaseous species by more extensive measurements.

Especially, the research work mentioned under points (1) and (3) will benefit by the developments in mass spectrometry which have the potential for increased sensitivity of the partial pressure determination in the Knudsen cell and for an enlarged dynamic range or the partial pressure measurements. The investigations in point (2) include the study of technical multicomponent systems and basic studies of comparatively simple systems. Since Knudsen effusion mass spectrometry is one of the most useful methods for thermodynamic investigations at high temperatures, further tasks referring to point (2) originate for example in the active fields of energy research, lamp chemistry, or high temperature materials research.

Acknowledgement. The author is indebted to Prof. K. G. Weil for reading the manuscript and helpful discussions. He also wishes to thank Prof. H. Nickel, director of the Institute of Reactor Materials, for his kind support and advice.

8 References

1. Margrave JL (ed) (1967) The characterization of high-temperature vapors, Wiley, New York
2. Hastie JW (ed) (1979) Characterization of high-temperature vapors and gases. National Bureau of Standards, Special Publication 561/1 and 561/2, U.S. Government Printing Office, Washington, vol 1, 2
3. Eyring L (ed) (1967) Advances in high-temperature chemistry, Academic, New York, vol 1, (1969) vol 2, (1971) vol 3, 4
4. Gole JL (1986) In: Moskovits M (ed) Metal clusters, Wiley, New York, p 131
5. Moskovits M (1986) In: Moskovits M (ed) Metal clusters, Wiley, New York, p 185
6. Chupka WA, Inghram MG (1953) J. Chem. Phys. 21: 371, 1313
7. Honig RE (1954) J. Chem. Phys. 22: 126
8. Inghram MG, Drowart J (1960) In: High temperature technology, McGraw-Hill, New York, p 219
9. Drowart J (1964) In: Rutner E, Goldfinger P, Hirtz JP (eds) Condensation and evaporation of solids, Gordon and Breach, New York p 255
10. Grimley RT (1967) In: Margrave JL (ed) The characterization of high-temperature vapors, Wiley, New York, p 195
11. Drowart J (1986) In: Todd JFJ (ed) Advances in mass spectrometry 1985, Part A, Wiley, New York, p 195
12. Chatillon C, Pattoret P, Drowart J (1979) High Temperatures – High Pressures 7: 119
13. Raychaudhuri PK, Stafford FE (1975) Mat. Sci. Eng. 20: 1
14. Gingerich KA (1966) Advances in mass spectrometry, Proc. 3rd Int. Mass Spectrometry Conf., Paris (Sept. 1964), vol 3, p 1009
15. Gilles PW, Conard BR, Sheldon RI, Bennet JE (1975) In: Thermodynamics of nuclear materials, IAEA, Vienna, vol 2, p 499
16. Tomiska J (1980) CALPHAD 4: 63
17. Berkowitz J, Chupka WA (1960) Annals of the New York Acad. of Sci. 79: 1073
18. Ionov NI (1948) Dokl. Akad. Nauk SSSR 59: 467
19. Gingerich KA (1972) Chimia 26: 619
20. Gingerich KA (1971) J. Cryst. Growth 9: 31
21. Gingerich KA (1980) In: Kaldis E (ed) Current topics of materials science, North Holland, Amsterdam, vol 6, p 345
22. Goldfinger P (1965) In: Reed RI (ed) Mass spectrometry, Academic, New York, p 265
23. Drowart J, Goldfinger P (1967) Angew. Chemie 79: 589

24. Drowart J (1971) In: Marsel J (ed) Proc. Int. School on Mass Spectrometry (1969), Ljubljana
25. Drowart J, Pattoret A, Smoes S (1967) Proc. British Ceramic Soc. 8: 67
26. Babeliowsky TPJH (1962) Physica 28: 1150
27. Drowart J, Goldfinger P (1962) Ann. Rev. Phys. Chem. 13: 459
28. Gorokhov LN, Semenov GA (1971) In: Quayle A (ed) Adv. in mass spectrometry, Elsevier, Amsterdam, vol 5, p 349
29. Büchler A, Berkowitz-Mattuk JB (1970) Techn. Metals Res. 4: 161
30. Boerboom AJH (1965) In: Reed RI (ed) Mass spectrometry, Academic, New York, p 251
31. Porter RF (1956) In: Proc. on high temperature – a tool for the future, Stanford Research Institute, Menlo Park, California, p 182
32. Stafford FE (1971) High Temperatures – High Pressures 3: 213
33. Drowart J (1982) Int. J. Mass Spectrom. Ion Phys. 45: 243
34. DeMaria G (1970) AGARD Conf. Proc. 52: 34
35. Clark NJ (1974) Proc. Royal Australian Chem. Inst. 41: 17
36. Avery DF, Cuthbert J, Prosser NJD, Silk C (1966) J. Sci. Instrum. 43: 436
37. Hastie JW (1984) Pure Appl. Chem. 56: 1583
38. Brittain RD, Hildenbrand DL (1983) J. Phys. Chem. 87: 3713
39. Marsel J, Popovic A, Susic R, Kaposi D (1981) Vestn. Slov. Kem. Drus. 28: 11
40. Popovic A, Marsel J (1980) In: Quayle A (ed) Advances in mass spectrometry, Hayden, London, vol 8A, p 402
41. Frisch MA, Reuter W (1979) J. Vac. Sci. Technol. 16: 1020
42. Frisch MA (1980) In: Quayle A (ed) Advances in mass spectrometry, Hayden, London, vol 8A, p 391
43. Grimley RT, Forsman JA (1979) NBS Spec. Publ. (U.S.) 561–1, p 211
44. Brittain RD, Lau KH, Hildenbrand DL (1982) J. Phys. Chem. 86: 5072
45. Hastie JW (1984) Proc. Fourth Int. Conf. High Temp. and Energy-Related Materials, Santa Fe, 1–6 Apr.
46. Platel PG (1965) J. Chim. Phys. 62: 1176
47. Franklin JL, Dillard JG, Rosenstock HM, Herron JT, Draxl K, Field FH (1969) Ionization potentials, appearance potentials and heats of formation of gaseous positive ions, NSRDS-NBS 26
48. Hilpert K, Gingerich KA (1983) Int. J. Mass Spectrom. Ion Phys. 47: 247
49. Moore CE (1970) Ionization potentials and ionization limits derived from the analysis of optical spectra, NSRDS-NBS 34
50. Viswanathan R, Hilpert K (1984) Ber. Bunsenges. Phys. Chem. 88: 125
51. Hilpert K (1984) Ber. Bunsenges. Phys. Chem. 88: 132
52. Hilpert K, Bencivenni L, Saha B (1985) Ber. Bunsenges. Phys. Chem. 89: 1292
53. Hilpert K, Bencivenni L, Saha B (1985) J. Chem. Phys. 83: 5227
54. Gurvich L (1974) Kondratiev VN (ed) Dissociation energies, ionization potentials, electron affinities, Nauka, Moscow
55. Rosenstock HM, Draxl K, Steiner BW, Herron JT, J. Phys. Chem. Ref. Data 1977: 6
56. Levin RD, Lias SG (1982) Ionization potentials and appearance potential measurements 1971–1981, NSRDS-NBS 71
57. Scheuring T, Weil KG (1985) Ber. Bunsenges. Phys. Chem. 89: 811
58. Odoj R, Hilpert K (1976) Z. Phys. Chem. Neue Folge 102: 191
59. Bauer SH, Porter RF (1964) In: Blander M (ed) Molten salt chemistry, Wiley (Interscience), New York, p 607
60. Hilpert K, Viswanathan R, Gingerich KA, Gerads H, Kobertz D (1985) J. Chem. Thermodynamics 17: 423
61. Hilpert K, Gerads H, Kobertz D, Miller M (1987) Ber. Bunsenges. Phys. Chem. 91: 200
62. Miller M, Hilpert K (1987) Ber. Bunsenges. Phys. Chem. 91: 642
63. Hilpert K, Miller M (1988) High Temp.-High Pressures 20: 231
64. Hilpert K, Miller M (1990) J. Electrochem. Soc. (in press)
65. Pittermann U, Weil KG (1980) Ber. Bunsenges. Phys. Chem. 84: 542
66. Kapala J, Skudlarski K (1983) Int. J. Mass Spectrom. Ion Phys. 47: 257
67. Kapala J, Skudlarski K (1984) Int. J. Mass Spectrom. Ion Phys. 55: 133
68. Miller M (1984) Int. J. Mass Spectrom. Ion Processes 61: 293
69. Otvos JW, Stevenson DP (1956) J. Amer. Chem. Soc. 78: 546
70. Mann JB (1967) J. Chem. Phys. 46: 1646

71. Mann JB (1970) In: Ogata K, Hayakawa T (eds) Recent developments in mass spectrometry, University of Tokyo Press, Tokyo, p 814
72. Mann JB, Personal communication and Ref. 71
73. Tawara H, Kato T (1987) Atomic Data and Nuclear Data Tables 36: 167
74. Electron impact ionization, Märk TD, Dunn GH (eds) (1985) Springer, Berlin Heidelberg New York
75. Wagner K, Schäfer H (1979) Z. anorg. allg. Chem. 451: 67
76. Feather DH, Searcy AW (1971) High Temp. Sci. 3: 155
77. Joyce TE, Rolinski EJ (1972) J. Phys. Chem. 76: 2310
78. Biefeld RM (1978) J. Chem. Thermodynamics 10: 907
79. Gingerich KA, Cocke DL, Miller F (1976) J. Chem. Phys. 64: 4027
80. Hilpert K, Gingerich KA (1980) Ber. Bunsenges. Phys. Chem. 84: 739
81. Neubert A (1978) High Temp. Sci. 10: 261
82. Kordis J, Gingerich KA (1973) J. Chem. Phys. 58: 5141
83. Rovner L, Drowart A, Drowart J (1967) Trans. Faraday Soc. 63: 2906
84. Kant A, Strauss BH (1966) J. Chem. Phys. 45: 822
85. Sullivan CL, Prusaczyk JE, Carlson KD (1972) High Temp. Sci. 4: 212
86. Hilpert K (1980) Habilitationsschrift Technische Hochschule Darmstadt, Darmstadt; (1981) Report from KFA Jülich, Jül–1744, Jülich, FRG
87. Freeman RD, Edwards J (1961) In: Margrave JL (ed) The characterization of high-temperature vapors, Wiley, New York, p 508
88. Odoj R, Hilpert K, Nürnberg HW (1973) Report from KFA Jülich, Jül-1460, Jülich, FRG
89. Wahlbeck PG (1986) High Temp. Sci. 21: 189
90. Chase MW, Jr, Davies CA, Downey JR, Jr, Frurip DJ, McDonald RA, Syverud AN (1986) JANAF Thermochemical Tables, Third Edition, Part I and II, American Institute of Physics, New York
91. Barin I, Knacke O (1973) Thermochemical properties of inorganic substances, Springer, Berlin Heidelberg New York
92. Barin I, Knacke O, Kubaschewski O (1977) Thermochemical properties of inorganic substances, Supplement, Springer, Berlin Heidelberg New York
93. Glushko VP et al. (1978) Thermodynamic data for individual substances, vol 1, vol 2 (1979), vol 3 (1981), High Temperature Institute, State Institute of Applied Chemistry, National Academy of Sciences of the U.S.S.R., Moscow
94. Hultgren R, Desai PD, Hawkins DT, Gleiser M, Kelley KK, Wagman DD (1973) Selected values of the thermodynamic properties of the elements, Am. Soc. Met., Metals Park, Ohio
95. Kelley KK (1960) Contributions to the data on theoretical metallurgy, XIII. High-temperature heat content, heat capacity and entropy data for the elements and inorganic compounds, Bureau of Mines, Bulletin 584, US Government Printing Office, Washington
96. Kelley KK, King EG (1961) Contributions to the data on theoretical metallurgy, XIV. Entropies of the elements and inorganic compounds, Bureau of Mines, Bulletin 592, US Governments Printing Office, Washington
97. Lamoreaux RH, Hildenbrand DL (1984) J. Phys. Chem. Ref. Data 13: 151
98. Lamoreaux RH, Hildenbrand DL, Brewer L (1987) J. Phys. Chem. Rev. Data 16: 419
99. Pedley JB, Marshall EM (1983) J. Phys. Chem. Ref. Data 12: 967
100. The oxide handbook, Samsonov GV (ed) (1982) 2nd ed., IFI/Plenum, New York
101. Brewer L, Rosenblatt G (1961) Chem. Rev. 61: 257
102. Brewer L, Rosenblatt G (1969) In: Eyring E (ed) Advances in high temperature chemistry, Academic, New York, vol 2, p 1
103. Brewer L (1953) Chem. Rev. 52: 1
104. Kubaschewski O, Alcock, CB (1979) Metallurgical thermochemistry, Pergamon, Oxford
105. Lewis GN, Randall M, Pitzer KS, Brewer L (1961) Thermodynamics, McGraw Hill, New York, p 516
106. Huber KP, Herzberg G (1979) Molecular spectra and structure, IV. Constants of diatomic molecules, Van Nostrand, New York
107. Mizushima M (1975) The theory of rotating diatomic molecules, Wiley, Chichester
108. Rosen B (1970) Tables internationales de constantes sélectionnées 17, données spectro-scopiques relatives aux molécules diatomiques, Pergamon, Oxford
109. Hilpert K, Bencivenni L (1985) Surf. Sci. 156: 436
110. Guggenheimer KM (1946) Proc. Phys. Soc. 58: 456

111. Gordy W (1946) J. Chem. Phys. 14: 305
112. Frurip DJ, Blander M (1980) J. Chem. Phys. 73: 509
113. Hilpert K (1984) Z. Metallkde. 75: 70
114. Hilpert K (1984) Ber. Bunsenges. Phys. Chem. 88: 37
115. Sidorov LN, Korobov MV (1981) Shitsuryo Bunseki 29: 199
116. Lyubimov AP, Zobens VYa, Rakhovski VI (1958) Zh. Fiz. Khim. 32: 1804
117. Belton GR, Fruehan RJ (1967) J. Phys. Chem. 71: 1403
118. Neckel A, Wagner S (1969) Ber. Bunsenges. Phys. Chem. 73: 210
119. Neckel A, Wagner S (1969) Monatsh. Chem. 100: 664
120. Hilpert K, Kobertz D, Venugopal V, Miller M, Gerads H, Bremer FJ, Nickel H (1987) Z. Naturforsch. 42a: 1327
121. Chastel R, Bergman C (1985) Journées de calorimetrie et analyse thermique, Paris 16: 322
122. Hildenbrand DL (1968) J. Chem. Phys. 48: 3657; Drowart J, Goldfinger P (1958) J. Chim. Phys. 55: 721
123. Hampson PJ, Gilles PW (1971) J. Chem. Phys. 55: 3712
124. Tomiska J (1984) J. Phys. E17: 1165
125. Hilpert K (1978) In: Daly NR (ed) Advances in Mass Spectrometry, Heyden, London, vol 7A, p 584
126. Hilpert K, Ruthardt K (1987) Ber. Bunsenges. Phys. Chem. 91: 724
127. Hilpert K (1984) Ber. Bunsenges. Phys. Chem. 88: 260
128. Petrov AA, Kazenas EK, Tagirov VK, Bol'shikh MA, Chadin AN, Nesterenko PA (1982) Instrum. Exp. Techn. 25, Part 2: 448
129. Novozhilov AF, Belousov VI, Murav'ev VV, Matveev IV (1981) Instrum. Exp. Techn. 24, Part 2: 752
130. Moore RH, Robinson D, Argent BB (1975) J. Phys. E 8: 641.
131. Hastie JW, Swingler D (1969) High Temp. Sci. 1: 46; Hastie JW (1966) Molten salts and their vapors, Ph.D. Thesis, University of Tasmania
132. Hilpert K (1980) Ber. Bunsenges. Phys. Chem. 84: 494
133. Hilpert K (1983) In: Hoinkis E (ed) Transport of fission products in matrix and graphite, HMI-B 372, Report from the Hahn-Meitner-Institut für Kernforschung, Berlin, p 139
134. Fraser DG, Rammensee W (1982) Geochim. Cosmochim. Acta 46: 549
135. Plante ER (1979) NBS Spec. Publ. (U.S.), vol 561-1, p 265
136. Ono K, Nishi S, Oishi T (1984) Trans. Jpn. Inst. Met. 25: 810
137. Kematick R, Anderegg J, Franzen H (1985) High Temp. Sci. 19: 17
138. Edwards JG, Starzynski JS, Peterson DE (1980) J. Chem. Phys. 73: 908
139. Maul JL (1974) doctoral thesis, Technische Universität München, München, p 29
140. Riekert G, Lamparter P, Steeb S (1981) Z. Naturforsch., Teil A36: 447
141. Neubert A, Zmbov KF (1974) High Temp. Sci. 6: 308
142. Timberg L, Toguri JM, Azakami T (1981) Metall. Trans. 12B: 275
143. White D, Walsh PN, Goldstein HW, Dever DF (1961) J. Phys. Chem. 65: 1405
144. Grotemeyer J, Schlag EW (1988) Angew. Chem. 100: 461
145. Gomez M, Chatillon C, Allibert M (1982) J. Chem. Thermodynamics 14: 447
146. Riekert G, Lamparter P, Steeb S (1981) Z. Metallkde. 72: 765
147. Neubert A, Zmbov KF (1983) Chem. Phys. 76: 469
148. Banon S, Chatillon C, Allibert M (1982) High Temp. Sci. 15: 105
149. Chatillon C, Allibert M, Pattoret A (1979) NBS Spec. Publ. (U.S.), vol 561-1, p 181
150. Chatillon C, Allibert M, Pattoret A (1978) In: Daly NR (ed) Advances in mass spectrometry, Heyden, London, vol 7, p 675
151. Chatillon C, Senillou C, Allibert M, Pattoret A (1976) Rev. Sci. Instrum. 47: 334
152. Zimmermann E (1986) doctoral thesis, Rheinisch-Westfälische Technische Hochschule Aachen, Aachen
153. Paulaitis ME, Eckert CA (1983) J. Chem. Thermodynamics 15: 55
154. Hackworth JV, Hoch M, Gegel HL (1971) Metall. Trans. 2: 1799
155. Gupta SK, Gingerich KA (1979) J. Chem. Phys. 70: 5350
156. Lukas W, Chatillon C, Allibert M (1979) J. Less Common Met. 66: 211
157. Rickert H (1964) In: Condensation and evaporation of solids, Gordon and Breach, New York, p 201
158. Detry D, Drowart J, Goldfinger P, Keller H, Rickert H (1967) Z. Phys. Chem. Neue Folge 55: 314
159. Alcock CB, Butler J (1980) Nippon Kinzoku Gakkaishi 44: 1239

160. Alcock CB, Butler J, Ichise E (1980) Solid State Ionics, vol 3/4 (Proc. 3rd Int. Meet. Solid Electrolytes – Solid State Ionics Galvanic Cells, Tokyo, Japan, 15–19 Sept. 1981), p 499
161. Chatillon C, Allibert M, Pattoret A (1976) High Temp. Sci. 8: 233
162. Neubert A, Guggi D (1978) J. Chem. Thermodynamics 10: 297
163. Ikeda Y, Tomaki M, Matsumoto G, Amioka K, Mizuno T (1982) Spectrochim. Acta, Part B, 37: 647
164. Wagner LC, Grimley RT (1974) Chem. Phys. Lett. 29: 594
165. Kohl FJ, Uy OM, Carlson KD (1967) J. Chem. Phys. 47: 2667
166. Honig RE (1954) J. Chem. Phys. 22: 1610
167. Drowart J, DeMaria G, Boerboom AJH, Inghram MG (1959) J. Chem. Phys. 30: 308
168. Gingerich KA, Desideri A, Cocke DL (1975) J. Chem. Phys. 62: 731
169. Gingerich KA (1980) Faraday Discussions Chem. Soc. 14: 109
170. Gingerich KA (1982) ACS Symp. Ser. 178: 109
171. Hartmann A, Weil KG (1989) Z. Phys. D 12: 11
172. Drowart J (1967) In: Rudman PS, Stringer J, Jaffee RI (eds) Phase stability in metals and alloys, McGraw, New York, N.Y.
173. Vargaftik NB (1975) Tables on the thermophysical properties of liquids and gases, 2nd edn., Wiley, New York
174. Pauling L (1960) The nature of the chemical bond, Cornell University Press, Ithaca, N.Y., 3rd edn.
175. Brewer L (1968) Science 161: 115
176. Gingerich KA (1973) Chem. Phys. Lett. 23: 270; (1974) J. Chem. Soc. Faraday II, 70: 471; (1978) Intern. J. Quantum Chem. Symp. 12: 489
177. Miedema AR, Gingerich KA (1979) J. Phys. B12: 2081; (1979) J. Phys. B12: 2258
178. Gingerich KA, Shim I, Gupta SK, Kingcade JE (1985) Surf. Science 156: 459
179. Gingerich KA, Ramakrishnan EA, Kingcade JE (1986) High Temp. Sci. 21: 1
180. Wu CH (1983) J. Phys. Chem. 87: 1534
181. Wu CH, Ihle HL (1980) Advances in Mass Spectrometry, vol 8, Proc. 8th Int. Mass Spectrometry Conf., Oslo (Aug. 1979) p 374
182. Wu CH, Ihle HR, Gingerich KA (1983) Int. J. Mass Spectrom. Ion Phys. 47: 235
183. Scheuring T, Weil KG (1983) Int. J. Mass Spectrom. Ion Phys. 47: 227
184. Wu CH, Ihle HR (1981) 29th Annu. Conf. Mass Spectrom. Allied Top., Minneapolis, Minn., USA 24–29 May 1981 (Abstr.), p 93
185. Neubert A, Ihle HR, Gingerich KA (1980) J. Chem. Phys. 73: 1406
186. Neubert A, Zmbov KF, Gingerich KA, Ihle HR (1982) J. Chem. Phys. 77: 5218
187. Neubert A, Ihle HR, Gingerich KA (1982) J. Chem. Phys. 76: 2687
188. Hilpert K, Kath D, High Temp. Sci. (in press)
189. Scheuring T, Weil KG (1985) Surf. Sci. 156: 457
190. Busse B, Weil KG (1981) Ber. Bunsenges. Phys. Chem. 85: 309
191. Busse B, Weil KG (1979) Angew. Chem. Int. Ed. Engl. 18: 629
192. Busse B, Weil KG (1982) Ber. Bunsenges. Phys. Chem. 86: 93
193. Hartmann A, Weil KG (1988) Angew. Chem. Int. Ed. Engl. 27: 1091
194. Haque R, Gingerich KA (1980) J. Chem. Thermodynamics 12: 439
195. Gingerich KA, Gupta SK, Haque R, Nappi BM, Pelino M, 28th Annu. Conf. Mass Spectrom. Allied Top., N.Y., USA 26–30 May 1980, p 199
196. Haque R, Pelino M, Gingerich KA (1980) J. Chem. Phys. 73: 4045
197. Sundaram Ramakrishnan E, Shim I, Gingerich KA (1984) J. Chem. Soc., Faraday Trans. 2, 80: 395
198. Gupta SK, Nappi BM, Gingerich KA (1981) Inorg. Chem. 20: 966
199. Kingcade JE, Gingerich KA (1986) J. Chem. Phys. 84: 4574
200. Haque R, Pelino M, Gingerich KA (1979) J. Chem. Phys. 71: 2929
201. Nappi BM, Gingerich KA (1981) Inorg. Chem. 20: 522
202. Gupta SK, Pelino M, Gingerich KA (1979) J. Chem. Phys. 83: 2335
203. Gupta SK, Pelino M, Gingerich KA (1979) J. Chem. Phys. 70: 2044
204. Ruthardt K, Hilpert K (1989) Ber. Bunsenges. Phys. Chem. 93: 1070; Ruthardt K, Hilpert K, Weil KG, Report from KFA Jülich, Jül-2014, Jülich, FRG
205. Kingcade JE, Gingerich KA, 33rd Annu. Conf. Mass Spectrom. Allied Top., San Diego, Calif., USA 26–31 May 1985, p 903
206. Shim I, Gingerich KA (1984) J. Chem. Phys. 80: 5107
207. Shim I, Kingcade JE, Gingerich KA (1987) Z. Phys. D. 7: 261

208. Shim I, Kingcade JE, Gingerich KA (1986) J. Chem. Phys. 85: 6629
209. Hilpert K (1979) Ber. Bunsenges. Phys. Chem. 83: 161
210. Gingerich KA, Haque R, Kingcade JE (1979) Thermochim. Acta 30: 61
211. Kingcade JE, Choudary UV, Gingerich KA (1979) Inorg. Chem. 18: 3094
212. Kingcade JE, Gingerich KA (1986) J. Chem. Phys. 84: 3432
213. Hilpert K (1982) J. Chem. Phys. 77: 1425
214. Riekert G, Rainer-Harbach G, Lamparter P, Steeb S (1981) Z. Metallkde. 72: 406
215. Srinivasa RS, Edwards JG (1984) J. Electrochem. Soc. 131: 2954
216. Piacente V, Gigli R (1982) J. Chem. Phys. 77: 4790
217. Balducci G, Ferro D, Piacente V (1981) High Temp. Sci. 14: 207
218. Balducci G, Piacente V (1980) J. Chem. Soc., Chem. Commun. 24: 1287
219. Kingcade JE, Nagarathna-Naik HM, Shim I, Gingerich KA (1986) J. Phys. Chem. 90: 2830
220. Zmbov KF, Neubert A, Ihle HR (1981) Z. Naturforsch., Teil A, 36: 913
221. Grimley RT, Grindstaff QG, DeMercurio TA, Forsman JA (1982) J. Phys. Chem. 86: 976
222. Viswanathan R, Saibaba M, Darwin D, Raj A, Balasubramanian R, Mathews CK (1986) In:
     Todd JFJ (ed) Advances in mass spectrometry 1985, Part B, Wiley, New York, p 1087
223. Wu CH (1976) J. Chem. Phys. 65: 3181
224. Koutecký J (1986) Chem. Rev. 6: 539
225. Hilpert K, Kath D (to be published)
226. Salahub DR (1987) Adv. Chem. Phys. 69: 447
227. Morse MD (1986) Chem. Rev. 86: 1049
228. Kant A, Strauss B (1966) J. Chem. Phys. 45: 3161
229. Neubert A (1979) J. Chem. Thermodynamics. 11: 971
230. Alikhanyan AS, Steblevskii AV, Leont'eva VA, Saidulaeva M, Marenkin SF (1985) Inorg.
     Mater. (USSR) (Engl. transl.) 21: 1116
231. Hilpert K (1980) In: Quayle A (ed) Advances in mass spectrometry, Heyden, London, vol 8A,
     p 383
232. Gordienko SP, Fenochka BV (1982) Inorg. Mater. (USSR) (Engl. transl.) 18: 1554
233. Pelino M, Gupta SK, Cornwell LR, Gingerich KA (1979) J. Less Common Met. 68: 31
234. Schmidt SR, Franzen HF (1986) J. Less Common Met. 116: 73
235. Myers CE, Kematick RJ (1987) J. Electrochem. Soc. 134: 720
236. Storms EK, Myers CE (1985) High Temp. Sci. 20: 87
237. Oforka NC, Argent BB (1985) J. Less Common Met. 114: 97
238. Timberg L, Toguri JM (1982) J. Chem. Thermodynamics 14: 193
239. Myers CE, Murray GA, Kematick RJ, Frisch MA (1986) In: Munir A, Cubicciotti D (eds) Proc.
     of the Symposium on High Temperature Materials Chemistry III, The Electrochemical
     Society, Pennington, vol 86–2, p 47
240. Schiffman RA, Franzen HF, Ziegler RJ (1982) High Temp. Sci. 15: 69
241. Rammensee W, Fraser DG (1981) Ber. Bunsenges. Phys. Chem. 85: 588
242. Tomiska J (1986) Z. Metallkde. 77: 97
243. Tomiska J (1985) Z. Metallkde. 76: 532
244. Saha B, Viswanathan R, Saibaba M, Darwin D, Raj A, Balasubramanian R, Karunasagur D,
     Mathews CK (1985) J. Nucl. Mater. 130: 316
245. Levin ES, Gel'd PV (1979) Russ. J. Phys. Chem. (Engl. Transl.) 53: 1628
246. Tomiska J, Erdelyi L, Nowotny H, Neckel A (1982) High Temp. Sci. 15: 41
247. Peterson DE (1985) J. Nucl. Mat. 131: 44
248. Peterson DE, Starzynski JS (1985) J. Less Common Met. 105: 273
249. Oishi T, Nishi S, Ono K (1986) Trans. Jpn. Inst. Met. 27: 288
250. Tomiska J, Neckel A (1984) Ber. Bunsenges. Phys. Chem. 88: 551
251. Tomiska J, Neckel A (1983) Int. J. Mass Spectrom. Ion Phys. 47: 223
252. Hilpert K, Miller M, Gerads H, Nickel H (1990) Ber. Bunsenges. Phys. Chem. 94: 40
253. Golonka J, Botor J, Dulat M (1979) Met. Technol. 6: 267
254. Bergman C, Chastel R, Mathieu JC (1986) J. Chem. Thermodynamics 18: 835
255. Martin-Garin L, Chatillon C, Allibert M (1979) J. Less Common Met. 63: P9
256. Erdelyi L, Tomiska J, Neckel A, Rose G, Ramakrishnan ES, Fabian DJ (1979) Metallurgical
     Transactions A10: 1437
257. Glazov VM, Lebedev VV, Molodyk AD, Pashinkin AS (1979) Inorg. Mater. (USSR) (Engl.
     Transl.) 15: 1363
258. Said H, Chastel R, Bergman C, Castanet R (1981) Z. Metallkde. 72: 360
259. Lamparter P, Cocke DL, Steeb S (1982) Z. Metallkde. 73: 149

260. Srinivasa RS, Edwards JG (1986) Monatsh. Chem. 117: 695
261. Belousov VI, Vendrikh NF, Novoshilov AF, Pashinkin AS (1981) Inorg. Mater. (USSR) (Engl. Transl.) 17: 880
262. Kalashnikov AA, Drowart J, Burdeinyi AN, Pashinkin AS (1982) Dokl. Phys. Chem. (Engl. transl.) 261: 1122
263. Frisch MA (1981) 29th Annu. Conf. Mass Spectrom. Allied Top., Minneapolis, Minn., USA 24–29 May
264. Hilpert K (1983) Ber. Bunsenges. Phys. Chem. 87: 818
265. Allibert M, Chatillon C (1980) Rev. int. hautes tempér. réfract. 17: 271
266. Kazenas E, Chizhikov DM, Tsvetkov YuV (1974) In: Kornilov II, Matveeva NM (eds) Str. Svoistva Primen. Metallid. (Mater. Simp.), 2nd 1972, Nauka, Moskow, USSR, p 158
267. Rauh EG, Ackermann RJ (1975) Can. Metall. Quart. 14: 205
268. Büchler A, Berkowitz-Mattuk JB (1967) In: Eyring L (ed) Adv. High Temperature Chem., Academic, New York, vol 1, p 95
269. Choudary UV, Gingerich KA, Kingcade JE (1975) J. Less Common Metals 42: 111
270. Ngai LH, Stafford FE (1971) In: Eyring L (ed) Adv. High Temperature Chemistry, Academic, New York, vol 3, p 213
271. Wagman DD, Evans WH, Parker VB, Halow I, Bailly SM, Schumm R (1968) Natl. Bur. Stand. (U.S.) Tech. Note 270-3
272. Wagman DD, Evans WH, Parker VB, Halow I, Bailly SM, Schumm R (1969) Natl. Bur. Stand. (U.S.) Tech. Note 270-4
273. Parker VB, Wagman DD, Evans WH (1971) Natl. Bur. Stand. (U.S.) Tech. Note 270-6
274. Wagman DD, Evans WH, Parker VB, Halow I, Bailly SM, Schumm RH, Churney KL (1971) Natl. Bur. Stand. (U.S.) Tech. Note 270-5
275. Schumm RH, Wagman DD, Bailly SM, Evans WH, Parker VB (1973) Natl. Bur. Stand. (U.S.) Tech. Note 270-7
276. Wagman DD, Evans WH, Parker VB, Schumm RH, Nuttall RL (1981) Natl. Bur. Stand. (U.S.) Tech. Note 270-8
277. Ackermann RJ, Thorn RJ (1961) Progress in Ceramic Science, Pergamon, New York, vol 1, p 39
278. Kimura H, Asano M, Kubo K (1980) J. Nucl. Mater. 92: 221
279. Ikeda Y, Ito H, Matsumoto G, Nasu S (1979) Shitsuryo Bunseki 27: 263
280. Kudo H, Wu CH, Ihle HR (1978) J. Nucl. Mat. 78: 380
281. Wu CH, Kudo H, Ihle HR (1979) J. Chem. Phys. 70: 1815
282. Wu CH (1987) Chem. Phys. Lett. 139: 357
283. Nakagawa H, Asano M, Kubo K (1982) J. Nucl. Mat. 110: 158
284. Ohmichi T, Takeshita H, Nasu S, Sasayama T, Maeda A, Miyake M, Sano T (1979) J. Nucl. Mat. 82: 214
285. Kuligina LA, Semenov GA (1985) Vestn. Leningr. Univ., Fiz. Khim. 3: 39
286. Ikeda Y, Ito H, Mizuno T, Matsumoto G (1982) J. Nucl. Mat. 105: 103
287. Mart PL, Clark NJ (1982) High Temp. Sci. 15: 1
288. Takeshita H, Ohmichi T, Nasu S, Watanabe H, Sasayama T, Maeda A, Miyake M, Sano T (1978) J. Nucl. Mat. 78: 281
289. Neubert A (1985) J. Chem. Phys. 82: 939
290. Ikeda Y, Ito H, Matsumoto G, Hayashi H (1981) J. Nucl. Mat. 97: 47
291. Ikeda Y, Ito H, Matsumoto G, Hayashi H (1980) J. Nucl. Sci. Technol. 17: 650
292. Wu CH, Ihle HR, Zmbov K (1980) J. Chem. Soc., Faraday Trans. 2, 76: 447
293. Brüning D, Guggi D, Ihle HR, Neubert A (1984) Proc. of the 13th Symposium Fusion Technology 1984, 24–28 Sept. 1984, Varese, Italy, vol 1, p 427
294. Nakagawa H, Asano M, Kubo K (1981) J. Nucl. Mat. 102: 292
295. Semenov GA, Smirnova LN (1986) Dokl. Phys. Chem. (Engl. Transl.) 284: 893
296. Spoliti M, Piacente V, Bencivenni L, Ferro D, Nunciante Cesaro S (1980) High Temp. Sci. 12: 215
297. Ermilova IO, Kazenas EK, Zviadadze GN (1976) Russ. J. Phys. Chem. 50: 1309
298. Ogden JS, Alexander CA, Litman AP (1977) 25th Annual Conf. on Mass Spectrometry and Allied Topics, Washington D.C., USA, 29 May – 3 June, Amer. Soc. for Mass Spect. N. WP-36, p 664
299. Byker HJ, Eliezer I, Howald RC, Ehlert TC (1979) High Temp. Sci. 11: 153
300. Ehlert TC (1977) High Temp. Sci. 9: 237
301. Farber M, Srivastava RD, Moyer JW (1982) J. Chem. Thermodynamics 14: 1103

302. Farber M, Srivastava RD, Moyer JW, Leeper JD (1987) J. Chem. Soc. Faraday Trans. 1, 83: 3229
303. Rudny EB, Sidorov LN, Kuligina LA, Semenov GA (1985) Int. J. Mass Spectrom. Ion Processes 64: 95
304. Farber M, Srivastava RD, Moyer JW, Leeper JD (1985) J. Chem. Soc., Faraday Trans. 1, 81: 913
305. Farber M, Srivastava RD, Moyer JW, Leeper JD (1986) High Temp. Sci. 21: 17
306. Odoj R, Hilpert K (1980) Z. Naturforsch. 35a: 9
307. Odoj R, Hilpert K (1980) High Temp.-High Pressures 12: 93
308. Semenov GA, Nikolaev EN, Ovchinnikov KV (1978) J. Gen. Chem. USSR, Part 1, 48: 1752
309. Kazenas EK, Samoilova IO, Zviadadze GN (1983) Russ. J. Phys. Chem. 57, 1571
310. Farber M, Srivastava RD (1976) High Temp. Sci. 8: 195
311. Valderraman J, Jacob KT (1978) J. inorg. nucl. Chem. 40: 993
312. Farber M, Srivastava RD (1976) High Temp. Sci. 8: 73
313. Murad E (1981) 29th Ann. Conf. Mass Spectrom. Allied Top., Minneapolis, Minn., USA, 24–29 May 1981 (Abstr.), p 81
314. Nosenko AE, Kozenchuk NM, Bilyi AI, Kravchishin VV (1986) Inorg. Mater. (USSR) (Engl. Transl.) 22: 1234
315. Farber M, Srivastava RD (1981) J. Chem. Phys. 74: 2160
316. Kudin LS, Balducci J, Jili G, Guido M (1982) Izv. Vyssh. Uchebn. Zaved., Khim. Khim. Tekhnol. 25: 3
317. Kudin LS (1981) Izv. Vyssh. Uchebn. Zaved., Khim. Khim. Tekhnol. 24: 837
318. Kudin LS, Balducci J, Jili G, Guido M (1982) Izv. Vyssh. Uchebn. Zaved., Khim. Khim. Tekhnol. 25: 259
319. Hirayama C, Kleinosky RL, Bhalla RS (1980) Thermochim. Acta 39: 187
320. Srivastava RD (1976) High Temp. Sci. 8: 225
321. Hilpert K, Gerads H (1975) High Temp. Sci. 7: 11
322. Hastie JW, Zmbov KF, Bonnell DW (1984) High Temp. Sci. 17: 333
323. Butman MF, Kudin LS, Burdukovskaya GG, Krasnov KS (1984) High Temp. (Engl. Transl.) 22: 551
324. Kudin LS, Butman MF, Krasnov KS (1986) High Temp. (Engl. Transl.) 24: 50
325. Ovchinnikov KV, Nikolaev EN, Semenov GA (1983) J. Gen. Chem. USSR, Part 1, 53: 852
326. Banon S, Chatillon C, Allibert M (1982) High Temp. Sci. 15: 17; (1982) High Temp.-High Pressures 14: 383
327. Granier B, Gilles PW (1981) Rev. int. hautes températures réfract. 18: 227
328. Heideman SA, Reed TB, Gilles PW (1980) High Temp. Sci. 13: 79
329. Sheldon RI, Gilles PW (1977) J. Chem. Phys. 66: 3705
330. Hildenbrand DL (1976) Chem. Phys. Lett. 44: 281
331. Balducci G, Gigli G, Guido M (1985) J. Chem. Phys. 83: 1913
332. Rauh EG, Garg SP (1981) High Temp. Sci. 14: 121
333. Belov AN, Semenov GA (1979) Russ. J. Phys. Chem. 53: 1752
334. Ackermann RJ, Garg SP, Rauh EG (1979) High Temp. Sci. 11: 199
335. Banchorndhevakul W, Matsui T, Naito K (1985) Thermochim. Acta 88: 301
336. Balducci G, Gigli G, Guido M (1983) J. Chem. Phys. 79: 5616
337. Balducci G, Gigli G, Guido M (1986) J. Chem. Phys. 85: 5955
338. Matsui T, Naito K (1983) Int. J. Mass Spectrom. Ion Phys. 47: 253
339. Matsui T, Naito K (1982) J. Nucl. Mat. 107: 83
340. Kamegashira N, Matsui T, Harada M (1981) J. Nucl. Mat. 101: 207
341. Matsui T, Naito K (1981) J. Nucl. Mat. 102: 227
342. Balducci G, Gigli G, Guido M (1981) J. Chem. Soc., Faraday Trans. 2, 77: 1107
343. Ikeda Y, Ito H, Mizuno T, Amioka K, Matsumoto G (1983) High Temp. Sci. 16: 1
344. Chizhikov DM, Kazenas EK, Ermilova IO (1976) Russ. J. Phys. Chem. 50: 878
345. Marushkin KN, Alikhanyan AS, Greenberg JH, Lazarev VB, Malyusov VA, Rozanova ON, Melekh BT, Gorgoraki VI (1985) J. Chem. Thermodynamics 17: 245
346. Aleshko-Ozhevskaya LA, Il'in MK, Makarov AV, Nikitin OT (1980) Vestn. Mosk. Univ., Ser. II. Khim. 21: 248
347. Chizhikov DM, Kazenas EK, Ermilova IO (1976) Russ. J. Phys. Chem. 50: 1859
348. Smoes S, Drowart J (1984) High Temp. Sci. 17: 31
349. Murad E (1980) J. Chem. Phys. 73: 1381

350. Frisch MA, Dai TT (1983) Proc. 31st Annu. Conf. Mass Spectrom. Allied Top., Boston, Mass., USA, 8–13 May 1983, p 783
351. Balducci G, Gigli G, Goldenberg MMF, Guido M (1984) J. Chem. Thermodynamics 16: 207
352. Skudlarski K, Kapala J, Kowalska M (1982) J. Less Common Met. 83: L39
353. Kazenas EK, Zviadadze GN, Bol'shikh MA (1984) Izv. Akad. Nauk SSSR, Met. N. 2: 67
354. Grade M, Hirschwald W (1982) Ber. Bunsenges. Phys. Chem. 86: 899
355. Behrens RG, Mason CFV (1981) J. Less Common Met. 77: 169
356. Grade M, Hirschwald W, Stolze F (1978) Ber. Bunsenges. Phys. Chem. 82: 152
357. Bagarat'yan NV, Nikitin OT, Gorokhov LN (1980) Vestn. Mosk. Univ., Ser. II. Khim. 21: 139
358. Chervonnyi AD, Piven VA, Kashireninov OE, Manelis GB (1977) High Temp. Sci. 9: 99
359. Paule RC (1976) High Temp. Sci. 8: 257
360. Milushin MI, Emel'yanov AM, Gorokhov LN (1986) High Temp. (Engl. Transl.) 24: 347
361. Ovchinnikov KV, Nikolaev EN, Semenov GA (1981) J. Gen. Chem. USSR, Part 1, 51: 207
362. Kaposi O, Lelik L, Semenov GA, Nikolajev EN (1985) Acta Chim. Hung. 120: 79
363. Kligina LA, Semenov GA, Stolyarova VL (1986) J. Gen. Chem. USSR, Part 1, 56: 1486
364. Kvande H, Wahlbeck PG (1976) High Temp.-High Pressures 8: 45
365. Sasamoto T, Kobayashi M, Sata T (1981) Shitsuryo Bunseki 29: 249
366. Sasamoto T, Kobayashi M, Sata T (1979) NBS Spec. Publ. (U.S.) 561: 283
367. Plies V (1982) Z. Anorg. Allg. Chem. 484: 165
368. Kazenas EK, Chizhikov DM, Vasyuta YuV (1976) Russ. J. Phys. Chem. 50: 309
369. Semenikhin VI, Rykov AN, Sidorov LN (1983) Russ. J. Phys. Chem. 57: 1008
370. Zmbov KF, Miletic M (1978) In: Daly NR (ed) Advances in mass spectrometry, Heyden, London, vol 7, p 753
371. Nikolaev EN, Ovchinnikov KV, Semenov GA (1984) J. Gen. Chem. USSR, Part 1, 54: 869
372. Drowart J (1984) Pure Appl. Chem. 56: 1569
373. Plies V, Jansen M (1983) Z. Anorg. Allg. Chem. 497: 185
374. Semenov GA, Frantseva KE, Aurova OA (1983) Izv. Vyssh. Uchebn. Zaved., Khim. Khim. Tekhnol. 26: 66
375. Sidorov LN, Minayeva II, Zasorin EZ, Sorokin ID, Borshchevskiy AYa (1980) High Temp. Sci. 12: 175
376. Smoes S, Drowart J (1984) J. Chem. Soc., Faraday Trans. 2, 80: 1171
377. Murad E, Michael I (1978) 26th Annu. Conf Mass Spectrom. Allied Top., St. Louis, Mo., USA, 28 May–2 June 1978, p 265
378. Murad E, Hildenbrand DL (1980) J. Chem. Phys. 73: 4005
379. Belov AN, Semenov GA, Vinokurov IV, Krasil'nikov MD (1979) Inorg. Mater. (USSR) 15: 1281
380. Murad E (1978) Chem. Phys. Lett. 59: 359
381. Hildenbrand DL (1977) Chem. Phys. Lett. 48: 340
382. Kordis J, Gingerich KA (1977) J. Chem. Phys. 66: 483
383. Balducci G, Gigli G, Guido M (1985) J. Chem. Phys. 83: 1909
384. Balducci G, Gigli G, Guido M (1983) J. Chem. Phys. 79: 5623
385. Balducci G, DeMaria G, Gigli G, Guido M (1986) High Temp. Sci. 22: 145
386. Balducci G, Gigli G, Guido M (1977) J. Chem. Phys. 67: 149
387. Balducci G, Gigli G, Guido M (1977) J. Chem. Phys. 67: 147
388. (1981) 29th Annu. Conf. Mass Spectrom. Allied Top., Minneapolis, Minn., USA 24–29 May 1981 (Abstr.), p 85
389. Storms EK (1985) J. Nucl. Mat. 132: 231
390. Belov AN, Lopatin SI, Semenov GA, Vinokurov IV (1984) Inorg. Mater. (USSR) 20: 384
391. Matsui T, Naito K (1985) J. Nucl. Mat. 136: 67
392. Younes C, Long Den Nguyen, Pattoret A (1981) High Temp.-High Pressures 13: 105
393. Kashireninov OE, Chervonnyi AD, Piven VA (1982) High Temp. Sci. 15: 79
394. Farber M, Srivastava RD (1979) High Temp. Sci. 11: 1
395. Drowart J, Colin R, Exteen G (1965) Trans. Faraday Soc. 61: 1376
396. Shul'ts MM, Stolyarova VL, Semenov GA (1980) J. Non-Cryst. Solids, Part 2, 38: 581
397. Shul'ts MM, Stolyarova VL, Semenov GA (1979) Fiz. Khim. Stekla 5: 42
398. Altemose VO, Tong SSC (1980) J. Non-Cryst. Solids 38, 39: 587
399. Shul'ts MM, Stolyarova VL, Semenov GA (1979) Dokl. Phys. Chem. (Engl. Transl.) 246: 387
400. Shul'ts MM, Semenov GA, Stolyarova VL (1979) Inorg. Mater. (USSR) (Engl. Transl.) 15: 788

401. Kambayashi S, Kato E (1984) J. Chem. Thermodynamics 16: 241
402. Kambayashi S, Kato E (1983) J. Chem. Thermodynamics 15: 701
403. Shul'ts MM, Ivanov GG, Stolyarova VL, Makhmatkin BA (1987) Sov. J. Glass Phys. and Chem. 12: 147
404. Boike M, Hilpert K, Müller F (in preparation)
405. Minaeva II, Karasev NM, Yurkov LF, Sharov SN, Sidorov LN (1981) Fiz. Khim. Stekla 7: 223
406. Sharkan IP, Mikulaninets SV, Firtsak YuYu, Dorgoshei NI, Chepar DV, Kutsenko YaP (1981) Fiz. Khim. Stekla 7: 380
407. Stolyarova VL, Semenov GA (1979) Fiz. Khim. Stekla 5: 107
408. Asano M, Yasue Y (1986) J. Nucl. Mat. 138: 65
409. Kowalska M, Skudlarski K, Botor J (1986) J. Chem. Thermodynamics 18: 997
410. Rammensee W, Fraser DG (1982) Geochim. Cosmochim. Acta 46: 2269
411. Rogez J, Chastel R, Bergman C, Brousse C, Castanet R, Mathieu JC (1983) Bull. Minéral. 106: 119
412. Fraser DG, Rammensee W, Hardwick A (1985) Geochim. Cosmochim. Acta 49: 349
413. Sasamoto T, Hara H, Sata T (1981) Bull. Chem. Soc. Jpn. 54: 3327
414. Allibert M, Chatillon C, Jacob KT, Lourtau R (1981) J. Am. Ceram. Soc 64: 307
415. Belov AN, Lopatin SI (1984) Inorg. Mater. (USSR) (Engl. Transl.) 19: 1337
416. Belov AN, Lopatin SI, Semenov GA (1981) Russ. J. Phys. Chem. 55: 932
417. Belov AN, Semenov GA (1980) Inorg. Mater. (USSR) (Engl. Transl.) 16: 1513
418. Lukas W, Kowalska M (1980) Polish J. Chem. 54: 893
419. Lukas W, Kowalska M, Skudlarski K (1980) J. Chem. Thermodynamics 12: 885
420. Allibert M, Chatillon C (1979) Can. Metall. Q. 18: 349
421. Rocket TJ, Foster WR (1965) J. Am. Ceram. Soc. 48: 75
422. Hervig PC, Navrotsky A (1985) J. Am. Ceram. Soc. 68: 314
423. Fraser DG, Bottinga Y (1985) Geochim. Cosmochim. Acta 49: 1377
424. McPhail DS, Hocking MG, Jeffes JHE (1985) J. Mat. Sci. 20: 449
425. Schäfer H (1983) Adv. Inorg. Chem. 26: 201
426. Schäfer H (1976) Angew. Chem. 88: 775
427. Øye HA, Gruen DM (1976) NBS Spec. Publ. (US) 561
428. Hastie JW (1975) High temperature vapors, Academic, New York
429. Hastie JW (1977) In: Braunstein J, Mamantov G, Smith GP (eds) Advances in molten salt chemistry, New York
430. Novikov GI, Gavryuchenkov FG (1967) Russ. Chem. Rev. 36: 156
431. Brooker MH, Papatheodorou GN (1983) In: Mamantov G (ed) Advances in molten salt chemistry, Elsevier, New York, vol 5, p 95
432. Papatheodorou GN (1982) In: Kaldis E (ed) Current topics in materials science, North Holland, New York
433. Martin TP (1983) Phys. Rept. 95: 169
434. Davidovits P, McFadden DL (1979) Alkali halide vapors, Academic, New York
435. Brewer L, Brackett E (1961) Chem. Rev. 61: 425
436. Brewer L, Somayajulu GR, Brackett E (1963) Chem. Rev. 63: 111
437. Schäfer H (1962) Chemische Transportreaktionen, Chemie, Weinheim/Bergstr.
438. Hilpert K (1989) J. Electrochem. Soc. 136: 2099
439. Yamawaki M, Yasumoto M, Hirai M, Kanno M (1982) J. Nucl. Sci. Technol. 19: 563
440. Reinhart GH, Behrens RG (1980) J. Less Common Met. 75: 65
441. Malkerova IP, Alikhanyan AS, Butskii VD, Pervov VS, Gorgoraki VI (1985) Russ. J. Inorg. Chem. 30: 1571
442. Gorokhov LN, Ryzkov MYu, Khodeev YuS (1985) Russ. J. Phys. Chem. 59: 1761
443. Khodeyev YuS, Ryzkov MYu (1986) In: Todd JFJ (ed) Advances in mass spectrometry 1985, Part B., J. Wiley, New York, p 1027
444. Korobov MV, Bondarenko AA, Sidorov LN, Nikulin VV (1983) High Temp. Sci. 16: 411
445. Korobov MV, Nikulin VV, Chilingarov NS, Sidorov LN (1986) J. Chem. Thermodynamics 18: 235
446. Chilingarov NS, Korobov MV, Rudometkin SV, Alikhanyan AS, Sidorov LN (1986) Int. J. Mass Spectrom. Ion Phys. 69: 175
447. Farber M, Srivastava RD (1984) J. Chem. Phys. 81: 241
448. Emons HH, Horlbeck W, Kiessling D (1981) Z. Chem. 21: 416
449. Emons HH, Kiessling D, Horlbeck W (1982) Z. anorg. allg. Chem. 488: 219

450. Makarov AV, Gamin VV, Troyanov SI, Nikitin OT (1985) Vestn. Mosk. Univ., Ser. II: Khim. 25: 219
451. Verkhoturov EN, Makarov AV, Nikitin OT (1981) Vestn. Mosk. Univ., Ser. II: Khim. 22: 439
452. Novikova LN, Ratkovskii IA (1980) Khim. Khim. Tekhnol. (Minsk) 15: 32
453. Schäfer H, Wiese U, Rabeneck H (1984) Z. anorg. allg. Chem. 513: 157
454. Skudlarski K, Dudek J, Kapala J (1987) J. Chem. Thermodynamics 19: 151
455. Skudlarski K, Dudek J, Kapala J (1987) J. Chem. Thermodynamics 19: 857
456. Gesenhues U, Wendt H (1986) Int. J. Mass Spectrom. Ion Processes 70: 225
457. Schäfer H, Flörke U (1980) Z. anorg. allg. Chem. 462: 173
458. Skudlarski K, Kapala J (1984) J. Chem. Thermodynamics 16: 91
459. O'Brien NW, Mucklejohn SA (1987) J. Chem. Thermodynamics 19: 1065
460. Imperatori P, Ferro D, Piacente V (1982) J. Chem. Thermodynamics 14: 461
461. Sapegin AM, Baluev AV, Evdokimov VI (1982) Russ. J. Phys. Chem. 56: 132
462. Hildenbrand DL, Lau KH (1982) Proc. 30th Annu. Conf. Mass Spectrom. Allied Top., Honolulu, Hawaii, USA, 6–10 June 1982, p 444
463. Hilpert K, Miller M (to be published)
464. Mucklejohn SA, O'Brien NW (1986) In: Todd JFJ (ed) Advances in mass spectrometry 1985, Part B., J. Wiley, New York, p 999
465. Farber M, Srivastava RD (1980) High Temp. Sci. 12: 21
466. Hilpert K, Miller M, Raimondo F (1990) J. Chem. Phys. (in press)
467. Lau KH, Hildenbrand DL (1987) J. Chem. Phys. 86: 2949
468. DeMercurio TA, Grimley RT (1986) In: Munir A, Cubicciotti D (eds) Proc. of the Symposium on High Temperature Materials Chemistry-III, The Electrochemical Society, Pennington, vol 86–2, p 59
469. Lelik L, Sajó G, Vass-Balthazár K, Kaposi O (1983) Acta Chim. Acad. Sci. Hung. 113: 61
470. Emons HH, Horlbeck W, Kiessling D (1982) Z. anorg. allg. Chem. 488: 212
471. Yamawaki M, Hirai M, Nakasaki T, Kanno M (1982) 30th Annu. Conf. Mass Spectrom. Allied Top., Honolulu, Hawaii, USA, 6–11 June 1982 (Abstr.), p 453
472. Saha B, Hilpert K, Bencivenni L (1986) In: Todd JFJ (ed) Advances in mass spectrometry 1985, Part B., J. Wiley, New York, p 1001
473. Dettingmeijer J-H, Dielis HR, DeMaagt BJ, Vermeulen PAM (1985) J. Less Common Met. 107: 11
474. Grade M, Rosinger W (1984) Ber. Bunsenges. Phys. Chem. 88: 767
475. Piechotka M, Kaldis E (1986) J. Less Common Met. 115: 315
476. Hirayama C, Kleinosky RL (1981) Thermochim. Acta 47: 355
477. Kaposi O, Lelik L, Balthazár K (1983) High Temp. Sci. 16: 299
478. Kaposi O, Ajtony Zs, Popovic A, Marsel J (1986) J. Less Common Met. 123: 199
479. Flesch M, Knacke O, Münstermann E (1986) Z. anorg. allg. Chem. 535: 123
480. Mohazzabi P, Searcy AW (1977) Int. J. Mass Spectrom. Ion Phys. 24: 469
481. Grimley RT, Forsman JA, Grindstaff QG (1978) J. Phys. Chem. 82: 632
482. Pauling L (1956) Proc. National Academy of Sciences India, vol. XXV, Section-A, Part I.
483. Welch DD, Lazareth OW, Dienes GJ (1976) J. Chem. Phys. 64: 835
484. Cordfunke EHP (1986) Thermochimica Acta 108: 45
485. Hilpert K, Seehawer J (1986) Technisch-wissenschaftliche Abhandlungen der Osram-Gesellschaft 12: 31
486. Miller RC, Kusch P (1956) J. Chem. Phys. 25: 860
487. Datz S, Smith WT (1959) J. Phys. Chem. 63: 938; Datz S, Smith WT, Taylor EH (1961) J. Chem. Phys. 34: 558
488. Murgulescu IG, Topor L (1970) Rev. Roum. Chim. 15: 997
489. Work DE (1981) J. Chem. Thermodynamics 13: 491
490. Mucklejohn SA, O'Brien NW (1984) J. Chem. Thermodynamics 16: 767
491. Kappes MM, Radi P, Schär M, Schumacher E (1985) Chem. Phys. Letters, 113: 243
492. Peterson KI, Dao PD, Castleman AW Jr (1983) J. Chem. Phys. 79: 777
493. Feather DH, Searcy AW (1971) High Temp. Sci. 3: 155
494. Schäfer H (1980) Z. anorg. allg. Chem. 461: 29
495. Gräber P, Weil KG (1972) Ber. Bunsenges. Phys. Chem. 76: 410
496. Gräber P, Weil KG (1973) Ber. Bunsenges. Phys. Chem. 77: 507
497. Bernauer O, Weil KG (1974) Ber. Bunsenges. Phys. Chem. 78: 1339
498. Bernauer O, Busse B, Weil KG (1979) Ber. Bunsenges. Phys. Chem. 83: 603

499. Ricart JM, Rubio J, Illas F (1986) Chem. Phys. Lett. 123: 528
500. Schäfer H, Hönes WJ (1956) Z. anorg. allg. Chem. 288: 62
501. Alikhanyan AS, Steblevaskii AV, Lazarev VB, Kalinnikov VT, Grinberg YaKh, Zhukov EG, Agamirova LM, Gorgoraki VI (1980) Inorg. Mater. (USSR) 16: 56
502. Sidorov LN, Gubarevich VD (1982) Sov. J. Coordination Chem. 8: 248
503. Nikitin MI, Sidorov LN (1980) Int. J. Mass Spectrom. Ion Phys. 35: 101
504. Sidorov LN, Zhuravleva LV, Varkov MV, Skokan EV, Sorokin ID, Korenev YuM, Akishin PA (1983) Int. J. Mass Spectrom. Ion Phys. 51: 291
505. Sidorov LN, Karasev NM, Korenev YuM (1981) J. Chem. Thermodynamics 13: 915
506. Korobov MV, Badtiev EB, Voronin GF, Bondarenko AA, Sidorov LN, (1980) Vestn. Mosk. Univ., Ser. II, Khim. 21: 200
507. Zhuravleva LV, Nikitin MI, Sorokin ID, Sidorov LN (1985) Int. J. Mass Spectrom. Ion Processes 65: 253
508. Skokan EV, Nikitin MI, Sorokin ID, Korenev YuM, Sidorov LN (1983) Russ. J. Phys. Chem. 57: 1325
509. Emons HH, Horlbeck W, Kiessling D (1983) Z. anorg. allg. Chem. 507: 142
510. Bloom H, Williams DJ (1981) J. Chem. Phys. 75: 4636
511. Williams DJ (1982) Aust. J. Chem. 35: 1531
512. Schäfer H, Boos D (1981) Z. anorg. allg. Chem. 477: 35
513. McPhail DS, Hocking MG, Jeffes JHE (1984) Int. J. Mass Spectrom. Ion Processes 59: 261
514. Sasamoto T, Itoh M, Sata T (1982) Bull. Chem. Soc. Jpn. 55: 3643 and literature quoted therein.
515. Gesenhues U, Wendt H (1984) Z. Phys. Chem. Neue Folge 142: 93
516. Wahlbeck PG (1987) J. Chem. Phys. 87: 654
517. Gesenhues U, Reuhl K, Wendt H (1983) Int. J. Mass Spectrom. Ion Phys. 47: 251
518. Hilpert K, Miller M, Venugopal V (in preparation)
519. Sasamoto T, Itoh M, Sata T (1983) Bull. Chem. Soc. Jpn. 56: 2415
520. Schäfer H, Flörke U (1981) Z. anorg. allg. Chem. 479: 89
521. Schäfer H, Flörke U (1981) Z. anorg. allg. Chem. 478: 57
522. Schäfer H, Flörke U (1981) Z. anorg. allg. Chem. 479: 84
523. Emons H-H, Horlbeck W (1984) In: Blander M (ed) Proceedings of the 4th Int. Symp. on Molten Salts, The Electrochemical Society, Pennington, vol 84–2, p 411
524. Lelik L, Sajó G, Vass-Balthazár K, Kaposi O (1983) Acta Chim. Acad. Sci. Hung. 113: 75
525. Kaposi O, Bencze L, Zhuravleva LV (1986) J. Chem. Thermodynamics 18: 635
526. Hilpert K, Kobertz D, Gerads H (1986) In: Todd JFJ (ed) Advances in mass spectrometry 1985, Part B., Wiley, New York, p 1045
527. Mucklejohn SA, O'Brien NW (1983) Int. J. Mass Spectrom. Ion Phys. 47: 231
528. Gavrilin EN, Chilingarov NS, Skokan EV, Sorokin ID, Kaposi O, Sidorov LN (1987) Russ. J. Phys. Chem. 61: 265
529. Kaposi O, Lelik L, Balthazár K (1983) High Temp. Sci. 16: 311
530. Hilpert K, Miller M, Gerads H, Saha B (1990) Ber. Bunsenges. Phys. Chem. 94: 35
531. Miller M, Skudlarski K, (1987) J. Chem. Thermodynamics 19: 565
532. Miller M, Hilpert K (to be published)
533. Hilpert K, Miller M (to be published)
534. Miller M, Skudlarski K (1985) Ber. Bunsenges. Phys. Chem. 89: 916
535. Miller M (1984) Int. J. Mass Spectrom. Ion Processes 55: 133
536. Kapala J, Skudlarski K (1987) J. Chem. Thermodynamics 19: 27
537. Kapala J, Skudlarski K (1983) J. Chem. Thermodynamics 15: 1025
538. Hilpert K, Miller M (in preparation)
539. Miller M, Skudlarski K (1986) In: Todd JFJ (ed) Advances in mass spectrometry 1985, Part B., Wiley, New York, p 1015
540. Kapala J, Skudlarski K (1981) J. Mass Spectrom. Ion Phys. 40: 255
541. Itoh M, Sasamoto T, Sata T (1981) Bull. Chem. Soc. Jpn. 54: 3391
542. Emons HH, Horlbeck W, Pohl D (1979) Ext. Abstr. Meet. Int. Soc. Electrochem., p 193
543. Hirayama C, Liu CS, Zollweg RJ (1978) In: Hildenbrand DL, Cubicciotti DD (eds) Proc. Symp. on high temperature metal halide lamp chemistry, The Electrochem. Soc., Princeton, vol 78–1, p 95
544. Hirayama C, Castle PM, Liu CS, Zollweg RJ (1977) J. Illuminating Engineering Society 6: 209
545. Skokan EV, Nikitin MI, Sorokin ID, Gusarov AV, Sidorov LN (1981) Russ. J. Phys. Chem. 55: 1062

546. Sidorov LN, Sorokin ID, Nikitin MI, Skokan EV (1981) Int. J. Mass Spectrom. Ion Phys. 39: 311
547. Pyatenko AT, Gusarov AV, Gorokhov LN (1981) Teplofiz. Vys. Temp. 19: 1167
548. Pyatenko AT, Gusarev AV, Gorokhov LN (1981) Teplofiz. Vys. Temp. 19: 329
549. Skokan EV, Sorokin ID, Sidorov LN, Nikitin MI (1982) Int. J. Mass Spectrom. Ion Phys. 43: 309
550. Sidorov LN, Borshchevsky AYa, Rudny EB, Butsky VD (1982) Chem. Phys. 71: 145
551. Chilingarov NS, Korobov MV, Sidorov LN, Mit'kin VN, Shipacher VA, Zemskov SV (1984) J. Chem. Thermodynamics 16: 965
552. Sorokin ID, Sidorov LN, Nikitin MI, Skokan EV (1981) Int. J. Mass Spectrom. Ion Phys. 41: 45
553. Sidorov LN, Borshchevsky AYa, Boltalina OV, Sorokin ID, Skokan EV (1986) Int. J. Mass Spectrom. Ion Processes 73: 1
554. Pyatenko AT, Gorokhov LN (1984) Chem. Phys. Lett. 105: 205
555. Sidorova IV, Pyatenko AT, Gorokhov LN, Smirnov VK (1984) High Temp. 22: 857
556. Lelik L, Kaposi O, Korobov MV, Sidorov LN (1984) Magy. Kém. Folyóirat 90: 74
557. Kaposi O, Lelik L, Korobov MV, Chilingarov NS, Gavrilov EN, Sidorov LN, Sorokin ID (1986) In: Todd JFJ (ed) Advances in mass spectrometry 1985, Part B., J. Wiley, New York, p 1003
558. Hastie JW, Plante ER (1981) Proc. Specialists meeting on coal fired MHD power generation, Sydney, Australia, 4–6 November 1981, p 3.8.1–3.8.10 (National conference publication no. 81/13).
559. Plante ER, Bonnell DW, Hastie JW (1983) Proc. 31st Annual conf. mass spectrometry allied topics, Boston, Mass., USA, 8–13 May 1983, p 830
560. Moon KA, Kant A, Croft WJ (1980) J. Am. Ceram. Soc. 63: 698
561. Semenov GA, Frantseva KE, Nikolaev EN, Pirogov YuA, Babkina LA (1976) Izv. Akad. Nauk SSSR Neorg. Mater. 12: 2223
562. Anikin IN, Pat'kovskii IA, Ashuiko VA, Asnovich EZ (1984) Dokl. Akad. Nauk SSSR 274: 84
563. Itoh N, Nakamura T (1987) Bull. Chem. Soc. Jpn. 60: 503
564. Hilpert K, Gerads H, Kobertz D (1985) Ber. Bunsenges. Phys. Chem. 89: 43
565. Hilpert K, Gerads H, Kath D, Kobertz D (1988) High Temp.-High Pressures 20: 157
566. Hilpert K, Ali-Khan I (1978) J. Nucl. Mat. 78: 265
567. Hilpert K, Gerads H, Lupton DF (1979) J. Nucl. Mat. 80: 126
568. Hilpert K (1980) Proc. Symp. Thermodynamics of nuclear materials 1979, Jülich, 1979, Report of the International Atomic Energy Agency Vienna, IAEA-SM-236, Vienna, vol 1, p 61
569. Hirai M, Kozura A, Yamawaki M, Kanno M (1983) J. Nucl. Sci. Technol. 20: 333
570. Huang JY-K, Gilles PW, Bennett JE (1984) High Temp. Sci. 17: 109
571. Neubert A, Zmbov KF (1974) Trans. Faraday Soc. 70: 2219
572. Sidorov LN, Zhuravleva LV, Sorokin ID (1986) Mass Spectrometry Reviews, Wiley, New York, vol 5, p 73
573. Ahlrichs R, Becherer R, Binnewies M, Borrmann H, Lakenbrink M, Schunck S, Schnöckel H (1986) J. Am. Chem. Soc. 108: 7905
574. Binnewies M, Lakenbrink M, Schnöckel H (1986) High Temp. Sci. 22: 83
575. Binnewies M (1983) Z. anorg. allg. Chem. 507: 66; Schnöckel H, Lakenbrink M (1983) Z. anorg. allg. Chem. 507: 70
576. Binnewies M (1983) Z. anorg. allg. Chem. 505: 32; Schnöckel H, Lakenbrink M, Lin Zhengyan (1983) J. Mol. Struct. 102: 234
577. Gingerich KA (1985) J. Less-Common Met. 110: 41
578. Gingerich KA, Choudary UV, Krishnan K, Hilpert K (1985) J. Chem. Phys. 83: 1237
579. Shim I, Finkbeiner HC, Gingerich KA (1987) J. Phys. Chem. 91: 3171
580. Shim I, Gingerich KA (1984) J. Chem. Phys. 81: 5937
581. Gupta SK, Gingerich KA (1981) J. Chem. Phys. 74: 3584
582. Gupta SK, Gingerich KA (1980) High Temp.-High Pressures 12: 273
583. Milne TA, Greene FT (1968) Advances Chem. Ser. 72: 68
584. Milne TA, Greene FT (1969) In: Eyring L (ed) Advances in high-temperature chemistry, vol 2, Academic, New York, p 107
585. Bonnell DW, Hastie JW In: Ref 2. p 357
586. Hastie JW, Bonnell DW (1983) In: Ribeiro da Silva MA (ed) Thermochemistry today and its role in the immediate future, Reidel, Boston

587. Olstad RA, Olander DR (1975) J. Appl. Phys. 46: 1499
588. Hastie JW, Bonnell DW, Schenck PK (1987) Report of the National Institute of Standards and Technology, USA, NBSIR 87-3561
589. Ohse RW, Babelot JF, Cercignani C, Hiernaut JP, Hoch M, Hyland GJ, Magill J (1985) J. Nucl. Mat. 130: 165
590. Ohse RW, Babelot JF, Cercignani C, Frezzotti A, Long KA, Magill J, Scotti A (1980) High Temp. Sci. 13: 35
591. Shim I, Gingerich KG (1989) Int. J. Quantum Chem.: Quantum Chem. Symp. No. 23: 409
592. Sapegin AM, Baluev AV, Evdokimov VI (1984) Russ. J. Phys. Chem. 58: 1792
593. Kleinschmidt PD, Ward JW (1986) J. Less-Common Met. 121: 61

Heavy Elements Synthesized in Supernovae and Detected in Peculiar A-Type Stars

Christian K. Jørgensen

Section de Chimie, Université de Genève, CH 1211 Geneva 4, Switzerland

Since two-digit Z elements are only known to be produced in supernovae, the chemically peculiar main-sequence stars (being 6 to 8 percent of Sirius-like objects within a rather narrow temperature interval) are enigmatic. The hypothesis of magnetic accretion is compared to other explanations, and it seems likely that the intensification of Fraunhofer lines (by factors 10 to a million) is indeed predominantly due to factual high abundances of a few elements including europium (and other lanthanides), uranium, bismuth, mercury, platinum, osmium, gallium and manganese in individual stellar photospheres. A given supernova seems to provide a quite specific selection of elements, and the accretion of the locally enriched interstellar gas (and probably protoplanetary grains) enhances it. Sidelights are thrown from silicon-rich stars on the striking difference between the Sun and (the Earth, its moon, and Mars), supernova dust contributing significantly to the inner planets, with 1300 ppm of weight having Z above 30 (but less than 0.6 ppm of solar matter). The lanthanides represent 120 and 0.04 ppm, respectively. A new parameter of atomic spectra is proposed as being relevant to magnetic accretion.

Structure and Bonding 73
© Springer-Verlag Berlin Heidelberg 1990

1 Spectral Types and the Hertzsprung-Russell Diagram

Inorganic chemists would hardly find life worthwhile, if they did not have at disposal elements with two-digit Z values, and living organisms would seem inconceivable without carbon, nitrogen and oxygen. The large majority of the three latter elements, and sofar we know, essentially all elements heavier than neon, are formed in the collapsing core of supernovae [1, 2] and dispersed in the shock-wave of the explosion. In a review in this volume [3] the rapid modifications the last 200 years of our ideas about "constituents" of matter, the agglutination of quarks (three at a time) to baryons 10 to 20 microseconds after the Big Bang (at a temperature close to 2×10^{12} K) and the reactions of baryons during the first four minutes [4] having as end products 0.78 (of the rest-mass) hydrogen ^{1}H, 0.22 helium ^{4}He, a few times 10^{-5} of each of the rarer ^{2}D and ^{3}He, around 10^{-9} lithium ^{7}Li and even much smaller amounts of other elements, were followed by processes of Stellar nucleosynthesis, probably starting after 10^8 years.

The purpose of the present review is to scrutinize a most peculiar group of unconventional stars. As already reviewed [3] the highly differing selections of elements detected by Fraunhofer absorption lines of stellar spectra were originally thought to represent, at face value, the chemical composition—the hottest stars (like most of the brilliant stars in Orion) mainly containing helium, and then, the less hot stars (like Vega and Sirius) predominantly hydrogen, and yellowish stars (like our Sun and α Centauri) iron, calcium and other metallic elements. About 60 years ago, this opinion was revised, mainly due to Cecilia Payne [5], to there being a roughly constant composition of the stellar *photosphere* (the surface layer, where the Fraunhofer lines are produced) but giving entirely different spectra, essentially as a function of the temperature T prevailing in the photosphere [and evaluated from the continuous background in the visible and near ultraviolet corresponding to the standard emission of an opaque body, peaking at a wavelength scale at 2897 nm time $(1000 \text{ K})/T$]. Minor spectral differences [6] are related to the surface gravitation (with a concomitant density gradient of the photosphere) and the small concentration of free electrons determining the distribution of a given element on neutral atoms and ions of differing charge.

It is useful to remember the spectral classification, of which the six regular types are B(25000 K), A(12000 K), F(8000 K), G(6000 K), K(5000 K) and M(3700 K) where the T in parentheses are representative values for B0, A0, ... Each interval is divided [6] into steps such as B0, B1, B2, ..., B8, B9, A0, A1, ... The spectral class O stars show higher T than the class B, up to well above 50000 K. A part of these (few, but spectacular) stars show emission lines, and are the strongest light-emitters (stable on a centennial time-scale) with absolute luminosities (the absolute magnitudes in literature refer to a distance of 10 parsec = 32.6 light-years) 2000 times that of Sirius. A related, "early" (i.e. high T) class is W, Wolf-Rayet stars which seem to be rather short-lived precursors

[7–10] to red supergiants (such as M1 type Betelgeuse and Antares) and subsequently (on a 10^6 y time-scale) to some of the supernovae [2, 7, 8]. Also the "late" types (parallel and subsequent to M) have been reorganized [6] to type S [characterized by band spectra of the diatomic molecule ZrO (and hence being a case of "chemical peculiarity") rather than TiO in the far more frequent type M and type C [band spectra of C_2, CH and CN with highly varying $^{13}C/^{12}C$ isotopic ratio]. Since C has replaced the previous classes R and N, it has removed the tail from the mnemotechnic phrase "O, Be A Fine Girl, Kiss Me Right Now, Smash" (which may be considered a monument to the pioneer contribution of American ladies, beginning 100 years ago, to stellar classification at the Harvard, Yerkes and other great observatories).

When absolute luminosities became available (from the parallax of stellar positions induced by the basis line 3×10^8 km by the annual motion of the Earth around the Sun) the Danish astronomer Ejnar Hertzsprung (1873–1967) and the US astrophysicist (also known from Russell-Saunders-coupling) Henry Norris Russell (1877–1957) established in 1905 the H-R diagram [6]. The *main-sequence* involves the large majority of stars (per volume unit; however, the white dwarfs of type A some 10^4 times less luminous than Sirius, but with comparable T, are usually detected in binary systems, but occur as single stars within the first 20 light-year distances, and are very difficult to detect and count, like the controversial low-T "brown dwarfs") with the absolute luminosity decreasing monotonically from type B2 to type M2 by a factor 10^5. A given star remains for most of its lifetime at a fairly invariant position in the Hertzsprung-Russell diagram, but then terminates with a rapid excursion as a red giant (types K and M, roughly as luminous as Sirius) before ending as a white dwarf (slowly cooling down) except for heavy stars terminating as neutron stars, conceivably higher-density [3] *uds*-matter, or black holes. Among the 5000 *visible* stars, which are discriminating strongly against low absolute luminosity, about half are main-sequence B, A, F, about 40 percent giant G, K, M and only about 5 percent G, K, M main-sequence stars (representing the large majority, if white and brown dwarfs are ignored).

2 Plausible Classifications of A_p Stars

Once it had been generally accepted [1, 5] that the highly different stellar spectra belonging to different main-sequence types (and red giants) can be rationalized by the photospheres having roughly the same relative abundance of elements, there was a tendency to forget evidence for different chemical composition, rather explained away as experimental uncertainty. However, in a few instances, overwhelming indications for *strong overabundance* (compared with the Sun) for one or several elements were discovered, as reviewed shortly [3]. Such cases are "CP" (chemically peculiar) and noted as a lower-case letter p after the spectral class, such as B9p for HR 465 (HD 9996), A0p for α^2 Canum Venaticorum (Cor Caroli), A4p for HR 4816 (HD 110066) and F0p for β Coronae Borealis. Preston [11] introduced four kinds of CP in the B, A and F classes:

CP-1: F_m and A_m stars ("metallic line" stars with exceptionally strong lines of iron and adjacent metallic elements, as compared to hydrogen and calcium)

CP-2: "Cool" A_p stars (with exceptionally to exorbitantly strong lines of chromium, strontium and/or europium)

CP-3: "Hot" A_p stars (with exceptionally to exorbitantly strong lines of mercury and/or manganese)

CP-4: B_w and A_w stars (with exceptionally *weak* helium lines).

On the whole, the photospheric T increases from CP-1 to CP-4. In a sense, there is a precursor to CP-1 in the form of G and K stars [6] with the Ba^+ line some 10 times stronger than usual. There is a general tendency for most CP-2 and CP-3 to show strong Si^+ lines (as discussed below, this has a great impact on the total abundance of two-digit Z elements, of which silicon and iron normally are the two best represented examples). It may be argued [12] that platinum can be at least as abundant as mercury, and gallium [13] at least as abundant as manganese, in the CP-3 category. This (highly individualistic) occurrence of lanthanides in CP-2 is discussed below. The only stars where uranium has been identified with a high level of confidence, also belong to CP-2. On the other hand, an enormous abundance of the last non-radioactive element, bismuth, has been found [13] in one CP-3 star. The category CP-4 might seem to fall outside our scope, but its quite incredible isotopic ratio $^3He/^4He$ of order 1 in some members has relevance to the diffusion theory of Michaud [14] discussed below, as has also been argued in the case [15] of the rather startling enrichment [16] in some CP-3 stars of the heaviest stable mercury isotope ^{204}Hg. Also, at least some CP-4 stars show high abundances of odd-Z elements such as phosphorus, manganese and gallium. It should immediately be brought to attention that the overabundances [17, 18] we speak about for Z above 50 sometimes reach factors from 10^4 up to 10^6 in CP-2 stars, but varying strongly from one lanthanide to the next. The overabundances relative to the Sun can also reach 50 for silicon and are typically 100 for chromium.

Although it will become evident below that the four categories CP (and in particular CP-2) are not at all homogeneous, it was a most worthwhile effort of North [19–21] to characterize the various CP stars by their parameters in the Vilnius-Geneva seven-color photometric system [22]. Since direct study of Fraunhofer lines is not easy beyond ninth magnitude stars, it is helpful for detecting CP candidates to use multi-color photometry of weaker stars, in particular in dense clusters in our Galaxy. Since the typical absolute luminosity of A_p stars is comparable to Sirius, most A_p stars are beyond eighth magnitude, if further away than 300 light-years, which is closer than from where we still perceive quite bright stars (like Deneb; Rigel and four other stars in Orion of about second magnitude). The perceived luminosity has the ratio $(2.512)^6 = 250$ for a difference of six magnitude indices, and 10000 for ten magnitudes, and is, under equal circumstances, inversely proportional to the square of the distance.

North [20] gives a catalog of 1028 CP stars. Though it is an old joke that one person classifies 300 plants or butterflies in 4 species, and another in 400, the criteria in the catalog have about 10 alternatives. Among the stars brighter than magnitude 7.0, the symbol Eu for europium occurs (either in the main table, or in added "remarks") 69 times. In the following, stars are characterized by their HR numbers (Harvard Revised Catalog, synonymous with BS, Catalog of Bright Stars, i.e. 9190 stars brighter than magnitude 6.5) and/or HD numbers (Henry Draper catalog of 225300 stars, mostly brighter than 8.3) besides the ancient notations of a Greek letter or (for weaker stars crowded in large constellations) the Flamsteed numbers followed by the genitive of the Latin name of the constellation (Canes Venatici for the "Hunting Dogs" being Canum Venaticorum in genitive). The 88 constellations (the "Serpent" cut in two non-contiguous areas, separated by "Ophiuchus") were organized by IAU (the International Astronomical Union, one of the most efficient and helpful in its kind) to have straight-line north-south and east-west segments (year 1900 positions) as borders, and a three-letter (of which the second or the third may be a second capital letter) symbol, such as CVn or (for Corona Borealis) CrB (the high chromium content in its next-brightest star is presumably accidental).

The difficulty (and great luck needed) of detecting CP stars beyond seventh magnitude decided North [20] to concentrate statistical attention on the 314 HR stars in his list of 1028 (more or less) CP stars. For the chemist, *europium* is the most fascinating element (first seen in α^2CVn, the second member of a fairly close optical pair, by Baxandall in 1913) and 38 cases "Eu" is explicitly given among the 314 HR stars, to which are added 19 cases in the list "remarks" as a result of new reports, debatable examples, and afterthought. The brightest case is indeed α^2CVn (2.85) where the magnitude is added in parenthesis, followed by β CrB (3.67). The reviewer is reluctant to accept γ TrA (Trianguli Australis) (2.87) cited as "Eu" but lacking corroborative properties. Five other "europium stars" have magnitudes between 4.6 and 5.2: HR 873 (21 Persei) 21 Per (5.11); HR 4327 (46 Centauri) 46 Cen (5.14); HR 7879 (73 Draconis) 73 Dra (5.19); HR 8151 (ϑ^1 Microscopii) ϑ^1 Mic (4.81); and HR 8949 (ι Phoenicis) ι Phe (4.68).

A slightly larger number, 65, HR stars are listed [20] with *mercury* (and hence CP-3). On the whole, there occur brighter (but perhaps a more heterogeneous set, as discussed below) mercury stars, such as: α Andromedae (α And, 2.07); φ Eridani (φ Eri, 3.55); μ Leporis (μ Lep, 3.29); β Tauri (β Tau, 1.65); γ Canis Majoris (γ CMa, 4.11); HR 4072 (HD 89822) 4.93; HR 5475 (π^1Boo) 4.9; HR 5971 (ι CrB) 4.97; HR 5982 (υ Her) 4.71; HR 6023 (φ Her) 4.23; and HR 8937 (β Sculptoris) β Scl (4.36). To this list of 11 stars brighter than 5.0 may be added HR 5883 (HD 141556) χ Lupi (χ Lup, 3.96) only [20] noted "Sr" but reported [16] to contain large amounts of the heaviest stable mercury isotope ^{204}Hg, and

HR 8911 (HD 220825) ϰ Piscium (ϰ Psc, 4.94) noted "Cr, Sr" but reported [23] to contain huge amounts of osmium, iridium and uranium.

North [20] elaborated a detection of CP stars using seven-color photometry and discussed in his Sect. 4.2 (page 209) the number of HR stars having a distinct parameter Y (of which a wide choice of critical limit hardly has any influence on the number of stars included). Thus, 11 among 13 Sr, Cr, Eu stars are detected, 12 among 22 Sr and Sr, Cr stars, 21 among 24 Cr and Cr, Sr stars (chromium lines more prominent than strontium), 17 among 21 Si, Cr stars, but only 50 among 81 Si stars. It is seen that this photometric detection of 111 out of 161 CP-2 stars (among HR stars) is more successful for some categories than for others. The main conclusion of North is that the detection of the CP-3 Hg, Mn stars is not feasible with the photometric color parameter Y, only 4 among the 89 HR stars of this category are detected. Then, it is not unexpected that only 7 such stars are recognized among the 29 HR cases of CP-4 (helium weak). It is more surprising that North only discusses 13 Sr, Cr, Eu stars of HR brightness, when his list clearly mentions 38 cases Eu (not including 19 "remarks"). Anyhow, North concludes that 10.7 percent of stars with masses between 3 and 5 solar masses are CP-2 (of Sr, Cr, Eu and the related types) and between 2 and 3 solar masses 7.4 percent. A related conclusion can be derived from the fact [24] that slices of half a unit magnitude width between 6.0 and 6.5; between 5.5 and 6.0; and between 5.0 and 5.5 all three contain a number of stars of spectral classes from A0 to A9 inclusive (1311, 794, and 465, respectively) which is remarkably close to 23 percent of all stars in the slice. The number of such A stars, 3105, between 7.5 and 8.0 only shows a moderate "thinning out" to 19 percent (whereas *all* stars closer than 40 light-years are only A stars to an extent of less than 2 percent). For our purposes, stars from A0 through A9 are a good model of "eligible" stars for CP-2 behavior (there are indeed a few B8 and B9 in the group [24] but they are compensated by a decreased incidence in A8 to F0). Hence, the 9190 HR stars are expected to include roughly 2000 "eligible" candidates, among which North [20] indicates 1.9 percent "first-class" and 1 percent "remark" europium stars, and 3.2 percent mercury stars.

Before discussing the many distinct (and highly unexpected) details of A_p stars, it must be emphasized that the Aristotelian principle of the excluded middle hardly applies to Eu, non-Eu, Nd, non-Nd, ... alternatives. After a well-timed IAU Colloquium in Vienna 1975 with 755 pages published of the proceedings [25] several opinions arose. For instance, Sidney C. Wolff and Richard J. Wolff, as well as L. Mestel, suggested *two* distinct CP sequences, covering essentially the same T range, one characterized by strong (and variable) magnetic fields (typically 1000 to 8000 Gauss) starting with the coolest CP-2 stars [roughly [26] in the order 10 Aql (HR 7167); γ Equ (HR8097); β CrB; HR 7575 (HD 188041); 73 Dra (HR 7879); 78 Vir (HR 5105); 52 Her (HR 6234); etc.]. In this list, the (supra-)peculiar HR 465 (HD 9996); HD 216533; and ϰ Psc (HR 8911) come at considerably higher T, though this quantity is much more difficult to define than in normal main-sequence stars (but there is no convincing evidence that CP-2 stars are marginal sub-giants). This *magnetic sequence* goes on with hot silicon stars such as HD 200311 and 49 Cnc (HR 3465) and seems to end with (a few of) the CP-4 helium-weak stars, such as HR 7129 (HD 175362) and 3 Scorpii (HD 142301). The *non-magnetic sequence* starts with the CP-1 category F_m and A_m stars (having unusually strong lines of iron-group elements), go on with CP-3 Hg, Mn stars and end with (most of) the CP-4 helium-weak stars. In all of these stars, the magnetic fields are too weak to be detected, and usually well below 200 Gauss (to be compared with roughly 1 Gauss on the Sun).

In view of the tentative and provisional character of many of the present-day rationalizations of the one (or two; or more?) A_p phenomena, it seems suitable to start with three sections on observations and numerical treatment before discussing the proposed mechanism in Sects. 6 to 8.

3 Degrees of Lanthanide Peculiarity in A_p Stars

The word *overabundance* refers to the ratio between the concentration of a given element (relative to hydrogen) in a given stellar photosphere and in the Sun [1] as given in the Table 1 of the review in this volume [3]. In most cases, such a value is not much better determined than within a factor 2 corresponding to 0.3 in the decadic logarithm. In analogy to "decibel", astrophysicists frequently use the unit 0.1 "dex" for 0.1 in the decadic logarithm, perhaps accustomed to one unit of stellar magnitude being (by definition) 0.4 dex. Most papers give $12 + \log_{10}(Z/H)$ of atomic abundances, relative to hydrogen having the fixed abundance 12.0 [though a system with 6.0 silicon has also been used, providing 10.4 for solar [1] hydrogen]. Most measurable abundances are above 3. They can be converted to 0.75 A times 1 mg/ton for $\log_{10}(Z/H) = -9$ [or 1 g/ton for $\log_{10}(Z/H) = -6$]. Two excellent reviews [17, 18] treat 21 sharp-lined "cool" A_p stars, and for comparison, the A1 star o Pegasi (HD 214994) of magnitude 4.79, and the Sun. The expression $-\log_{10}(Z/H)$ used by Adelman [17, 18] is the well-known p(Z/H) for a chemist. We also call the decadic overabundance $D(Z)$ [sometimes called [Z/H] in literature]. It should not be construed to be identical (even taking experimental uncertainty into account) for most of the 21 A_p stars, but there are quite clear-cut trends. Nearly all $D(Mg^+)$ are between zero and +0.3. It is helpful for comparison to indicate the ionic charge producing the Fraunhofer lines studied, but they have been corrected (as well as it can be done) for the relative concentration of Mg^+ and of all the other monatomic species (Mg^0 and Mg^{+2}) expected at the conditions prevailing. $D(Si^+)$ is roughly 0.5. $D(Ti)$ varies between 0.5 and 2.8, rather wildly. The chromium abundance is 8 times (0.9 unit) higher in the Sun than that of titanium. $D(Cr)$ varies between 1.2 and 3.6, but is typically close to 2. $D(Mn)$ varies between 1 and 2.8, and $D(Fe)$ between 0.4 and 1.8. It may be noted that $\log_{10}(Fe/Cr)$ is uniformly smaller than 2.0 [or [1] 1.6] in the Sun, varying between 0.2 and 1.2. $D(Ni)$ is -0.3 to 1.2, and

$D(Sr^+)$ 1.8 to 3.3. The values of $D(Y^+)$ are much lower, 0.2 to 0.8. The relatively few values determined for $\log_{10}(Zr/Y)$ fall between 0.5 and 1.0.

In the twelve A_p stars where La^+ is detected [17, 18] $D(La^+)$ is quite low (0.6 to 1.7) and $D(Ce^+)$ 1 to 2.1. It turns out that most A_p stars contain more neodymium than cerium, $D(Nd^+)$ being 1.2 to 3.6. Similarly, $D(Sm^+)$ is 2.0 to 3.4, but $D(Eu^+)$ 2.8 to 5.2, the two highest values in HD 2453 and HD 5797, justifying the name "europium stars". As discussed in Sect. 6, it may be argued that Eu^+ is similar [29] to alkaline-earth ions Ca^+, Sr^+ and Ba^+ (all playing an important rôle in astrophysics) but it may be noted that $\log_{10}(Eu/Sr)$ varies from -1.86 in the Sun [or [1] -2.2] to $+0.87$ in γ Equulei. The analogs $\log_{10}(Eu/Gd)$ varies from -0.17 in the Sun [or [1] -0.4] to $+1.7$ in β Coronae Borealis. We return in Sect. 7 to the corresponding $D(U^+)$ estimated by Adelman [17, 18] to vary between 1.6 and 4.2, the most reliable value perhaps being 3.4 in HD 2453. For our purpose, uranium is a rather specific problem.

There are at least three A_p stars each showing a quite individualistic distribution of lanthanides: α^2CVn; HR 465 (HD 9996) of magnitude 6.38; and Przybylski's star [30] HD 101065 with magnitude 8.02. The brightest (and earliest known) α^2CVn has attracted attention for many reasons, including its magnetic field, varying in a complicated way with a reproducible period 5.47 days, and its absorption spectrum in the far-ultraviolet (measured with spectrometric equipment on a satellite down to 1025 Å) further discussed in Sect. 8. It is still not a resolved problem, to what extent these unusual physical properties are related to the observed europium overabundance (it must be emphasized that it refers to the layer of the surface where the Fraunhofer lines originate; $D(Eu^+)$ is unlikely to be invariant at larger depths). The abundances [31] given for α^2CVn can be translated to direct weight concentrations [like Table 1 for the Sun [3]] producing (per ton) 600 g Ti, 1600 g Cr, 60 kg Fe, 160 g Nd, 1 kg Eu and 60 g Gd, probably not far from 2 kg lanthanides. Even if some or all of these values are overestimated by a factor 3, they are still baffling. $D(Eu^+)$ is suggested to be 6 and $D(Fe)$ about 2. The mean values [17] for 21 sharp-lined "cool" A_p stars are 3.8 and 1.06, respectively, or (with the solar iron abundance assumed [17] there) 7 kg Fe and 6 g Eu per ton, to be compared with 3 kg Si, 2.5 g Nd, 4 g Gd and 0.12 g U. The maximum $D(Ln)$ for lanthanides (the list did not include α^2CVn) give 50 g Nd, 400 g Eu, 25 g Gd, suggesting roughly 1 kg lanthanides per ton.

HR 465 (HD 9996) is perhaps, in many ways, the most exorbitantly peculiar A_p system. Preston and Wolff [32] suggested it to be a long-periodic variable (22 to 24 years) with a rare-earth maximum (RE-max) around 1960. In the interval August 1965 to October 1969, the Eu^+ line 4335.58 Å went markedly weaker, and the chromium line strengthened. At the RE-min, the spectrum is quite characteristic for a sharp-lined, magnetic A_p. Cowley and Rice [33] noted in 1981 that the RE-max is returning. There is a strong U^+ line at 3859.6 Å (to be discussed in Sect. 7), especially strong Dy^{+2} lines, as well as Ho^+ and Er^+. Good evidence is also obtained for osmium, platinum and thorium. That the drastic variation is, at least, roughly periodic can be seen [32] from Mount Wilson spectrograms from 1948 and 1949. Further on, HR 465 is a single-line spectro-scopic binary (mutual distance somewhat above 10^8 km) with the period 273 days. Cowley [34] indicates $\log_{10}(Ln/H) + 12$ values for the RE-max, La 4.3; Ce 5.2; Nd 6.0; Sm 6.0; Eu 5.6; Gd 5.1; Tb 5.4; Dy 5.2; Ho 4.8; Er 5.3; Yb 3.9; and Hf 4.2. This means in g/ton 120 Nd and Sm, between 8 and 50 of Ho, Gd, Ce, Dy, Tb and Eu, probably altogether 500 g/t of lanthanides. Tb^+ has not been observed in any other star, strong Dy^+ lines are known also from γ Equ and HR 6958 (HD 170973), they are moderately strong in HD 25354, β CrB and 10 Aquilae. Ho^+ lines are only known with certainty from HR 465 (RE-max), HD 51418, HD 101065 and HD 137909. Er^+ has mainly been identified [34] in HR 465. More recently Dy^{+2} was observed [35] in 16 chemically peculiar B and A stars, in order of decreasing strength HR 465 (RE-max), α^2CVn, HD 192913, HD 200311, HD 51418, HD 25354 (Sr, Cr, Eu star [20] of magnitude 7.84), ...

In addition to the rather astounding variation of huge lanthanide concentra-tions over time intervals of decades, HR 465 (RE-max) also shows [36] an enormous amount of tellurium, and, what is of interest in Sect. 4, also large abundances of palladium, osmium, platinum and mercury. The $\log_{10}(Z/H) + 12$ are 6.4 for Pd, above 8.0 for Te, below 3.0 for Ba, 5.3 for Os, 5.8 for Pt and 5.1 for Hg. This corresponds (with the solar values [1] given by Trimble) to overabund-ances $D(Z) = 5.1$ for Pd, probably above 6 for Te, below 1.0 for Ba, but 4.5 for Os, 3.7 for Pt and about 3 for Hg. It is perhaps even more impressive to transform the results [36] to weight concentrations in g/t, yielding 200 g pallad-ium, more than 10 kg tellurium, less than 0.1 g barium, 30 g osmium, 90 g platinum, and 15 g mercury. These measurements were related to the r-process [36] in the raw material, absorbing neutrons at a time-scale much shorter than the half-lifes of β-radioactivity, as needed to provide any element with Z above 83, as further discussed in Sect. 7.

Przybylski [30] discovered the so-called *holmium star* HD 101065 which had previously been classified [24] as B5 (a rather high T) without any conspicuous peculiarity. However, high-resolution spectra of this eighth magnitude star showed almost all lanthanides present at a concentration higher than strontium and more than 50 times that of calcium. In one way, Przybylski's star is at the opposite extreme of HR 465 (RE-max). The sum [37] of $12 + \log_{10}(Z/H)$ for all metallic elements with Z above 19 is 5.0 (roughly corresponding to 20 g/t) well below 7.4 for iron [1] in the solar photosphere, but the concentration 0.16 g/t strontium, 0.4 g/t barium, and the lanthanides between 0.4 g/t and 1.3 g/t (neo-dymium), all the twelve detected having the sum 10 g/t. Warner [38] demon-strated the presence of lithium (among all elements) to the extent of 10 mg/t (250 times the solar photosphere), similar to calcium, and also detected Lu^+, but no ytterbium.

At the 1975 colloquium in Vienna [37] Przybylski pointed out that HD 101065 has a visible spectrum extremely rich in Fraunhofer lines [38], though the star is exceptionally poor in metals. He suggested a quite low T of the order 6200 K (similar to the spectral type F8) in sharp contrast to the previous B5 (implied by weak metallic lines). Although he had not detected [37] iron lines, S. N. Shore said at the discussion that he had found a few Fe^+ lines by comparison with β CrB and γ Equ. This problem was taken up again [39] and the underabundance $D(Fe)$ found to be more negative than -2.4 (less than 4 g/t). It remains tenable that HD 101065 is the only known star with less iron than the sum of all lanthanides. There are, however, other iron-group elements detected. There seems to be as much cobalt as iron, but no evidence was found for Ni^0 and Ni^+. Neutral yttrium atoms were not detected, and Y^+ is less abundant than the typical Ln^+. Zirconium Zr^+ is comparable to Ln^+, but very little Te^+ (cf. HR 465) and some Nd^{+2} was found.

Cowley has done much to dissipate the belief that the Sr, Cr, Eu stars constitute a fairly homogeneous class (with the exception of the three maverick stars discussed above). The number of detectable spectral lines, and their intensities, of yttrium, barium and lanthanides [40] in 29 A_p and A_m stars were carefully compared. In most of these stars, yttrium and lanthanum are unexpectedly scarce. The relative solar abundances of 12 lanthanides (altogether 0.041 g/t) are known [1], not including the radioactive promethium discussed in Sect. 7, terbium and holmium. The geological values (in ppm or g/t) are: La 18, Ce 46, Pr 5.5, Nd 24, Sm 6.5, Eu 1.1, Gd 6.4, Tb 0.9, Dy 4.5, Ho 1.1, Er 2.5, Tm 0.4, Yb 2.7, Lu 0.5 (and the subsequent element Hf 4.5). The general alternation between even and odd Z values (comprehensible by the fact that odd Z present only one or two non-radioactive isotopes) is, ironically enough, attenuated in holmium. In the Earth's crust, cerium is distinctly more abundant than neodymium. The situation was known to be the other way [17, 18] in the A_p stars. Cowley [40] describes the differing lanthanide abundances as a superposition of two components, [A] with an alternating curve like the terrestrial (and chondritic meteorite) Ln, but, in practice, *not* having Tm, Yb and Lu detected in stars [with exception of HR 465(RE-max) and Przylbylski's star discussed above] and usually [18] (in some cases more than 20 times) more Nd^+ than Ce^+. The other component [B], only predominant in a smaller sample of the 29 stars shows very high abundances of europium ($Z = 63$) and gadolinium ($Z = 64$), surrounded by samarium, terbium and dysprosium in a symmetric, narrow peak having its maximum between $Z = 63$ and 64.

Cowley and Henry [41] wrote "Numerical taxonomy of (40) A_p and A_m stars" using the techniques familiar to zoologists of optimizing an attempt of objective classification. Without sharply clear-cut borders, a class "A" is defined, which is lanthanide-rich, and where Dy^+ and Er^+ frequently are identified. Prominent members are HR 465 (RE-max); HR 6958 (HD 170973); HD 25354; HD 51418; HD 101065; HD 192913; HD 200311; and HD 221568. The class "B" stars show strong La^+ and Ce^+ lines, usually Eu^+ and Gd^+ too, but Nd^+ and Sm^+ are weak or absent, such as 10 Aquilae; β CrB; 52 Her (HD 152107); HR 7575 (HD 188041); HD 71866; and the variable star RR Lyncis (HD 44691). The class "C_1" is iron group-rich, the best represented Ln^+ usually being Ce^+ and Nd^+. This class is probably the most common A_p category, and comprises, for instance, 53 Camelopardalis (HD 65339); 49 Cancri (HD 74521); $α^2$CVn; 45 Leonis (HD 90569); 41 Tauri (HD 25823); 78 Virginis (HD 118022); HR 465

(RE-min); HR 4816 (HD 110066); HR 4854 (HD 111133) and HD 2453. Although α^2CVn was the earliest detected europium star, it has also very high titanium, chromium, manganese and iron abundances [17, 31]. The class "C_2" is a more miscellaneous category, the Ln^+ spectra may even be rather weak. Among its members are 73 Draconis (HD 196502); HR 8216 (HD 204411); HD 8441 [only Gd^+]; HD 42616; and HD 216533. Class "D": cooler stars of this class show light and intermediate Ln, probably without extreme overabundances. In the hotter objects, Ln^+ lines can be difficult to detect. Among the members are 32 Aquarii (HD 209625); γ Equulei; 33 Libri (HD 137949); 63 Tauri (HD 27749); τ Ursae Majoris (HD 78362); ζ^1 Lyrae (HD 173648); τ^7 Serpentis (HD 140232); 68 Tauri (HD 27962); and HR 4751. The correlations of the various properties of these 40 stars are studied in clustering dendrograms, and seven classes of iron-group spectra determined [41]. However, only quite slight correlations were found between the lanthanide and the iron-group categories. For our purpose (in Sect. 4) it is useful to note the category "IE" of silicon stars with very strong lines of Ti^+, Cr^+ and Fe^+ consisting of HR 6958 (HD 170973); 45 Leonis and 41 Tauri (HD 25823).

Cowley and Downs [42] studied spectra of *barium stars* detected via the strong Ba^+ line at 4554.04 Å. Barium isotopes are formed by the s-process, the capture of neutrons (the starting nuclei being ^{56}Fe and adjacent iron-group nuclei) on a time-scale *slow* compared to the half-lifes of β-radioactive nuclei (much like a moderated fission reactor) which is supposed to provide the major part of elements [1] with Z above 28 both in the solar system and in most stars. The highest absolute luminosity (220 times that of Sirius) known for a Ba^+ star is the yellow supergiant ζ Capricorni (HD 204075 of magnitude 3.74) probably having about 20 solar masses. The overabundances [6] relative to the Sun are not dramatic, 8 times (0.9 logarithmic unit) for barium, and 2 to 4 times for Sr, Y, La, Ce, Nd, Sm and Gd. The typical Ba^+ stars [42] are low in europium, but (presumably due to small cross-sections for neutron capture) high for Sr, La, Ce and Pr. Spectroscopically speaking, Ba^+ stars are rather similar to A_m and F_m stars, which are, however, main-sequence stars. They all [1] lack scandium. We now turn to other stars which exhibit huge overabundances of elements other than lanthanides and alkaline-earths.

4 Mercury, Osmium, Platinum, Gallium, Manganese and Silicon Stars

Next to europium, the most familiar element characterizing a sub-set of A_p stars is mercury ($Z = 80$). Since this sub-set clearly tends to have a considerably higher T than the typical Sr, Cr, Eu stars (and usually gets classified [24] as B5, B8 or B9 rather than A0 and "later" classes) the only important line in practice belongs to Hg^+ at 3983.98 Å. As carefully analyzed by White, Vaughan, Preston and

Swings [16] and by Cowley and Arnold [26, 43] both the Fe^0 line coinciding at 3983.96 Å, a weak Fe^+ line at 3984.06 Å, and worst of all at high chromium abundance, a strong Cr^0 line at 3983.906 Å, can mask the Hg^+ line. Hence, spectrograms at lower dispersion taken of weaker stars tend to suggest doubtful mercury occurrence. This has a pernicious effect on the remarkable phenomenon [16] that the isotopic abundance in 30 mercury stars is distributed between approximately the terrestrial mix (very little $A=196$; 10 to 23 percent each of 198, 199, 200, and 201; 29.8 percent $A=202$, and 6.85 percent of 204) in ϰ Cancri (HR 3623 = HD 78316) to comparable amounts of ^{202}Hg and ^{204}Hg in ı Coronae Borealis (HR 5971 = HD 143807) and to pronounced preponderance of ^{204}Hg in χ Lupi (HR 5883 = HD 141556). Taken at face value, this enrichment of the heaviest stable isotopes [the wavelength increases 0.07 Å of the line (decreasing the energy 0.44 cm^{-1}) between each pair of A and $A+2$] indicates a very high neutron concentration, wherever the mercury nucleosynthesis took place.

The estimated values of $12+\log_{10}(Hg/H)$ are typically 4 to 5 (corresponding to from 1 g/t to 20 g/t) or an overabundance $D(Hg)$ some 2 or 3 units of logarithm. The characterization as a mercury star is slightly less clear-cut in direction of smaller $D(Hg)$ which can be 1 or 0.7 unit only. North [20] only notes "Sr" for χ Lupi. Quite early, several other heavy metals were associated with A_p and B_p stars. Jaschek and Malaroda [44] studied 73 Draconis of magnitude 5.2 (HR 7879 = HD 196502) noted [20] as Sr, Cr, Eu. Not only are lines of Cr, Mn, Fe, Co, Ni and Sr strengthened, but 32 lines of Os^0 and 3 lines of Os^+ were identified, indicating a quite exceptional *osmium* abundance, several Pt^+ lines are certain, gold is probable (four strongest lines of Au^0 are present, but three of those are blended), and five U^+ lines seem sure. Iridium, mercury, lead and bismuth were not detected. Curiously enough (for $T=9000$ K of 73 Dra) about hundred lines of the molecular spectra of CN and of CH were found in the violet. This similarity to a much cooler star like Procyon (α Canis Minoris) may suggest a huge "sunspot".

Galeotti and Lovera [23] studied a somewhat related case, ϰ Piscium (HR 8911 = HD 220825) of magnitude 4.94 noted [20] Cr, Sr. Strong lines were found of Os^0, U^+ (as discussed in Sect. 7, this result was not confirmed [45] in another observatory) and the presence of Ir^0 and Pt^+ considered highly likely. No Au, Hg, Pb and Bi were detected. Ten Pt^+ lines [12] were found both in "manganese stars" without perceptible magnetic field, such as HR 4072 (HD 89822) of magnitude 4.93 noted "Hg" and 46 Draconis B (a component of a spectroscopic double star) HD 173524 of (combined) magnitude 5.04, and in several "magnetic A_p stars" such as HR 4854 (HD 111133) noted [20] Sr, Cr, Eu (magnitude 6.31) and HR 7575 (HD 188041) also noted Sr, Cr, Eu (magnitude 5.63). These (simultaneous) *platinum* stars [12] have overabundances $D(Pt^+)$ up to 4 or 5 logarithmic units, meaning [1] some 100 to 1000 g/t Pt at face value. An extreme case is HR 465 (RE-max) mentioned above said [36] to contain per ton 200 g Pd, above 10 kg Te, 30 g Os, 90 g Pt and 15 g Hg, in addition to 500 g lanthanides.

A quite individualistic case [13] is the *bismuth star* HR 7775 (HD 193452) of magnitude 6.2 and [24] spectral type B9, having [20] the unique noted Hg, Pt, Sr, Y. Jacobs and Dworetsky [13] measured also the far-ultraviolet spectrum (outside our atmosphere) and found not only gallium as Ga^+ but also Hg^+ and a huge amount of Bi^+ a million times the solar abundance, close to 2 kg/t, twice the solar iron content by weight. This, hotter, star approached the standard notation of Hg, Mn, Si stars as a major part (CP-3) of chemically peculiar stars. North [20] noted 65 among the 314 bright (HR) stars "Hg" (alone, or with other symbols added) constituting 0.7 percent of all HR stars. If the "eligible candidates" are the A0 to A9 stars, it would be about 3.2 percent actually being A_p. However, the Hg, Mn stars are mainly recruited between B5 and A0, a far more restricted community, and their percentage (rendered fuzzy by the question of how small $D(Hg)$ is still sufficient for inclusion) could easily be 6 or 10.

In Vienna 1975, Dworetsky was chairman of an "Open Discussion on Mercury-Manganese and Related Stars" reported on pp 725–736 of the proceedings [25]. As a minor correction to the dichotomy between "non-magnetic" [all A_m stars, Hg, Mn, Si and (nearly all) helium-weak stars] on one hand, and "magnetic" Eu, Sr, Cr stars on the other hand (defended by Sidney and Richard Wolff), Cowley pointed out that intermediate objects exist, such as HR 465, HD 25354 and HD 200311 containing Hg^+ and Pt^+. One may inquire how much Ln is needed in a manganese star (main sequence 11000 K) to be detected; cerium would need 100 to 1000 times the solar abundance, and europium 10^3 to 10^4 times. Also, there *are* manganese stars without Hg^+, of which 53 Tauri (HD 27295) is a classical case. Then, Wolff wanted to say there is never simultaneously Mn *and* Si overabundance. There are certainly some silicon stars with Hg. Hack said that HD 36916 is a Si, Mn star with very strong lines of Mn^+, and the Hg^+ line.

The great novelty in manganese and silicon stars is that a modest overabundance makes a great impact on the overall weight composition, whereas huge overabundances of lanthanides, osmium, platinum and mercury still are in the 100 g/t class. $12 + \log_{10}(Mn/H) = 5.4$ in the solar photosphere means 10 g/t. Hence, $D(Mn)$ of one or two units correspond to 0.1 or 1 kg/t. This is a fortiori true for silicon known [17, 18] in many A_p stars to have an average $D(Si) = 0.5$ bringing the silicon content from 0.9 up to 3 kg/t, and several stars are reported well above 20 kg/t. This is not a universal rule for all two-digit Z elements, magnesium is much less influenced. We return in Sect. 8 to some of the corollaries of high silicon content.

The hotter Si stars also develop specific peculiarities, such as high abundance of *phosphorus, gallium* and yttrium (e.g. compared to the preceding element silicon, zinc and strontium, or to the following, such as sulfur). Cowley [46] pointed out that chromium lines tend to be weak, when phosphorus and gallium are abundant. This has an important bearing on the credibility of the strong Ga^+ line 4262.00 Å quite close to the Cr^+ line at 4261.92 Å. These questions are related to the helium-weak stars in Sect. 5. Kodaira [47] evaluated abundances in α Andromedae and in ten other Mn stars, finding $12 + \log_{10}(Mn^+)$ in the former star 6.29 (with silicon 6.50) corresponding to 80 g/t, and usually 5.5 to 5.9 (i.e. 30 g/t), for the other 10 stars. By the way, Ga^+ has practically the same $\log_{10}(Ga/H)$ close to 5.5 (or 16 g/t) making it equally justified to speak about gallium ($Z = 31$) as manganese stars. α And also shows a phosphorus concentration close to 800 g/t, and mercury 40 g/t. Cr^+ is low (8 g/t), Fe^+ moderate (250 g/t). In most of these 11 stars, the amount of phosphorus is larger than, or comparable to, the amount of Si, which is almost unheard of in other stars. Actually, an enormous ephemeric excess of neutrons during the nucleosynthetic process may increase A of ^{28}Si to 29, 30, and β-active (half-life 150 s) 31 forming ^{31}P; or the s-processed ^{88}Sr via ^{89}Sr (52 days) producing ^{89}Y.

5 Helium-weak B_p Stars

This category (CP-4) of Preston [11] is dramatically different from the other categories discussed in this review, by having a pronounced *under*-abundance of the next-most frequent element. Nevertheless, it has been twisted into contact with our subject of lanthanide, U, Sr, Cr, Hg, Os, Pt, Bi, Ga, Mn and Si overabundances by several authors. There are only about 100 helium-weak stars known [6] and among the 314 HR stars included in the 1028 B_p, A_p and F_p listed by North [20] only 30 are noted helium-weak (actually, there are even fewer noted among the less brilliant non-HR stars). Their concentration per unit volume of our galactic vicinity is very low, since most have distances [24] of order 500 light-years. The typical helium-weakness corresponds to 50 kg/t of helium, and say, 10 to 30 kg/t of "metals", i.e. $Z = 6$ and higher values, and the rest hydrogen. Conceptually, this atomic ratio (He/H) decreased from its usual value 0.08 to around 0.013 can *also* be described as a hydrogen overabundance, but it does not answer the real question, how helium *can* be depleted by a factor of order 5.

However, about 20 or a-fifth of the known helium-weak B_p stars (they might also be called B_w stars) have an additional [48, 49] peculiarity: a monumental isotopic ratio $^3He/^4He$. The primordial Weinberg mixture [4] has this ratio close to 10^{-4}. Atmospheric air shows 1.3×10^{-6} (out of only 5 cm^3 helium per m^3 air). However, the terrestrial gravitation is not able to retain any of the two helium isotopes for a long time, and the ^4He originating as α-particles from uranium and thorium minerals (and the source of helium in many wells of natural gas) also disappears rapidly on a geological time-scale. Anyhow, ^3He stars are known [48] with the ratio $^3He/^4He$ between 0.47 and 2.7 (corresponding to 32 to 73 percent ^3He in the helium) and 5 to 20 times less total helium than normal B stars. The best known ^3He star is 3 Cen A, the A-component of a spectroscopic binary 3 Centauri (HR 5210 = HD 120709) of type B5 and magnitude 4.32 and an absolute luminosity approximately 16 times that of Sirius. Not only does it contain 2.5 kg/t of ^3He, but some of the other elements have astonishing abundances [50] such as 300 g/t carbon, 600 g/t nitrogen, 80 g/t oxygen, 20 g/t magnesium, 30 g/t silicon, 25 g/t phosphorus, 15 g/t iron, 2 g/t gallium and 1.3 g/t krypton. Though the elements heavier than helium apparently add up to only 1 kg/t (an all-time record of low "metallicity") the abundance of phosphorus and gallium relative to other elements reminds one of B-type silicon stars. The gallium concentration is 60 times larger than in the solar photosphere, and the ratio (Ga/Fe) is 4000 times larger. Hack and Stalio [51] report on the far-ultraviolet observations of 3 Cen A by means of the Princeton spectrometer aboard the Copernicus satellite. T is probably above 20000 K. The He^+ line at 1084.9 Å is absent, but was detected in the normal B3 type ι Herculis (HD 160762) and B2 type γ Pegasi (HD 886). Boron, among all things, was detected as the B^+ line 1081.1 Å in 3 Cen A. The intensities of various lines of C^0, C^+, C^{+2}, N^0 and N^+ are comparable in this star and in ι Her. Cl^+, Ar^0 and Ca^{+2} lines are, at the best, weak, but Mn^{+2} much stronger in the ^3He star. Both Ga^+ 1118.8 and Ga^{+2}1295.45 Å are much stronger in 3 Cen A than in ι Her. Also As^+, Se^+ and Y^{+2} seem present in 3 Cen A. Several Fe^+, Fe^{+2} and Ni^+ lines suggest abundances of the order 1 kg/t, two orders of magnitude higher than the previous iron values [50].

The anomalous abundances in helium-weak, and in particular ^3He, stars have nourished expectations [50] of nuclear reactions in the stellar surface layers, e.g. a strong magnetic field accelerating ^1H and ^4He nuclei to form ^2D and ^3He. This costs nearly as much energy as evolved by forming one ^4He in normal stars, and the efficiency of the process is probably very low. Also reactions removing a proton from ^{32}S and ^{56}Fe to form ^{31}P and ^{55}Mn are not too easy to defend quantitatively. Anyhow, we are now approaching the enigmatic area of the physical mechanism of chemical peculiarity in B, A and F stars.

6 Magnetic Accretion and Competing Models

The first idea one got to explain the extravagant composition of just a few CP stars was nuclear reactions in the interior, or on the stellar surface. We may have become a little too accustomed since 1943 to make great amounts of isotopes by neutron capture essentially at room temperature, and we tend to forget that the dominating competition in a star is neutron capture by the three-quarters of the mass being protons. Even the astrophysical description of s (slow) process neutron capture [1, 42] tends to consider cross-sections for 30 keV thermal energy, corresponding roughly to 350 megakelvin, a temperature 23 times higher than supposed to occur in the solar center. Actually, even at 110000 K, exceeding the surface T of all persistant O and Wolf-Rayet stars, kT is only 10 eV, and one would have huge effects of resonances in the capture cross-section. The main reason that kT around 0.03 eV usually is more efficient is that some (but not all) neutron cross-sections get extremely large because of the velocity-dependent De Broglie wavelength h/mv of the neutron. All of these arguments cannot circumvent the protons absorbing almost all available neutrons. They work better in a (science-fiction) nearly pure helium photosphere of ^4He.

There is one point where it may be worthwhile to keeping an eye open for the "alchemistic" style of nuclear reactions catalyzed by trace quantities of WIMP (weakly interacting massive particles) such as a negatively charged X^- with an atomic weight [52–54] somewhere between 100 and 10^6. Our major chance of observing such species in stars is the isotopic shift factor of all wave-numbers $5.5 \times 10^{-4} (A_1^{-1} - A_2^{-1})$ between A_1 and A_2 spectral lines due to the weak influence of the electrons on the center of gravity of the species with one nucleus. This shift [38, 52] is $5.5 \times 10^{-4}/42$ between ^6Li and ^7Li (a significant example because the tightly bound adduct of X^- and ^8Be may be stable, in contrast to ^8Be instantaneously split into two ^4He nuclei) but is 7 times larger between ^6Li and a very heavy lithium isotope. Nuclei with A well above 100 show isotope shifts of a different origin [16] connected with their charge distribution of finite size. Adducts like the manganese isotope XFe may show such effects to an extent well above 2×10^{-5} (that may otherwise be explained as a Doppler shift of $v = 6$ km/s) and stellar-spectra can sometimes show unexpected shifts from the laboratory values in the 0.1 Å class. The center of gravity effect would shift a line of a very heavy manganese isotope to higher energy 10^{-5} (in the opposite direction). Hyperfine and electroweak effects may also be perceptible in WIMP adducts. The propensity toward fission [55] may also be much smaller for a given Z value combined with one WIMP, since it may have a higher affinity for a nucleus with larger A, and hence, it does not exclude the possibility of finding long-lived WIMP species with three-digit Z values.

With the exception of such exotic catalysts, there is a general consensus today that the CP-2 and CP-3 stars do not perform the nucleosynthesis of two-digit Z elements internally. We recognize today that this would be as oxymoronic an idea as an orchid illuminated from the inside by a Roman candle. Then, the specific minority (small, but indisputable, like colorblindness in a human population) of CP stars must be rationalized in two steps: where and when were the rare elements formed under drastic conditions such as in supernovae [1, 2] and how did the elements enter and concentrate in the CP stars? Admittedly, there is one way of shortcircuiting this question, to believe that the star has some specific trick at disposal, fractionating standard material at hand, and then exhibiting a concentrate of a few, quite unexpected, elements in a thin, but *the* observable, zone of the photosphere.

Many astrophysical textbooks accept the hypothesis of Michaud [14] of efficient stratification in the photosphere of selected elements, and even isotopes, by a delicate balance between the radiative pressure on the atoms and monatomic cations absorbing the ambient photons, and the fall due to gravitational attraction. This theory still needs to be refined in many elaborate details. If it applies to the ^3He stars (Sect. 5) with the elusive helium isotope enriched by a factor 5000, the discrimination originates in the gravitational force being 25 percent smaller than in ^{4}He (which would be an appealing explanation on the surface of white dwarfs). The best established case of isotope fractionation in heavy atoms [16] is ^{204}Hg, needing pushing out in space of the lighter isotopes. Already in 1976, this diffusion hypothesis was considered [15] a conceivable, but very crude estimate of the overabundance (by a factor 10^4) in A_p mercury stars. The main problem was lack of data for highly ionized species (such as Hg^{+9} and Hg^{+11}) deeper down in the star. Everybody agrees that the diffusion hypothesis of Michaud necessitates a very calm photosphere, where diffusion takes place unperturbed by any variable directional flows (in spite of the turbulent wind around 5 km/s observed [30, 38] in Przybylski's star HD 101065), a scenario which is indeed more likely in A main-sequence stars than in almost all other stars.

The Vienna colloquium [25] included an Open Discussion (chairman: W. Deinzer): Controversial Theories-Accretion vs Diffusion theory; Dynamo vs Fossil Fields (pp 713-724 in the Proceedings). The theory of *magnetic accretion* was proposed by Ove Havnes and Peter Conti [56]. Magnetic A_p stars have selectively captured atoms from the interstellar medium by its interaction with the rotating stellar magnetic field. Heavier cations can penetrate deeper into the field (closer to the star) and are transferred to the stellar surface around the magnetic poles (α^2CVn and many other europium stars show a spectral variation within the rotational period, clearly indicating systematic concentration differences oriented by the magnetic field). A braking mechanism on a time scale 10^7 to 10^8 years slows down the rotation of magnetic A and B stars, due to their magnetic field interacting with the ionized interstellar gas. [At the Vienna colloquium [25] many lecturers treated the magnetic field as being due to an oblique rotator [6] and discussed the (not too obvious) correlations with rotation periods in normal and CP stars]. There is no stellar wind, and no evidence of hot chromospheres and coronas. Ionized material streams *toward* the A or B star. Convective zones develop only at F2 and cooler main-sequence stars [57].

Havnes and Conti [56] estimate the typical factors of abundance anomaly in A_p stars to be 1000 for lanthanides, 10 for silicon and iron-group elements, 1 for magnesium and 0.1 for helium and oxygen. Accretion from a diameter of 10^9 km is sufficient to change completely the observable photospheric composition, smaller than the effective radius 2×10^9 km of the stellar magnetic field. Nearly all elements, except helium, are ionized at the boundary of the rotating stellar field. The model of Mihalas [58] with $T_e = 10000$ K and log g $= 4$ gives 4×10^{21} atoms/cm^2 (roughly 200 g/m^2, perhaps for the whole stellar surface 10^{25} g) above optical depth 0.1 [where the *action is* of the accreted elements] which is about a tenth of the number of atoms in the photosphere.

Overabundance by a factor 1000 can be accreted in less than 10^7 years if the effective accretion radius is 3×10^8 km, smaller than the magnetosphere and much smaller than the ionization radius R_0 of the Strömgren sphere (being for A stars about 3 light-years or 3×10^{13} km). Inside R_0 the hydrogen atoms are nearly completely (at 10^4 K, 99 percent) ionized, and the concentration of free electrons and free protons [H II region] comparable. Elements with the first ionization energy I_1 below a rydberg will be M^+, but not helium, nitrogen and neon.

Elements with the second ionization energy I_2 below 1 rydberg (such as calcium, scandium, strontium, yttrium, zirconium, barium and all elements from lanthanum to ytterbium) can be M^{+2} to some extent, but the high density of free electrons tend to promote M^+, e.g. Ca^+. Havnes and Conti [56] note that a larger percentage of the lanthanides will be Ln^+ because of a much higher recombination coefficient [59] of Ln^{+2} with electrons. Unless there are in-homogeneities in the surface magnetic field, it is not possible [56] to understand why iron-group elements (such as chromium) appear in spots [60] around the magnetic equator. Table 1 compares the three consecutive ionization energies in eV (mainly from Charlotte Moore [61] and some from the lanthanide volume [62] of "Atomic Energy Levels", when known, and below 50 eV) with $I_1 = 13.598$ eV of hydrogen, 24.587 eV of helium, and the energy $H^* = 10.20$ eV of the first excited ($2s$ & $2p$) level of hydrogen atoms. It is perhaps significant that the nine elements having I_1 in the 3.4 eV wide interval between H^* and (I_1 of H), i.e. C, P, S, Cl, Br, I, Xe, Hg and Rn, and the twenty atoms with I_2 in the same interval, viz. Ca, Sc, Ti, Sr, Y, Zr, La and the subsequent elements through Yb [and very likely, also Th and U] provide many of the elements characterizing the A_p spectra, perhaps because of some buffer effect when their ionization is initiated by protons (before the magnetic accretion) or some specific property in the outer stellar atmosphere (much like the helium-containing diatomic HeX^+ and helium-organic cations in gaseous state reviewed [63] in this volume).

Michaud answered in Vienna [25] that his diffusion model can provide abundance anomalies of order 10^7 in 10^7 years, but accretion at most can give 10^3 or 10^4. Havnes said that Michaud's model does not work if the mixing velocity is larger than 10^{-3} cm/s. Today, there seems to be general agreement that the "magnetic" Eu, Sr, Cr type of A_p stars lose most of their angular momentum during their initial entrance in the main sequence, that is on a time-scale of 10^6 years. Non-magnetic silicon stars [64] are still losing angular momentum a long time after they have reached the main sequence. On the other hand, as will be discussed in Sect. 8 (and is really quite clear from the multifarious kinds of chemical peculiarity) the interstellar matter encountered as raw material for magnetic accretion probably depends dramatically on the kind of supernova and related events having fed it with two-digit (if not three-digit) Z values. The original model of Havnes and Conti [56] assumed a standard composition of interstellar matter, but it seems likely that although silicon and iron may come from a standardized pool, the rarer elements show quite "personalized" behavior. Havnes [25] also pointed out that a radiation belt around the magnetic poles sometimes may splash down particles into the photosphere, rationalizing both the irregular variation with time of CP element abundances, and their spotwise distribution on the stellar surface. The strong magnetic field at the surface is, to some degree, an epiphenomenon, many stars have probably very strong fields hidden just below their visible surface [65].

Table 1. Relative order of the first three ionization energies (in eV) of most atoms, compared to H* (the first excited state of hydrogen) and the first ionization energy of hydrogen and helium atoms

6 C	H* 11.260 H 24.383 He 47.887	46 Pd 8.34 H* H 19.43 He 32.93
7 N	H* H 14.534 He 29.601	47 Ag 7.576 H* H 21.49 He 34.83
8 O	H* H 13.618 He 35.116	48 Cd 8.993 H* H 16.908 He 37.48
9 F	H* H 17.422 He 34.970	49 In 5.786 H* H 18.869 He 28.03
10 Ne	H* H 21.564 He 40.962	50 Sn 7.344 H* H 14.632 He 30.502
11 Na	5.139 H* H He 47.286	51 Sb 8.641 H* H 16.53 He 25.3
12 Mg	7.646 H* H 15.035 He	52 Te 9.009 H* H 18.6 He 27.96
13 Al	5.986 H* H 18.828 He 28.447	53 I H* 10.451 H 19.131 He 33
14 Si	8.151 H* H 16.345 He 33.492	54 Xe H* 12.130 H 21.21 He 32.1
15 P	H* 10.486 H 19.725 He 30.18	55 Cs 3.894 H* H 23.17 He 33.38
16 S	H* 10.360 H 23.33 He 34.83	56 Ba 5.212 10.004 H* H He 35.79
17 Cl	H* 12.967 H 23.81 He 39.61	57 La 5.577 H* 11.06 H 19.175 He 45.95
18 Ar	H* H 15.759 He 27.629 40.74	58 Ce 5.47 H* 10.85 H 20.20 He 36.76
19 K	4.341 H* H He 31.625 45.72	59 Pr 5.42 H* 10.55 H 21.62 He 38.98
20 Ca	6.113 H* 11.871 H He	60 Nd 5.49 H* 10.72 H 22.14 He 40.41
21 Sc	6.54 H* 12.80 H He 24.76	61 Pm 5.50 H* 10.90 H 22.3 He 41.1
22 Ti	6.82 H* 13.58 H He 27.491	62 Sm 5.63 H* 11.07 H 23.43 He 41.37
23 V	6.74 H* H 14.65 He 29.310	63 Eu 5.67 H* 11.25 H He 24.70 42.65
24 Cr	6.766 H* H 16.50 He 30.96	64 Gd 6.14 H* 12.09 H 20.64 He 44.01
25 Mn	7.435 H* H 15.640 He 33.667	65 Tb 5.85 H* 11.52 H 21.91 He 39.37
26 Fe	7.870 H* H 16.18 He 30.651	66 Dy 5.93 H* 11.67 H 22.79 He 41.47
27 Co	7.86 H* H 17.06 He 33.50	67 Ho 6.02 H* 11.80 H 22.84 He 42.48
28 Ni	7.635 H* H 18.168 He 35.17	68 Er 6.10 H* 11.93 H 22.74 He 42.65
29 Cu	7.726 H* H 20.292 He 36.83	69 Tm 6.18 H* 12.05 H 23.68 He 42.69
30 Zn	9.394 H* H 17.964 He 39.722	70 Yb 6.254 H* 12.184 H He 25.03 43.74
31 Ga	5.999 H* H 20.51 He 30.71	71 Lu 5.426 H* H 13.9 20.96 He 45.25
32 Ge	7.899 H* H 15.934 He 34.22	72 Hf 7.0 H* H 14.9 23.3 He
33 As	9.81 H* H 18.633 He 28.351	73 Ta 7.89
34 Se	9.752 H* H 21.19 He 30.820	74 W 7.98
35 Br	H* 11.814 H 21.8 He 36	75 Re 7.88
36 Kr	H* H 13.999 24.359 He 36.95	76 Os 8.7
37 Rb	4.177 H* H He 27.98	77 Ir 9.1
38 Sr	5.695 H* 11.030 H He 41.64	78 Pt 9.0 H* H 18.563 He
39 Y	6.38 H* 12.24 H 20.46 He	79 Au 9.225 H* H 20.5 He
40 Zr	6.84 H* 13.13 H 22.99 He	80 Hg H* 10.437 H 18.756 He 34.2
41 Nb	6.88 H* H 14.32 He 25.04	81 Tl 6.108 H* H 20.428 He 29.83
42 Mo	7.099 H* H 16.15 He 27.16	82 Pb 7.416 H* H 15.032 He 31.937
43 Tc	7.28 H* H 15.26 He 29.54	83 Bi 7.289 H* H 16.692 He 25.56
44 Ru	7.37 H* H 16.76 He 28.47	86 Rn H* 10.748 H
45 Rh	7.46 H* H 18.08 He 31.06	88 Ra 5.279 10.147 H* H He

It should perhaps be added that the magnetic accretion proposed by Havnes [56, 66] has nothing to do with paramagnetism or non-closed shells of the ions most efficiently collected by the star; it is a question of ionization under appropriate conditions, and subsequent ion transport determined by the magnetic field lines. Especially in the case of heterogeneous interstellar material already enriched in certain elements, magnetic fields less than a tenth as intense as of "magnetic" A_p stars, and hence difficult for us to detect, may be more than sufficient for magnetic accretion [66] to Hg, Pt, Mn, Si stars of the CP-3 category.

7 Uranium and Rapid Neutron Capture

The nucleosynthesis in normal stellar central regions is the transmutation (in our Sun, on the average, 600 million tons per second) of ^1H to ^4He (with a few, alternative, intermediate reactions) between 10 and 30 megakelvin. At higher temperatures, collision of (moderately) positive nuclei of hydrogen or helium with nuclei of carbon (we speak here of a "modern" Population I star; in their absence in pronounced "ancient" Population II stars, the resonant collision of three ^4He forming an excited state of ^{12}C subsequently de-excited to the ground state, is a second choice), nitrogen, oxygen, form higher Z values, such as neon, magnesium and silicon [1, 2]. These elements tend to build concentric shells in very massive stars [7, 9, 67]. By sufficient heating, this process goes beyond argon and calcium, forming significant amounts of iron and nickel. The reactions between two positive nuclei do not go much beyond ^{60}Ni at this point. However, some isotopes (such as ^{21}Ne, ^{22}Ne, ^{25}Mg, ^{26}Mg) colliding with ^4He nuclei provide a copious source of *neutrons*. Since the build up of elements heavier than ^{60}Ni is strongly uphill energywise (the thermodynamical equilibrium below 1 or 2 gigakelvin makes matter most stable as a narrow distribution of isotopes around ^{56}Fe, unless there exist other, more stable forms [3] such as *uds*-matter, and if A is not well above 10^{55} where gravitational stabilization is significant) the synthesis of heavier nuclei is a non-equilibrium process (much like feeding of growing animals) consuming neutrons. It may be noted that an upper limit of nuclei with A above 70 is about one percent of the neutrons provided. This, optimized, situation gives a quantity not very different from the Earth's crust [3] where 1.2 kg/t of such elements occur (having needed about 0.5 kg/t of neutrons added to ^{56}Fe as basis material, and representing 10 kg/t of processed, unfrequent, neon and magnesium isotopes).

The elements with Z above 83 (bismuth) seem to have an awesome problem: we do not know any isotopes of $Z = 85$, 86 and 87 with a half-life of more than 9 hours, with the exception of 3.8 days for ^{222}Rn. The gap between $A = 211$ and 226 can only be bridged by exceedingly *rapid* (*r*-process) inflow of neutrons. The scenario is re-cycling [68] of $^{56}Fe \rightarrow$ fissile isotopes $\rightarrow$ fission products in supernovae with a time scale 4 to 16 seconds, until exhausting the neutron flux after 100 s. When the quantum number N of (Z, N) isotopes achieve [69] the Maria Goeppert-Mayer closed-shell value 126, they may have Z around 70, and after subsequent β-activity, they wind up around ^{196}Pt. However, a sufficiently high neutron concentration may go far beyond, close to the next stabilized $N = 184$. The isotopes known today with Z above 100 are highly neutron-deficient, the β-stability line passing $A = 253$ for $Z = 99$ is expected to have a slope (dA/dZ) at least 2.8, crossing $Z = 114$ (a protonic closed-shell) close to $A = 300$. It is somewhat frustrating [70, 71] that $^{298}114$ (or preceding cases of $N = 184$, such as $^{296}112$, $^{295}111$ or $^{294}110$) may have half-lifes in the 1 to 10^6 y class (the shorter half-lifes are most favorable for detection, the specific activity being inversely proportional to the half-life) since we have no known possibility of increasing N so much (actually, the Bohr-Wheeler parameter Z^2/A stabilizes the A as high as feasible without a prohibitively short β-half-life). Anyhow, species [71] such as ^{276}U, ^{280}Cm and ^{284}Fm are likely to disappear more rapidly by spontaneous fission than by β-activity.

Judged from evidence in stellar spectra, uranium (and all other trans-bismuth elements) seems indeed to be a scarce commodity. Only one bone fide Th$^+$ line has been identified in the solar spectrum, previously providing [1] the estimate $12 + \log_{10}(\text{Th/H}) = 0.9$. However, the radiative life-time of this Th$^+$ line is so short [102] that the corresponding intensity is 4 times higher than had been assumed, bringing $\log_{10}(\text{Th/H})$ down to -11.7 or 0.3 mg/t. From the (Th/U) ratio in meteorites, $\log_{10}(\text{U/H})$ has been estimated [17] to -12.3 or 10^{-4} g/t. For comparison, the terrestrial values [3] are in g/t: 16 Pb, 12 Th, 4 U, and the average value 8 for each of the 15 lanthanides. It must be noted in all fairness that potassium and uranium must have considerably lower abundance in the interior of the Earth (otherwise, the geothermal heat would make the crust incandescent). Nevertheless, the (U/Eu) ratio is at least 2 in the crust, compared with the solar value 0.1, and 0.001 to 0.01 in various A_p stars [17]. The new thorium solar value brings the (terrestrial/solar) ratio R up to 40000, much higher than 3000 for the lanthanide group, 1400 for tungsten and lead, and 500 for calcium. R is (in another context) 1.6 million for lithium. Though the R values obviously are influenced by chemical enhancement (say, of rubidium and barium in the outer crust) or depletion (by volatility or siderophilic extraction into the iron core) the very large R values (higher even than 300 for silicon, compared with 35 for magnesium) demonstrate a systematic contribution of specific supernova products (perhaps cooked at a higher temperature) to the inner planets. Several solar values were revised [103] downwards. Concomitant R for lithium and total Ln are 10^7, and 7000 (17 mg/t). Thorough studies [104] of Sm$^+$ restored (the high) 1.5 mg/t. Chondrite meteorites do not seem perfect to give "correct" Ln concentrations.

As early as 1970, five distinct U$^+$ lines were detected [44] in 73 Draconis (HD 196502). Seven years later, Cowley, Aikman and Fischer [72] published a 36-page review (with many spectral tracings) on *uranium stars*, extending previous reports [39, 73–75]. The U$^+$ line of ^{238}U in the laboratory is situated at 3859.57 Å, with ^{235}U$^+$ having three hyperfine components centered at about 0.07 Å *lower* wavelength. The stellar line (such as 3859.599 Å in HR 7575 = HD 188041) usually is shifted a trifle to higher wavelengths, such as β CrB showing a blend of the true U$^+$ line 3859.566 Å with an unidentified line 3859.72 Å (solar spectrum 3859.741 Å), on the tail of a strong Fe^0 line at 3859.91 Å. Also the feature at 3854.7 Å (stellar 3854.63 to 3854.89 Å) is a blend of U$^+$ 3854.66 and Cr^0 3854.22 Å, ^{235}U$^+$ is shifted 0.04 Å to lower wavelength. The weaker ^{238}U$^+$ line at 4241.66 Å has ^{235}U$^+$ at 4241.74 Å. Two strong U$^+$ lines shift by 0.2 Å, but 4244.37 Å is hopelessly confused with Mn$^+$, and 4116.0 Å with Fe^0 4115.89 Å. Here, it helps that Przybylski's star HD 101065 has so weak iron-group lines; 4244.300 as well as 4244.324 and/or 4244.364 Å and 4115.974 and 4116.01 Å (^{235}U$^+$ shows 4244.11 and 4115.93 Å). The ultra-peculiar HR 465 (HD 9996) in its rare-earth maximum [33] has several strong U$^+$ lines. The spectral variable, Osawa's star HD 221568, gives 98.5 percent confidence levels for 35 lines of $^{235}U^+$, but much less for $^{238}U^+$. Also in HD 101065 and HD 2453, the evidence for $^{235}U^+$ is the better one. Running the risk of discouraging tourism in the vicinity of these stars, taking their isotope shifts at face value has important bearings on the details of the r-process neutron capture, and the time evolved since.

The life-time of a type A main-sequence star is believed to be 10^8 to 10^9 y. If a supernova produced uranium at the beginning of the A career of the star, more or less comparable amounts of the ^{232}Th, ^{235}U and ^{238}U are to be expected. If production is more recent, one may also expect (half-lifes in parentheses): ^{233}U (162000 y); ^{234}U (247000 y); ^{236}U (23.9 My); ^{237}Np (2.14 My); ^{244}Pu (83 My); and ^{247}Cm (16 My). One of the enigmatic sides [72] of the "dedicated" uranium stars is that (Th/U) is well below 0.2, and actually, thorium is quite difficult to detect, except in the inimitable HR 465 (RE-max) where Th^{+2} and Th^+ lines *are* found [33] but indicating much less Th than U (7 g/t uranium [26]). Although it is quite unlikely that the stellar shifts are due to XNp adducts [52–54] we cannot entirely ignore that uranium "isotopes" may occur that we do not find in our minerals.

The search for stellar transuranium elements has not been particularly fruitful. Kuchowicz [69] looked for plutonium lines in 73 Draconis and a few other A_p stars, and went on [70] with more extensive comparisons. However, the fairly definitive study [74] by Cowley, Aikman and Hartoog of 51 A_p, A_m and normal A stars did not reveal any trace of Np^+, Pu^+ and Cm^+. The incapability of the sudden neutron burst in underground nuclear explosions to produce elements [78] beyond fermium ($Z = 100$) does not exclude that higher Z values (e.g. 164 which might form by droplet fractionation in the splash of objects falling down on a neutron or *uds*-star) or their characteristic fission products may occur in A_p stars. The fission (discovered 1939) gives typically two fragments with A fairly close to 0.4 and 0.6 times the original A value. However, abruptly between $A = 256$ and 258, the spontaneous fission becomes symmetric [78] with some $0.49A$ (taking into account the emissions of some neutrons). The special "non-chondritic" component [B] (with A close to 155, mainly europium and gadolinium) suggested by Cowley [40] to constitute the main part of lanthanides in a minority of A_p stars, might be a symmetric fission product of some isotope with A close to 310 in r-capture (a remote guess would be $^{310}126$ if it was uniquely stabilized by the closed-shell [69] quantum numbers $Z = 126$ and $N = 184$). The narrow range of observed A values in the [B] product becomes a kinetic question of the competitive rates of spontaneous fission and neutron capture.

A related problem is the conspicuous *xenon* isotope anomalies in various meteorites. The heavy fraction [79–87] enriched in ^{136}Xe was originally [79, 80] ascribed to γ-ray or neutron-induced fission of ^{242}Pu (half-life 83 million years, only 0.08 percent probability of spontaneous fission; otherwise, ^4He emitter) and a controversy maintained about plutonium(IV) occupying sites in (quite scarce mineral fractions) and forming xenon; or the noble gas directly incorporated from the protoplanetary cloud. Some of the most remarkable components of the

Allende meteorite were then suggested [85] to contain (still, or previously) a fairly volatile, superheavy element such as $Z = 115$, 114 or 113. The status today is not far from the remarks at the end of this review, that the rarer elements may show a strong disparity in differing CP stars, because the origin in supernova-like events suffers quite a varied dispersion of detailed reaction rates.

As long r- or s-process [36] neutron captures from scratch (i.e. iron) were plausible processes in a (not very steady-state) A_p star, the question of Pm^+ in HR 465 (RE-max) was considered important potential evidence for the r-process. *Promethium* [29] is best known as the fission product ^{147}Pm (half-life 2.5 year, β-emitter) but the longest half-life (17.7 y) known is of ^{145}Pm (which cannot be formed from neutrons and neodymium, ^{145}Nd being stable). *If HR 465 shows Pm^+ lines* [88] (this problem seems caught in statistical quicksand [89, 90], and has today lost some of the urgent priority, as seen in the next paragraph), and in particular, if they oscillate with time, they were produced locally (in or outside the star) less than 80 years ago. As in uranium minerals [80] it seems today likely (simply because of the shorter half-life) that a given star contains 10^5 times less Pm than technetium.

The S-type stars [6] are red giants with low T (around 3000 K) showing very conspicuous absorption bands of the diatomic molecule ZrO (similar to TiO in the far more frequent M-type). When Merrill [91] detected spectral lines of technetium atoms, it was generally considered strong evidence for production of ^{99}Tc (half-life 212000 y) [^{97}Tc (2.6 My) and ^{98}Tc (1.5 My) are less accessible, because ^{97}Mo and ^{98}Mo are stable] by s-process neutron capture [36]. Among 100 such *technetium* stars known [92–94], abundances are available for 15, mostly with the ratio (Tc/Si) between 10^{-11} and 10^{-8}, but two, χ Cyg (the K0 type [24] HD 187796) and the S6 type R And (HD 1967) in the range 10^{-7} to 10^{-6} (probably close to a few mg/t of technetium). Rather than assuming the consecutive capture of roughly 40 neutrons, Malaney [95] suggested photo-fission in the star of some isotopes with A close to 240. In the quiescent period, the temperature (if at all defined; this is not a perfectly trivial question) close to 50 megakelvin determines the CNO cycle, where proton capture in ^{13}C and ^{14}N produce γ-rays with 7.55 and 7.293 MeV energy, respectively. That means that almost half of the energy evolved by transmuting ^1H to ^4He is available as these photons with energy well above fission threshold. Actually, ^{231}Pa and ^{237}Np produced by the γ-ray knocking out a proton, or a neutron followed by rapid β-decay (some days) of ^{231}Th and ^{237}U, are also susceptible to photo-fission. The total yield of photo-fission is expected [95] to be about 20 times the technetium yield. This may not be sufficient to explain the Tc concentration in the two richest S stars.

By the way, deuterium (which may be depleted in the relevant region), ^{13}C, ^{17}O and ^{21}Ne satisfy the condition for being able (energy-wise) to lose a neutron, when absorbing a 7.00 MeV photon (0.00752 amu) that the difference between the atomic weight of (Z, A) and $(Z, A-1)$ is above 1.00115 amu, and these isotopes might compete efficiently with photo-fission, in spite of their moderate abundance (g/t in the solar photosphere: 30, 40, 3 and 1.5, respectively). ^{25}Mg

(60 g/t) is a marginal case, it could only react with photons above 7.33 MeV. Otherwise, it is true that very few isotopes (strongly correlated [69] with the nucleonic shell model) such as ^{91}Zr, ^{143}Nd, ^{199}Hg, ^{201}Hg and ^{207}Pb would be able to eject a neutron with 7.3 MeV photons. ^{209}Bi has an exceptionally low threshold (3.8 MeV) for protons being ejected by a photon.

The new model [95] of technetium stars is not really a return to violent nuclear reactions (consecutive neutron capture needs copious production of neutrons) in stellar interiors, but is rather of the same kind as α-emission of uranium (supposed to have minor effects even in A_p stars, but supplying a major part of geothermal heat, and of great historical significance [3] the first four decades after the discovery of radium in 1898) as would also be spontaneous fission [e.g. of the californium isotope ^{254}Cf with half-life 61 days once held responsible [2] for the exponential decrease of supernova (type II) luminosity, now ascribed to β-decay of ^{56}Ni (half-life 6.1 days) forming β-active ^{56}Co (77 days) finally producing the common iron isotope]. The chemist may perhaps have a small qualitative argument in favor of the s-process: the S stars might be specializing in palladium-group elements (Zr) like the M stars in iron-group elements. Again, one might think naïvely that in a place containing so horren-dous amounts of neodymium (10^{21} g?) as the photosphere of HR 465 (RE-max) it might seem easy to produce a few hundred million tons of promethium. But it seems that disregarding slowed-down neutrons (from fission or fusion reactors) we have very few places cooler than 100 megakelvin manufacturing elements with Z above 30.

8 Silicon and Heterogeneity of Supernova Dust

When comparing elements in different stellar photospheres, even an element like silicon frequently shows an overabundance relative to the Sun by a factor 11, that is 10 kg/t rather than 0.9 kg/t. This may not seem extreme, when compared to an overabundance factor 35000 for europium or 1000 for platinum, in both cases bringing the enhanced element up to 20 g/t. But 10 kg/t is still 500 times more than 20 g/t. If we look at our soil or rocks, there is a most definite trend [29] toward enhanced silicon, the ratio between terrestrial and solar abundance (Table 1 of Ref. 3) R being 300, and not 70, the average value for the five elements Mg, Al, P, S and Cl. On Earth, silicon is the second most common element. For the most abundant, oxygen, $R = 58$. It is becoming clear that supernovae do not provide a standard composition of "supernova dust" and that already the silicon content in the interstellar medium and in protoplanetary entities is higher than in the material which previously has formed normal stars.

Virginia Trimble [96] reviews extensively the Supernova event seen 23. February 1987 in the Great Magellanic Cloud (at a distance about 170000 light-years, or about 10 times closer than the Andromeda galaxy). It has confirmed itself as a supernova of type II; after some 130 days, its exponential decay of total luminosity (not easily integrated over the whole spectrum) indicates the presence of 0.07 solar masses (1.4×10^{32} g) of ^{56}Co (formed from ^{56}Ni with $Z = N = 28$, and decaying to ^{56}Fe) as corroborated from observation of the corresponding nuclear γ-rays. However, this supernova behaved quite differently from earlier theoretical expectations, at several points. For instance, it has formed much more nitrogen, and even oxygen, than iron, and if this is a general property of type II supernovae, it may be that elements with Z above 14 are much more extensively formed in type I supernovae [2]. The central object has still not been clearly observed, and hints at paradoxical properties [96].

The CP stars discussed in this review are most likely to accrete material of a composition rather dependent on the detailed behavior and frequency of the supernovae providing it locally (with time), and starting with elements around silicon, there may not be a "standard interstellar material" even when specifying our distance from the galactic center. For instance, mm Wave studies [97] of the relative abundances of $A = 28$, 29 and 30 in the diatomic molecule SiO shows that the interstellar medium has (^{29}Si/^{28}Si) and (^{30}Si/^{28}Si) higher than the Earth, but (^{30}Si/^{29}Si) comparable.

A related question [56] is whether the magnetic accretion is restricted to gaseous ions, or also includes heterogeneous interstellar grains. The material arriving (and evaporating) at surfaces of A$_p$ stars may very well be similar to protoplanetary clouds, perhaps having already segregated mixed oxides of lanthanides; platinoid metals; and mercury. Recently, the impact of cometary material on magnetic white dwarfs has been suggested [98] to be the source of SGR-events, soft γ-ray [40000 eV] burst repeaters. As a plausible complement to the atomic ionization energies in Table 1, it is striking that most elements characterizing CP stars had not been dramatically depleted by volatility before the Earth was formed. The ratio R between the terrestrial and solar abundances has the quite ordinary value 120 for phosphorus, but is smaller (close to 30) for mercury (though not 1.3 as for sulfur).

Returning to A$_p$ stars, Leckrone [99] reviewed far-ultraviolet spectra (1025 to 3000 Å) obtained by satellite spectrometers. α^2CVn is closely similar to A0 type Vega (α Lyrae) down to 1280 Å. Below, it is much brighter, because Vega shows strong hydrogen Lyman lines, and carbon atom line and continuous absorption below 1120 Å. At no wavelength, does α^2CVn approach Regulus (α Leonis) in spite of similar UBV photometric colors. The C^0 ionization edge at 1104 Å seems to be completely absent in α^2CVn and in α Leo (spectral type B7) as also true for Si^0 (1520 Å). Carbon is perhaps underabundant in α^2CVn. Since most europium in α^2CVn should be Eu^{+2}, one would expect strong lines [100] of Eu^{+2} at 2375.46, 2444.99 and 2513.76 Å. They are, at the best, very weak. These measurements with the Copernicus satellite [99] have disclosed several new ways in which A$_p$ stars can be peculiar.

A persistent argument by many authors is that A$_p$ stars do not really have unusual composition; the observed strong spectral lines are due to some subtle effect in the tenuous outer region of the photosphere, overrepresenting some lines of a given element by factors of 10 to 10^5. It should be noted that this "second generation" Fraunhofer problem is the opposite of the classical [5] problem: how can a quarter of the solar photosphere be helium, and not having any influence in the spectrum (above the terrestrial ozone edge at 3000 Å) and the hydrogen Balmer lines be so weak, when they are so strong in Vega, Deneb and Sirius. It may be an attractive model that some stellar atmospheres remain so calm that turbulent mixing does not interiorize the monatomic or solid matter arriving on the outskirts. But it seems rather astounding to the reviewer, if the diffusion model is capable of segregating the elements (and isotopes) with such a variety of choices. The diversified CP stars may rather reflect the highly different mixtures of supernova products that a definite star (or an Earth-like planet during formation) happens to encounter.

Thirty years ago, the idea was frequently expressed that all A_p stars had been members of a double (or multiple) star system together with a progenitor that later exploded as a supernova, providing the unusual elements to the A star. It was then realized that there are roughly 10^9 A stars in our galaxy, among which a few times 10^7 are A_p stars. Since the beginning of their (quite recent) main-sequence existence, at the most 10^7 supernovae have occurred. Further on, it would be highly surprising if each progenitor systematically had a normal A-type companion. This does not prevent that a few spectacular A_p stars (such as HR 465 and/or α^2CVn) may have had this experience. It should be interesting for future astrophysicists to study the two blue (B-type) companions [96] that have quietly remained together, after the third blue star exploded in the Large Magellanic Cloud. The incidence of A_p character would be informative for "Blue Stragglers" of which some may have acquired their high velocities [101] by ejection from a binary system broken up by a progenitor. It is finally worth noting how strange things happen to the archaic Population II stars. The ^{3}He stars have the *other* elements almost indicating pristine Population III stars sprinkled with a bit of supernova dust, as is also true for HD 101065. In such cases, the dust on your shoes may tell where you have been walking.

Acknowledgement: I am grateful to Professor Virginia Trimble (University of California, Irvine, CA 92717 & University of Maryland, College Park, MD 20742) for the inspiration leading to this review, and for the most helpful information. I also thank Professor Ove Havnes, Aurora Borealis Observatory, Tromsø, Norway and Professor Torkild Andersen, Institute of Physics, University of Aarhus, Denmark for helpful correspondence and for kindly pointing out some important, recent literature.

9 References

1. Trimble V (1975) Rev. Mod. Phys. 47: 877
2. Trimble V (1982) Rev. Mod. Phys. 54: 1183; (1983) 55: 511

3. Jørgensen CK, Kauffman GB (this volume)
4. Weinberg S (1979) The first three minutes, Bantam Books, New York
5. Payne CH (1927) Stellar Atmospheres, Harvard College Press, Cambridge MA
6. Landolt-Börnstein Tables (1982) (New Series), Group VI, vol 2 (Astronomy and Astrophysics), subvol 2b: Stars and star clusters, Springer, Berlin Heidelberg New York
7. Maeder A (1981) Astron. Astrophys. 99: 97; 101: 385; 102: 401
8. Maeder A, Lequeux J (1982) Astron. Astrophys. 114: 409
9. Maeder A (1983) Astron. Astrophys. 120: 113 and 130
10. Taylor RJ (1989) Nature 337: 307
11. Preston GW (1974) Annu. Rev. Astron. Astrophys. 12: 257
12. Cowley CR (1977) Astrophys. J. 213: 451
13. Jacobs JM, Dworetsky MM (1982) Nature 299: 535
14. Michaud G (1978) Astrophys. J. 220: 592
15. Cowley CR, Day CA (1976) Astrophys. J. 205: 440
16. White RE, Vaughan AH, Preston GW, Swings JP (1976) Astrophys. J. 204: 131
17. Adelman SJ (1973) Astrophys J. 183: 95
18. Adelman SJ (1973) Astrophys. J. Suppl. 26: 1
19. Hauck B, North P (1982) Astron. Astrophys. 114: 23
20. North P (1984) Photométrie des Etoiles A_p. Thèse no. 2117, Observatoire, Université, Geneva
21. North P (1985) Astron. Astrophys. 148: 165
22. North P, Hauck B, Straizys V (1982) Astron. Astrophys. 108: 373
23. Galeotti P, Lovera E (1974) Nature 249: 130
24. Hirshfeld A, Sinnott RW (1982) Sky Catalogue 2000.0 (vol 1: Stars to magnitude 8.0), Sky Publishing, Cambridge MA
25. Weiss WW, Jenkner H, Wood HJ (1976) (eds) Physics of A_p stars. Proceedings of IAU Colloquium no. 32, Universitätssternwarte Wien, Vienna
26. Cowley CR: Ref. 25, p 289
27. Wolff RJ, Wolff SC (1976) J. Astrophys. 203: 171
28. Landstreet JD, Borra EF: Ref. 25, p 449
29. Jørgensen CK (1988) In: Gschneidner KA, Eyring L (eds) Handbook on the physics and chemistry of rare earths, North Holland, Amsterdam, vol 11, p 197
30. Przybylski A (1966) Nature 210: 20
31. Cohen JG (1970) Astrophys. J. 159: 473
32. Preston GW, Wolff SC (1970) Astrophys. J. 160: 1071
33. Cowley CR, Rice JB (1981) Nature 294: 636
34. Cowley CR: Ref. 25, p 294
35. Aikman GCL, Cowley CR, Crosswhite HM (1979) Astrophys. J. 232: 812
36. Cowley CR, Hartoog MR, Aller MF, Cowley AP (1973) Astrophys. J. 183: 127
37. Przybylski A: Ref. 25, p 351
38. Warner B (1966) Nature 211: 55
39. Cowley CR, Cowley AP, Aikman GCL, Crosswhite HM (1977) Astrophys. J. 216: 37
40. Cowley CR (1976) Astrophys. J. Suppl. 32: 631
41. Cowley CR, Henry R (1979) Astrophys. J. 233: 633
42. Cowley CR, Downs PL (1980) Astrophys. J. 236: 648
43. Cowley CR, Arnold CN (1978) Astrophys. J. 226: 420
44. Jaschek M, Malaroda S (1970) Nature 225: 246
45. Cowley CR, Aikman GCL, Hartoog MR (1976) Astrophys. J. 206: 196
46. Cowley CR: Ref. 25, p 731
47. Kodaira K: Ref. 25, p 521
48. Hartoog MR, Cowley AP (1979) Astrophys. J. 228: 229
49. Hartoog MR (1979) Astrophys. J. 231: 161
50. Wallerstein G (1962) Phys. Rev. Lett. 9: 143
51. Hack M, Stalio R: Ref. 25, p 555
52. Jørgensen CK (1981) Nature 292: 41
53. Cahn RN, Glashow SL (1981) Science 213: 607
54. Jørgensen CK (1989) Chemistry of systems containing unsaturated quarks. In: Liebman JF, Greenberg A (eds) From atoms to polymers, isoelectronic analogies, VCH Publishers, New York (Molecular structure and energetics)
55. Jørgensen CK (1983) Nature 305: 787
56. Havnes O, Conti PS (1971) Astron. Astrophys. 14: 1

57. Kraft RP (1967) Astrophys. J. 150: 551
58. Mihalas D (1967) Astrophys. J. 149: 169; (1968) 153: 317
59. Burgess A (1964) Astrophys. J. 139: 776; (1965) 141: 1588
60. Pyper DM (1969) Astrophys. J. Suppl. 18: 347
61. Moore CE (1970) Ionization potentials and ionization limits derived from the analyses of optical spectra, NSRDS-NBS, vol 34, National Bureau of Standards, Washington DC
62. Martin WC, Zalubas R, Hagan L (1978) Atomic energy levels, the rare earth elements, NSRDS-NBS, vol 60, National Bureau of Standards, Washington DC
63. Frenking G, Cremer D (this volume)
64. Wolff SC (1981) Astrophys. J. 244: 221
65. Mestel L: Ref. 25, p 722
66. Havnes O: Ref. 25, p 135
67. Maeder A (1987) Astron. Astrophys. 173: 247
68. Steinberg EP, Wilkins BD (1978) Astrophys. J. 223: 1000
69. Jørgensen CK (1981) Structure and Bonding 43: 1
70. Bolsterli M, Fiset EO, Nix JR, Norton JL (1972) Phys. Rev. C5: 1050
71. Howard WM, Nix JR (1974) Nature 247: 17
72. Cowley CR, Aikman GCL, Fisher W (1977) Publ. Dominion Astrophysical Observatory (Canada) 15: 37
73. Cowley CR, Adelman SJ (1974) Astrophys. Lett. 16: 5
74. Cowley CR, Aikman GCL, Hartoog MR (1976) Astrophys. J. 206: 196
75. Cowley CR, Allen MS, Aikman GCL (1977) Astrophys. Lett. 18: 83
76. Kuchowicz B (1970) Nature 227: 156
77. Kuchowicz B: Ref. 25, p 169
78. Hoffman DC (1984) Accounts Chem. Res. 17: 235
79. Kuroda PK (1960) Nature 187: 36
80. Kuroda PK (1982) The origin of the chemical elements, Springer, Berlin Heidelberg New York
81. Alexander EC, Lewis RS, Reynolds JH, Michel MC (1971) Science 172: 837
82. Manuel OK, Hennecke EW, Sabu DD (1972) Nature (Phys. Science) 240: 99
83. Black DC (1975) Nature 253: 417
84. Lewis RS, Srinivasan B, Anders E (1975) Science 190: 1251
85. Anders E, Higuchi H, Gros J, Takahashi H, Morgan JW (1975) Science 190: 1262
86. Sabu DD, Manuel OK (1976) Nature 262: 28
87. Blake JB, Schramm DN (1976) Nature 263: 707
88. Cowley CR, Aller MF (1972) Astrophys. J. 175: 477
89. Wolff SC, Morrison ND (1972) Astrophys. J. 175: 473
90. Havnes O, Van den Heuvel EPJ (1972) Astron. Astrophys. 19: 283
91. Merrill PW (1952) Science 115: 479
92. Smith VV, Lambert DL (1986) Astrophys. J. 311: 843; (1988) 333: 219
93. Dominy JF, Wallerstein G (1986) Astrophys. J. 310: 371; (1987) 317: 810
94. Wallerstein G, Dominy JF (1988) Astrophys. J. 330: 937
95. Malaney RA (1989) Nature 337: 718
96. Trimble V (1988) Rev. Mod. Phys. 60: 859
97. Wolff SC (1980) Astrophys. J. 242: 1005
98. Boer M, Hameury JM, Losota JP (1989) Nature 337: 716
99. Leckrone DS: Ref. 25, p 465
100. Sugar J, Spector N (1974) J. Opt. Soc. Am. 64: 1484
101. Trimble V (1988) Nature 336: 111
102. Andersen T, Petkov AP (1975) Astron. Astrophys. 45: 237
103. Grevesse N (1984) Physica Scripta T8: 49
104. Vogel O, Edvardsson B, Wännström A, Arnesen A, Hallin R (1988) Physica Scripta 38: 567

Crookes and Marignac – A Centennial of an Intuitive and Pragmatic Appraisal of "Chemical Elements" and the Present Astrophysical Status of Nucleosynthesis and "Dark Matter"

Christian K. Jørgensen[1] and George B. Kauffman[2]

[1] Section de Chimie, Université de Genève, CH 1211 Geneva 4, Switzerland
[2] Department of Chemistry, School of Natural Sciences, California State University, Fresno, CA 93740–0070, USA

The question, in what sense matter (with positive rest-mass) has constituents, is of great concern to the chemist. A decade before the Lavoisier paradigm was invalidated (by the characterization of the electron in 1897 and by the discovery of radioactive isotopes), Crookes suggested that a primordial material first formed hydrogen and then all the heavier elements by consecutive polymerization at exceedingly high temperatures, in remarkable agreement with recent discoveries in astrophysics. As early as 1860 Marignac had suggested that the heavier elements have slight negative deviations from additive mass, as quantitatively elaborated by Harkins and Wilson in 1915. The Z protons and $N = (A - Z)$ neutrons assumed in 1932–1964 to occur in nuclei were superseded by d, u, s, c, b (and perhaps t) quarks, the nuclear ground state being $u^{2Z+N}d^{Z+2N}$. Though single quarks may be inconceivable, the vacuum may "dress" them with (a fluctuating or invariant number) x additional quarks and x anti-quarks; and unsaturated systems (likely to contain comparable numbers of u, d, and s quarks) of $(3A+2)$ quarks have $Z = (Z_0 + \frac{1}{3})$ whose chemical properties can be predicted. Astrophysical arguments about nonbaryonic matter and the hot quark soup (2 terakelvin) coagulating 20 microseconds after the Big Bang are closely related to this problem.

1 Elements and Principles According to Lavoisier

Nearly all civilizations, having flourished in a given area during a given time interval, developed astronomy and zoology to a remarkable degree, based on careful observations and the inherent human propensity for plausible rationalization. This is much less the case with the concept of *chemical elements*, which is essentially a European one-time idea. This concept originated in the recognition (more than 2000 years ago) of the seven classical *Metals* known to the ancients: gold, silver, mercury, copper, iron, tin, and lead. With the exception of iron, the six other metals were known to form alloys in nearly all proportions, possessing additive mass to a high precision, but, of course, not pretending to describe all other kinds of minerals or materials prepared by chemical treatment. Many such materials had aspects closely similar to compounds, e.g., vitriols (sulfates) or nitrates of the classical metals, but their reduction to metals was not feasible (or was not attempted by appropriate means). The more general hypothesis that *all* matter consists of a limited number of elements, including the seven metals, began to take shape with Robert Boyle (1627–1691) in his book "The Sceptical Chymist", published in 1661, and was elaborated by Antoine-Laurent Lavoisier (1743–1794) during his last years before he was guillotined because he was a tax collector.

According to Lavoisier, approximately 30 elements [among which Mg, Ca, Ba, Al, Si, B, F, and Cl (chlorine was believed to be a higher oxide of murium) were characterized in compounds, but had not yet been isolated [1] as elements] are *nontransmutable* and have *strictly additive masses* (ensuring an invariant mass of closed systems). Lavoisier also lists a (very small) number of weightless *principles*, such as heat ("caloric", in the sense of ignoring transformation of kinetic energy to heat), light, and one (excess or deficit possible) or two (negative and positive) electricities. It should be noted that the experimental evidence then available was all on the side of Lavoisier; the profuse descriptions of alchemical transmutations had not been satisfactorily reproduced, and we know that the rest-mass changes less than one part in 10^9 in chemical reactions (except for the combination of two hydrogen atoms used in the Langmuir torch), just beyond the reach of the most sensitive balances yet constructed.

It should also be noted that the list of elements is not dogmatically invariant; Lavoisier pointed out that classification as an element is always provisional; tomorrow, an "element" may be found to be a mixture or a compound. These two possibilities were frequently realized during the last century; the rare earths [2–4] slowly revealed the presence of 16 elements, and for many years VO and UO_2 [characterized by Martin Heinrich Klaproth (1743–1817) in 1789] were considered elements because nearly all vanadium and uranium compounds contain constitutional oxygen.

In several places Lavoisier contemplated the question of whether some elements are more equal than others (to paraphrase George Orwell's "Animal Farm"). The four candidates for supreme elements were hydrogen, oxygen, nitrogen, and carbon (making both organic chemists and astrophysicists happy). For the next hundred years, nearly all textbooks on inorganic chemistry began with chapters on atmospheric air as a physical mixture and on water as a compound. This is an area where Lavoisier was less fortunate than in his theory of elements and principles. All acids *had* to contain oxygen, including muriatic acid (the earlier name for hydrochloric acid still used in very old drugstores and for adjusting the pH of swimming pools), and Lavoisier had nagging suspicions that all bases (including soda and potash) contain nitrogen (like ammonia and the amines). This is the reason why he did not predict metallic sodium or potassium [obtained by Humphry Davy (1778–1829) in 1807 by the electrolysis of molten salts], whereas he emphasized the likelihood that magnesia, lime, baryta (and strontia, characterized in 1794) are oxides of highly reactive metals (as confirmed by Davy in 1808) as is also true of alumina. Lavoisier missed a glorious opportunity for naming the lightest element oxygen (being the Brønsted definition of an acid) and the other (responsible for 88 percent of the mass of water) hydrogen or "water-former", although security committees today are violently opposed to exchanging the two names on gas cylinders.

One of the earliest scientists still remembered today, Thales (ca. 625–ca. 547 B.C.) of the Ionian colony of Miletus in Asia Minor proposed that matter consists of the universal element water. This statement would be even more impressive in German (*Wasserstoff*) or Swedish (*väte*), referring to hydrogen. In 1815 the English physiological chemist William Prout (1785–1850) pointed out [5] that all other elements may be polymers, H_n, of the lightest element, hydrogen. The major reason for this hypothesis (in addition to the ancient desire of philosophers for a unified external world, thus single-handedly creating all the complications by their deep, and sometimes wise, comments) was that additive masses, which had been the basis for the quantitative analysis of alloys for a millenium, were promoted by Lavoisier to be the essential parameter of a given element in compounds. A great effort was involved in determining *equivalent weights* (e.g., the amount of a given element occurring for each 35.45 g of chlorine in a binary chloride), which became *atomic weights* after multiplication by the oxidation number [6]. For elements known in gaseous compounds, it was very important that Avogadro's hypothesis (i.e., the coefficient n in $PV = nRT$) was widely accepted at the Chemical Congress in Karlsruhe in 1860, despite the unpalatable corollary of diatomic H_2, N_2, O_2, Br_2, etc. and oligo-atomic P_4 and S_8 (and ozone, O_3) molecules. The various forms of periodic systems (based entirely on the atomic weight as the parameter characterizing an element) then developed rapidly [1, 7, 8].

2 Marignac on Atomic Weights and Additive Mass of Constituents

Together with the Belgian Jean-Servais Stas (1813–1891) and one or two other Europeans, Jean Charles Gallisard de Marignac (1817–1894) was the major expert in determining the atomic weights (of 28 elements [9, 10] between 1842 and 1883), and he acquired a rather ambivalent opinion concerning Prout's hypothesis. Like several other chemists [5] he compared minor deviations in atomic weights from integers A (now called *mass numbers*) to the small deviations of the observed behavior of real gases from the ideal relation between P, V, and T. In 1860, in an answer and comment [9] to a paper by Stas, Marignac posed[1] the rather startling question: "Is it conceivable that the unknown cause (probably different from the physical and chemical interactions known to us), which has determined the formation of definite atomic structures from the unique primordial matter, giving each of the elements specific properties of their atoms, at the same time, may influence the modalities of the universal ['gravitational'] attraction in such a way that the weight of each atom is not exactly the sum of the weights of the primordial constituents of such an atom?"

It is clear that the emphasis on additive masses led Marignac to imagine attenuation of the observable manifestations of mass [inertia and gravitational weight, later shown by Roland Eötvös (1848–1919) to be precisely proportional] because of the enormous binding force between subatomic constituents (such as hydrogen in Prout's hypothesis). This idea was clearly incorporated into Albert Einstein's (1879–1955) treatment of general relativity (1915). On the other hand, a major stumbling block for Marignac was the obviously nonintegral atomic weights of chlorine (35.45), copper (63.55), zinc (65.4), strontium (87.6), tellurium (127.6), and barium (137.4).

To bring the discussion closer to the facts, we may consider some present-day atomic weights on the scale $^{12}C = 12.0000000$ amu (atomic mass units; chemical bonding influences the next digit) although chemists for almost a century set the natural mixture of oxygen isotopes (with low concentrations of ^{17}O ànd ^{18}O) at 16 amu. The most striking quasi-confirmations of Prout's hypothesis occur for elements having only one nonradioactive isotope such that the mass number A (the integer quite close to the atomic weight) is odd. Such elements are beryllium ($A = 9$; our only example with even atomic number Z), fluorine (19), sodium (23), aluminum (27), phosphorus (31), scandium (45), manganese (55), cobalt (59), arsenic (75), yttrium (89), niobium (93), rhodium (103), iodine (127), cesium (133), praseodymium (141), terbium (159), holmium (165), thulium (169), gold (197), and bismuth (209).

In this series the observed atomic weights 9.0122, 18.9984, and 22.9898 for the first three elements increasingly show more pronounced negative deviations,

---

[1] "Ne pourrait-on pas supposer que la cause inconnue, mais probablement différente des agents physiques et chimiques que nous connaissons, qui a déterminé certains groupements des atomes de la matière primordiale unique, de manière à donner naissance à nos atomes chimiques simples et à imprimer à chacun de ces groupes un caractère spécial et des propriétés particulières, a pu en même temps exercer une influence sur la manière suivant laquelle ces groupes d'atomes obéissent à la loi de l'attraction universelle, de telle sorte que le poids de chacun d'eux n'est pas exactement la somme des poids des atomes primordiaux qui le constituent?"

such as 44.956 for ^{45}Sc and 58.933 for ^{59}Co. The deviation (below A) is remarkably constant from 88.905 for yttrium to 140.907 for praseodymium, and then, monotonically, becomes less important, 168.934 for thulium, 196.967 for gold to 208.980 for bismuth. The atomic weight [11] must cross A close to 216, is 232.038 for the (almost isotopically pure) thorium, ^{232}Th, and is 241.057 for the americium isotope, ^{241}Am. The fission of heavy nuclei into two fragments (e.g., krypton and barium), is seen to decrease the rest-mass by 0.23 amu; this 0.1 percent difference is released as conventional energy. Beginning in 1920, the deviations of the atomic weight from the integer A were studied primarily by Francis William Aston (1877–1945) by means of his precise mass spectrograph.

Chemists have especially minimized the difference from A by choosing a typical atom like ^{12}C as a standard. The deviations from A vary very regularly [11] in mass-spectrum, but at the same time Prout's hypothesis is not nearly as perfect as is sometimes believed. The atomic weight of a bare proton is 1.00727647 amu, of an electron 0.00054858 amu, and hence of a hydrogen atom 1.0078250. The neutron weighs 1.0086650 amu. If the atomic weights of 90 hydrogen atoms and 142 neutrons are added, a total of 234.00086 amu is obtained, *two* units more than $A = 232$ of ^{232}Th.

Eighty-one of the elements (all with Z values through 83, except 43 and 61) have at least one nonradioactive isotope. Among the following elements, we add thorium and uranium because the α-radioactive half-life of ^{232}Th is longer than, and that of ^{238}U is comparable to the age of the Earth. If we select the intervals $(A - 0.10)$ to $(A + 0.05)$ we would expect about 0.15 times 83, or some 12 or 13 elements, to fall within these weight ranges. The actual number [12] is 39. Among these examples, the 20 odd A values mentioned above are a major part. Two elements show two odd A values with so similar an abundance that bromine and europium have the apparent $A = 80$ and 152, respectively. A related case is that of thallium, for which Gustavus Detlef Hinrichs (1836–1923) maintained a virulent polemic [7] to make the U.S. National Bureau of Standards admit that A is exactly 204. Twelve of the 83 elements have one highly predominant isotope, and only five (lithium, chromium, selenium, molybdenum, and ytterbium) are "genuine accidents". The probability [12] of 39 elements out of 83 falling within the intervals $(A - 0.10)$ to $(A + 0.05)$ is $(83!) (0.15)^{39}$ $(0.85)^{44}/(44!) (39!) = 4.2 \times 10^{-12}$ or perhaps larger by a factor of 7, if one takes the freedom of choosing the position of each "window" (selecting 0.15 of the totally available interval) into account. As early as 1901, Strutt [13] calculated that the probability of the then 9 best known atomic weights being a statistical result of equally probable deviations from A is 0.0012; and that of a larger sample of 27 elements is 0.0017. It is, of course, very difficult to draw absolute conclusions from a single set of observations which cannot be extended. It may be worthwhile to ponder the question of why we say it is accidental that the digit group 1828 is repeated among the first ten digits of $e = 2.718281828459045$ followed by 45, then 45 doubled and then 45 again.

3 Genesis of the Elements According to Crookes

Returning to 1887, Sir William Crookes (1832–1919) gave a lecture [14, 15] on 18 February 1887 to the Royal Institution in London (of which Davy and Faraday had been earlier presidents) with the somewhat provocative title "Genesis of the Elements". Crookes was the antithesis of a neophytic amateur; he had discovered thallium in 1861 [using lines in the flame spectrum, as Robert Wilhelm Bunsen (1811–1899) and Gustav Robert Kirchhoff (1824–1887) had done in discovering rubidium and cesium the previous year], and from 1878 he had worked intensively on the Crookes cathode ray tube and, in particular, on *cathodoluminescence*, the spectral lines emitted by the surface bombarded by high-energy rays [characterized as electrons in 1897 by Sir Joseph John Thomson (1856–1940)] as we know from color television, where the red lines are due to transitions [4, 16, 17] in $4f^6$ europium(III). The description of these lines passed through various stages [18, 19]; in 1909, the element europium, which had been discovered [3, 4] in 1901, was shown by Georges Urbain (1872–1938) to be the emitter [20], and the television screen [21, 22], containing syncrystallized $Eu_xY_{1-x}VO_4$ and more recently [23] $Eu_xY_{2-x}O_2S$, was later developed. The Crookes tube also allowed Wilhelm Conrad Röntgen (1845–1923) to discover X-rays in 1895. As far as our subject is concerned, the important break with the Lavoisier paradigm was Crookes' feeling, expressed on 2 September 1886, that elements may undergo [24] an evolution "like plants and animals of our globe according to Lamarck, Darwin and Wallace". In his lecture to the Royal Institution in 1887 Crookes suggested that

1. There is a primordial material (almost a precursor to matter) *prothyle*, a word already used [5] by Prout in 1816, though frequently spelled "protyle" with a less transparent etymology.
2. At a very early time most of the prothyle condensed to the lightest element hydrogen (having $A = 1$).
3. Under extreme physical conditions (very high temperature, possibly very high pressure and/or exceptionally intense electromagnetic fields) hydrogen is capable of polymerizing to heavier elements. [Although spectral lines of helium [4] had been characterized in 1868, the element was not found in minerals until 1895; it is a noble gas [25] like argon [26], isolated from air in 1894. Hence, the first oligomer known to Crookes was lithium ($A = 7$)].
4. The subsequent reactions would provide varying amounts of beryllium ($A = 9$), boron, carbon, etc. in close analogy to reactions of gaseous compounds induced by heating under laboratory conditions. Because of a decrease in oligomerization rate as A increases or because of the reversible equilibria of endothermic reactions (or a combination of both), only quite small amounts of elements were formed beyond nickel ($A = 59$), but the polymerization is manifestly feasible up to bismuth (209), thorium (232), and uranium (238).

5. The Periodic Table is viewed as a pendulum-like evolution of chemical properties as a function of increasing A and was designed as concentric curves (having a shape like the number 8) with extrema (the top and bottom turning-point of 8) C–Si–Ti–Ge–Zr–Sn–Ce–Ho– [the predecessor of Ta] –Pb–Th–.

6. A given chemical element may actually be a mixture of *meta-elements* with *almost* identical chemical behavior, but the observed atomic weight is a weighted average of the A values of the meta-elements.

For the chemist, the most appealing alternative of meta-elements was A values being differing integers (much like our *isotopes*). Small deviations of the atomic weights from an integer may correspond to one predominant meta-element accompanied by small concentrations of adjacent meta-elements, as we would say in the case of He, C, O, S, Ar, Ca, etc., whereas large deviations, such as B, Cl, Ni, Cu, Zn, etc. suggest a wide dispersion of meta-element A values. Another alternative would be varying amounts of incorporated prothyle, modifying A by only a small fraction, e.g. calcium 39.8; 39.9; 40.0; 40.1; 40.2; and 40.3. Crookes argued that the lanthanides, distributed in the interval A between 140 and 175, are meta-elements with unusually large mutual chemical differences, similar (for us) to deuterium, ^2D, compared to ^1H. Urbain [27] violently objected to the classification of lanthanides as meta-elements, rather than Lavoisier elements, but the past 40 years have seen astrophysical evidence for stellar nucleosynthesis surprisingly similar to "Genesis of the Elements", as discussed below.

It may seem unexpected that so radical a proposal by Crookes without his providing any specific evidence did not stir up more rejection in Victorian Britain. However, if the case is compared [28] with that of George G. Stokes (1819–1903) and William Thomson (later Lord Kelvin) (1824–1907) half a generation before, several physicists are seen to be far from narrow-minded. The Lavoisier paradigm may have seemed somewhat senescent in 1887 (although it was first broken ten years later with the discovery of the electron and radioactivity) and was preserved in the same schools commanding "Thou shalt not extract the square-root of a negative number", which co-existed with scientists working with complex numbers as a useful tool and William Rowan Hamilton (1805–1865) defending the utility of $(1, i, j, k)$ quaternions in physics. At any rate, Marignac found Crookes' suggestion so interesting that (despite [12, 29] having quite bad health after his discovery of gadolinium in 1880) he immediately took the time in 1887 to comment on it in a 12-page article "Quelques réflexions sur le groupe des terres rares à propos de la théorie de M. Crookes sur la genèse des éléments" (p 811 to the end of vol 2 of Ref. 10; cf. also the excellent biography by his collaborator and son-in-law E. Ador [29]). Marignac was reluctant to accept the hypothesis that each meta-element emitting one or a few sharp lines by cathodoluminescence can be identified with definite elements such as samarium, gadolinium, etc., but at the same time he was quite sympathetic to a primordial formation of elements and (despite chemists' not being able to detect their chemical differences) perhaps of meta-elements as well. Marignac reminds us that the cathodoluminescence of trace impurities can be far more intense – and vary quite unpredictably – than that of the major constituents of a solid.

In 1915 William Draper Harkins (1873–1951) and Ernest D. Wilson published three papers (the first [30] of which is entitled "The Changes of Mass and Weight Involved in the Formation of Complex Atoms") based on Prout's hypothesis. On p 1370 they used the word "packing effect" (later adopted by Aston as "packing fraction"), accepted the idea of isotopes in cases like magnesium, silicon, and chlorine (in analogy to the known existence of ^{20}Ne and ^{22}Ne), and attempted to evaluate the actual mass decrease by electromagnetic interaction between electrons and the positive constituents of the nucleus. Harkins and Wilson were very impressed by Marignac's comments of 1860 (Ref. 9 and p 693, vol 1 of Ref. 10), and they translated into English on pp 1372–1373 of their article a page (the last quarter of which we translated above). They then stated: "It has usually been assumed, and without any really logical basis for the assumption, that if a complex atom is made up by the union of simple atoms, the mass of the complex atom must be exactly equal to the masses of the simple atoms entering into its structure". This was certainly not conventional wisdom in 1915, although Einstein proposed his general theory of relativity in that same year.

The Crookes cathode ray tube provided the first clear-cut crack in the Lavoisier paradigm. In 1897 J. J. Thomson [31] demonstrated that cathodes made of several different metallic elements emit the same negatively charged particle (with an atomic weight soon determined to be 0.00055 amu). This *electron* [a word [5] coined by George Johnstone Stoney (1826–1911) in 1894] was the first Lavoisier principle shown to have a positive, but small, mass, and it combined (as a warning that earlier stratifications of chemical theory are frequently pushed up into the daylight again by some insensible mole) the chemically acceptable aspects of phlogiston (metallic elements and alloys containing free electrons; redox reactions being electron transfer) with the rationalization of the large majority of currents (except in molten salts and in salt solutions). Today, the charge 96485 coulomb of a mol of electrons (with opposite sign) is called a faraday in memory of Faraday's electrochemical laws.

4 Radioactive Isotopes from Thorium and Uranium Minerals

In 1896, when 76 Lavoisier elements were known [1], the discovery of the *radioactivity* of uranium by Henri Becquerel (1852–1908) [the first positive result in March 1896, after a series of negative experiments by other people studying fluorescent materials, subsequent to a rather far-fetched remark in December 1895 by the mathematician Henri Poincaré (1854–1912) that since X-rays induce a strong fluorescence in $BaPt(CN)_4 \cdot 4H_2O$, fluorescent solids may emit X-rays] was a much more dramatic foray against the Lavoisier classification. However, this was not immediately realized. As discussed by Abraham Pais [32], radioactivity was originally considered a serious problem for the conservation of energy (once neither illumination nor fluorescent properties were needed and the Sun, being almost at zenith or screened at night by 13000 km of terrestrial material, had no perceptible influence). The 1898 and 1899 publications by Maria Sklodowska Curie (1867–1934) and Pierre Curie (1859–1906) emphasized that radioactivity is an invariant property [103] of a given element (in all its compounds), thorium being 3 times less radioactive than uranium and radium three million times more radioactive than uranium. This is actually valid for a definite isotope, but examples accumulated between 1900 and 1905 which finally proved *spontaneous transmutation* (of what appeared to be meta-elements).

The first example [33] was thorium emanation, ^{220}Rn (diffusing from thorium compounds now known to contain ^{224}Ra in approximate equilibrium with the ancestor "radiothorium", ^{228}Th), having exponential decay of its radioactivity with a *half-life* $t_{1/2}$ close to one minute. The weak point about the three emanations and their radioactive deposit (such as the polonium isotopes "radium A", ^{218}Po, and "thorium A", ^{216}Po) was the fact that their character as elements, rather than some kind of mobile ephemeral contamination, was only firmly established a few years later.

A more convincing experiment performed by Crookes [34] involved Fe(OH)$_3$ as a carrier of "uranium X" (having the more convenient $t_{1/2} = 24$ days) precipitated from the soluble uranyl carbonato complex in excess (NH$_4$)$_2$CO$_3$. These observations show an insoluble hydroxide, but they did not establish the thorium(IV) chemistry of this ^{234}Th isotope, the first descendant of ^{238}U. Sir Ernest Rutherford (1871–1937) and Frederick Soddy (1877–1956) then showed [35] that $t_{1/2} = 3.6$ days of "thorium X" (^{224}Ra), soluble in aqueous ammonia, is replaced by radio-elements precipitating as hydroxides. In 1907 Rutherford demonstrated [36] that "ionium" (^{230}Th with $t_{1/2} = 80000$ years, the immediate ancestor of ^{226}Ra with $t_{1/2} = 1600$ years in uranium minerals), ^{232}Th ($t_{1/2} = 14100$ million years), and "radiothorium", ^{228}Th ($t_{1/2} = 1.9$ years) cannot be separated by chemical means.

For our purpose the most decisive argument for spontaneous transmutation was the proof [37] that α-particles (originating from various radio-elements) form helium on losing their initially high kinetic energy by travelling a few cm in air or 10^{-3} cm in condensed matter. A book [38] written in Pisa in March 1909 and translated into German depicts the situation just before 1911 when the atomic nucleus [39] was detected by the help of a violent scattering of a minor percentage of α-particles traversing matter. As expected, a decade of study of spontaneous transmutation of consecutive chains of radio-elements involved a few mistakes. Sir William Ramsay (1852–1916), who discovered neon, krypton, and xenon [26] in 1898 by the fractional distillation of liquid air, and A. T. Cameron argued that not only helium but also traces of lithium (whose detection by spectral lines is so sensitive) are formed by radon in aqueous copper(II) solutions, but Marie Curie and Ellen Gleditsch [40] showed that the lithium originated in the walls of glass containers attacked by the radioactive solution.

Nuclei [39] with densities of 10^{14} g/cm^3 were the end of the scenario (still plausible in 1910) of subatomic catenation (similar to CH_2 links adding 14 amu to alkane chains), and several authors had earlier interpreted the linear A increases in Döbereiner triads (Ca, Sr, Ba); (S, Se, Te); (Cl, Br, I); etc. as adducts of similar subunits, typically 47 ± 2 amu, but again, Marignac had too rigid a respect for precise atomic weights to accept this idea.

Since electrons are emitted from some β-radioactive nuclei, they became an obvious ingredient in nuclear models (although pre-existence is more compelling in obstetrics than in quantum mechanics [41]), awaiting a future rationalization of A protons and $(A - Z)$ electrons compressed (for whatever reason) within a tiny volume. The fission of ^{235}U was not discovered until 1939, although the interprotonic repulsion was known to be enormous (and roughly [11] proportional to $Z^{5/3}$). The (still largely continuing) wait for a rationalization was only mitigated by the idea that intranuclear attractions are an order of magnitude stronger than the interprotonic repulsion and allow (A/Z) to increase from close to 2 in light elements to around 2.6 in the heaviest known elements. For chemists, the most important conclusion was that chemical properties are almost exclusively determined by the positive charge Ze, throwing light on Crookes' thesis no. 6. Chemists were much less aware that Lavoisier mass-additivity shows irregularities up to almost 1 percent [11, 30].

It may be mentioned that rare radioactive processes involving expelled nuclei considerably heavier than ^4He have been detected recently such as ^{14}C emission [42] from ^{223}Ra (by itself β-radioactive), ^{24}Ne from ^{230}Th, ^{231}Pa, ^{232}U, ^{233}U and ^{234}U, which may also emit ^{28}Mg (the corresponding [43] probabilities of forming ^{210}Pb and ^{206}Hg are 6×10^{-13} and 2×10^{-13}, compared to α-emission producing ^{230}Th).

It was known [38] before 1910 that the longest-lived (still radioactive) descendant of ^{222}Rn ("radium D"), $^{210}Pb(t_{1/2} = 21$ years), cannot be chemically separated from nonradioactive lead, but it was first shown [44, 45] in 1914 that lead isolated from uranium minerals can be mainly ^{206}Pb and lead from thorium minerals can be predominantly ^{208}Pb, compared to the usual isotope mixture with 207.2 amu. Less than ten years later, it was established that isotopy is not restricted to A values above 205 but that the majority of elements show two or more stable isotopes in mass-spectra.

5 "Elementary" Particles and Astrophysics

The nuclear model of $(A - Z)$ confined electrons and A protons was very gratifying to chemists because only two elementary particles (with positive rest-mass; another rebuke to the Lavoisier mass-invariance is the fact that the electron kinetic energy after a potential difference of 511000 V is as large as the rest-mass of 0.00055 amu and was an early confirmation of Einstein's special theory (1905) of relativity, with the celebrated equation $E = mc^2$) correspond to two major types of chemical reactions; exchange of electrons being redox reactions, and exchange of protons (at least for Brønsted) being acid-base reactions. Nevertheless, this model broke down in around 1928, that is, before Sir James Chadwick (1891–1974) discovered the neutron in 1932.

The rather sophisticated argument derived from quantum mechanics, which, since 1925, has produced an avalanche of counterintuitive conclusions (mainly connected with the *absolute* identity of identical elementary particles) having the poor taste to be confirmed by experiments. Microscopic systems have to choose between Fermi-Dirac statistics (being *fermions* with the total spin quantum number I being an odd positive integer divided by 2) or Bose-Einstein statistics (being *bosons* with I a nonnegative integer). Experiments clearly show that electrons, protons (and neutrons) are fermions. An aggregate of an odd number of fermions is itself a fermion, and an even number of fermions constitutes a boson. Fine details of the molecular spectrum of N_2^+ (in electric discharges) show beyond any doubt that ^{14}N (containing 21 fermions, if 7 electrons are confined) is indeed a boson. As soon as the neutron was discovered, the prevailing nuclear model changed to Z protons and $N = (A - Z)$ neutrons, a total of A fermions.

Another complication was that Wolfgang Pauli (1900–1958) and Enrico Fermi (1901–1954) had to invent the *neutrino*, a neutral fermion with no or negliglible rest-mass (below 10^{-8} amu) and almost impossible to detect (it *was* finally detected in 1956). β-radioactivity is the simultaneous emission of a (not preexisting) electron and a neutrino, explaining both why the boson or fermion (odd A) character of the nucleus is not modified and why the kinetic energy of the detected electron is a continuous distribution between zero and the maximum energy available, the (strictly speaking anti-) neutrino running away with the complement. Today, inside CERN, close to Geneva, one highway makes a bump (at a road sign "Neutrino Tunnel") like a crossing above a railway.

Although, in hindsight, the 1932 model of Z protons and $N = (A - Z)$ neutrons became obsolete in 1964 (as seen below), it is still the colloquial way of thinking in nuclear physics, and it served to describe the stellar and the major part of the prestellar (primordial) nucleosynthesis. Since geologists pointed out their dissatisfaction with the maximum life-time of 50 million years for our Sun, if powered by gravitational contraction, as first calculated by Lord Kelvin [William Thomson (1824–1907)] one had looked to something like radioactivity, but it was found that the only viable reaction is

$$4\,^1H^+ + 2e^- \longrightarrow {}^4He^{++} + 2\,\text{neutrinos} \tag{1}$$

(disregarding intermediate species), decreasing the rest-mass by 0.0276 amu. This is eight times less than the fission of ^{235}U, but, since A is 59 times smaller, the exothermic effect per kilogram of helium formed is 7 times larger than that per kilogram of uranium. The present energy output of the Sun is 3.86×10^{26}

Watt (disregarding the 2×10^{25} W carried away by the elusive [104] neutrinos). Transforming one percent of the solar mass of 2×10^{30} kg according to Eq. (1) gives 1.3×10^{26} kg of energy as photons, or 1.2×10^{43} Joule, sufficient to maintain the present-day output for close to 10^9 years.

Rutherford was quite interested in determining the age of the Earth (obviously a lower limit to the age of the solar system). This can be done with surprisingly good precision, knowing that $t_{1/2} = 713$ million years of ^{235}U, having today an abundance of 0.7 percent in uranium from minerals. There are good reasons [46] to assume that the primordial ^{235}U abundance cannot have been much larger than that of ^{238}U and was probably somewhat smaller. This gives an age of 5200 ± 800 million years, very close, incidentally, to $t_{1/2} = 4468$ million years, of ^{238}U, of which half has now decayed to ^{206}Pb, if all of it was imported at the formation of the solar system. Hence 1500 million years ago, the ^{235}U abundance was close to 2.5 percent, which is considered an enrichment fit for water-cooled reactors; there is one case [47] known of a large natural reactor that functioned for about 1 million years in a river bed at Oklo, Gabon (which was detected by the extremely low residual ^{235}U abundance and the isotopic abundances of the stable end-products of fission).

Whereas it is the business of normal stars (like our Sun) to transmute hydrogen into helium according to Eq. (1), much more dramatic conditions are needed to form heavier elements, actually with Z at least 6, since the isotopes ^6Li, ^7Li, ^9Be, ^{10}B, and ^{11}B readily undergo thermonuclear reactions and are very scarce in stars. The production of the heavier elements (beginning with carbon) takes place (almost in agreement with Crookes' thesis no. 4) in *supernovae* [46, 48]. The four last supernovae in our galaxy were observed in 1006, 1054, 1572, and 1604. One was seen in the Andromeda galaxy in 1885. On 23 February 1987 one event was detected in the large Magellanic Cloud, from which both neutrinos and photons left some 170000 years ago [49–52]. Whereas T in the solar center is only 15 megakelvin (the continuous background spectrum emitted at the surface corresponds to 5800 K), supernovae come close to (if not well above) 1 gigakelvin. At this T the thermodynamic equilibrium is still largely in favor of ^{56}Fe and the other iron isotopes, with some nickel, cobalt, manganese, and chromium admixed. The formation of smaller amounts of higher Z values is due to a high concentration of neutrons expelled by collisions of rapid ^4He with ^{18}O, ^{22}Ne, ^{25}Mg, and heavier isotopes.

Astrophysicists have always been interested in the Fraunhofer lines in absorption (or, under exceptional conditions, emission) spectra, a unique source of information on chemical elements at the stellar surface. The predominance of the iron lines and the lines of Na, Mg, Ca, Ca$^+$, Ti, V, Cr, Mn, Co, Ni, Sr, and Sr$^+$ in the solar spectrum, when compared with the hydrogen Balmer lines in Sirius and other stars with T close to 11000 K or the helium, carbon, nitrogen, and oxygen lines in Orion-type stars with T between 20000 and 30000 K, originally suggested essential differences in chemical composition and hinted at a Crookes evolution [53]. The question (quite legitimate before the small, inaccessible nuclei were established in 1911) of whether T above 6000 K or laboratory spark spectra (as contrasted to flame and arc spectra) induce dissociation of atoms into subunits [5] beginning to approach the prothyle stage is to some extent a matter of definition, reversible removal of electrons from gaseous atoms to form M^+ and higher ionic charges being a (rather superficial) kind of atomic dissociation. Careful analysis of (mainly absorption) lines of stellar spectra and (usually emission) lines of laboratory atomic spectra drove the astrophysicists to conclude [46] that *most* stellar atmospheres have almost the same elemental abundances (relative to H) and that the spectral types (O, B, A, F, G, K, M, etc.) are essentially determined by T of the photosphere and, for minor details, by the pressure and the concentration of free electrons prevailing [112].

Table 1 compares the weight compositions (derived from atomic [46] abundances compiled by Virginia Trimble) of the solar photosphere in ppm (or g/ton) with the well-known geological table for the outermost few kilometers of the Earth's crust. Many of the latter values are established with great accuracy, whereas (with the exception of the major constituents hydrogen and helium)

Table 1. Weight abundances in parts per million (g/t) of elements in the solar photosphere (where the Fraunhofer absorption lines originate) and in the outermost few km of the Earth's crust

$Z =$	Sun	Earth	$Z =$	Sun	Earth
1 H	740000	1400	31 Ga	0.03	15
2 He	240000	—	32–36	0.2	14
3 Li	0.00004	65	37 Rb	0.03	310
4 Be	0.00005	6	38 Sr	0.05	150
5 B	<0.001	3	39 Y	0.004	28
6 C	3500	320	40 Zr	0.04	220
7 N	1300	—	41 Nb	0.01	20
8 O	8000	466000	42 Mo	0.02	15
9 F	0.5	300	43–46	0.012	0.02
10 Ne	600?	—	47 Ag	0.0007	0.02
11 Na	35	28300	48 Cd	0.008	0.2
12 Mg	600	20900	49 In	0.004	0.1
13 Al	70	81300	50 Sn	?	40
14 Si	900	277200	51–54	?	2
15 P	10	1200	55 Cs	<0.01	7
16 S	400	520	56 Ba	0.012	250
17 Cl	8	480	57 La	0.007	18
18 Ar	30	—	58 Ce	0.009	46
19 K	10	25900	59–71	0.025	56
20 Ca	70	36300	72 Hf	0.001	4
21 Sc	0.04	5	73 Ta	?	3
22 Ti	2.5	4400	74 W	0.05	70
23 V	0.4	150	75–78	0.04	0.01
24 Cr	25	200	79 Au	0.001	0.005
25 Mn	10	1000	80 Hg	<0.02	0.5
26 Fe	1000	50000	81 Tl	0.0015	0.3
27 Co	1.5	30	82 Pb	0.012	16
28 Ni	150	100	83 Bi	0.0015	0.2
29 Cu	0.8	70	90 Th	0.0003	12
30 Zn	1.2	80	92 U	0.0001?	4

many solar values are still uncertain within a factor of 2 or 3. The most striking difference in going from the order (H > He ≫ O > C > N > Fe > Si > ...) in our closest star to (O > Si > Al > Fe > Ca > Na > K > Mg ≫ Ti > ...) in the Earth's crust can be rationalized by the loss of volatile materials (nearly all the hydrogen, helium, carbon, nitrogen, neon, and about 80 percent of the oxygen) during the formation of the small, inner planets (Mercury, Venus, Earth and its moon, and Mars), whereas the large planets (Jupiter, Saturn, Uranus, and Neptune) still contain huge amounts of H_2, He, CH_4, and NH_3. A smaller difference is the unexpectedly high concentration of potassium, rubidium, and uranium in the terrestrial surface, compared to the denser minerals with more magnesium, titanium, and chromium found in lunar samples and in deep drilling holes in the Earth. Nevertheless, there is an overall enrichment of elements with Z above 30 (except for platinoid and other noble metals, presumably extracted into the hot core consisting of stainless steel, similar to iron meteorites) in the Earth, giving strong evidence for a far higher concentration of supernova dust in the raw material for the planet, which seems to have been imported [47, 54] within a narrow time interval (0 to 30 million years) about 4700 million years ago. Radioactive dating of lunar and terrestrial samples with the highest ages known since solidification indicate 4500 to 4000 million years, and recently, evidence has been found [55] for life 3800 million years ago. Convincing evidence for higher concentration of supernova dust (gratifying to inorganic chemists) is the (rare earths/silicon) weight ratio of 54×10^{-5} in the Earth's crust, but only 5×10^{-5} in the solar photosphere. It may be noted [46] that α Centauri (at only 4.3 light-years distance, half that of Sirius) has all elements with $Z = 6$ to 60, about twice as abundant as in the Sun.

Despite the strong reluctance of Harlow Shapley (1885–1972) and several other astronomers, Edwin P. Hubble (1889–1953) succeeded in proving before 1930 that not only are there many, many millions of galaxies like ours and the closest standard-size galaxy seen behind the stars in the constellation of Andromeda (the two Magellanic Clouds are relatively small satellite nebulae), but they all recede with huge speeds, roughly proportional to their distance. This is not a case of geophobia; it is a velocity field (compatible with Einstein's equations) making the Universe look much the same to all observers. Unfortunately, the proportionality constant (given in $km\,s^{-1}Mpc^{-1}$ where one megaparsec = 3.086×10^{19} km = 3.26 million light-years) is still not better known than within [56] a factor of 1.5 (values between 50 and 90 are still defended in the literature).

If the expansion is a straightforward Einstein explosion originating at a point (or a tiny volume), allowing for a bit of deceleration due to gravitational attraction, the reciprocal value of the Hubble constant (having the dimension of a time) gives, to a first approximation, the age of 10^{10} years since the Big Bang. Thirty years ago there was an alternative model of the steady-state continuous formation of matter, making the Universe conceptually eternal. However, according to a private communication from Fred Hoyle (Bologna, May 1988), it is no longer consistent to assume an invasion by hydrogen atoms or neutrons, although an alternative with an in-flow of 10^{45}kg-objects (able to form 10^{15} stars) can be defended. There is one sense in which the scenario can be considered steady-state in several recent theories: there may be a multitude [105] of universes (without direct intercommunication) forming in a foam-bath in a many-dimensional space, each soap-bubble (in the local Minkowski space) in the 10^{57} kg class (or above) technically being a black hole and looking like a familiar Universe in the interior. At any rate, when the cosmic isotropic microwave radiation corresponding to an actual $T = 2.7$ K was found by Arno Allan Penzias and Robert Woodrow Wilson in 1965, there was achieved an almost general consensus that there was a Big Bang (before which there was no local time) with the Universe having the average T not far from 10 gigakelvin (10^{10} K) divided by the square-root of the running time in seconds.

This large-scale dynamic upheaval has decoupled the question of stellar life-times. It is not significant that our Sun could continue its energy production at the present rate for 10^{10} years at the expense of forming helium out of one-tenth of its mass. The theory for main-sequence stars [112] in the Hertz-sprung-Russell diagram has been well established from the careful study of instantaneous pictures available for stars of differing mass and (photospheric) T. Stars like our Sun finally develop into red giants and then collapse to white dwarfs (like the Sirius companion) with a moderate density of 10 tons/cm^3. During their last 500 million years, they have emitted much more energy, and our Sun is assured 4000 ± 1000 million years of further activity. Main-sequence stars with T below 5000 K are scheduled for 2×10^{10} years, twice the time since the Big Bang. On the other hand, we know of a large number of hotter, profligate stars (e.g., of the spectral class B, the Orion type, roughly 30000 K) that are visible from afar (with absolute luminosities [112] thousands of times that of Sirius or 10^5 times that of the Sun) and shine for only 10^8 or 10^7 years. We are not aware of any really good theory for the formation, first of the 10^{12} observable galaxies, and then, on the average, of the 10^{11} stars in each galaxy, which must have taken place in the time interval between 10^8 to 10^9 years after the Big Bang.

It has been firmly established [46] that the elements were not formed in a one-pot operation immediately after the beginning of time, as was first believed (in 1946) by George Gamow (1904–1968), who ignored the consequences of ^{8}Be and all isotopes with $A = 5$ dissociating instantaneously. In his best-seller, "The First Three Minutes", Steven Weinberg [57] explained the evolution of protons and neutrons forming ^{4}He to the extent of 22 percent of the total mass, leaving, after several minutes, a few times 10^{-5} each of deuterium and of the rare helium isotope ^{3}He (including its slow formation from β-radioactive tritium ^{3}T) and close to 10^{-9} of the mass as lithium, ^{7}Li. At the densities and temperatures prevailing, any heavier conventional (Z, N) isotopes would be far scarcer.

It is now recognized that only very massive stars (typically 5 to 10 times the solar mass) have the opportunity to form the larger amounts of carbon, nitrogen, and oxygen that are dispersed in violent stellar winds at the end of their rather short career and that the major source of these and the heavier elements are supernova explosions [46, 48]. Astronomers speak of the "metallicity" of a star as the weight proportion of $Z \geq 6$. As seen in Table 1, the solar photosphere has 1.7 percent "metals", of which only one-fifth have Z above 10, and hence form metallic elements in the laboratory. These ratios are quite typical of the large majority of stars, called "Population I" in this connection, but many stars are "Population II" with metallicities 10 or 20 times lower. This is evidence that they have had very little occasion to scoop up supernova products, and there is still a search for almost pristine "Population III stars" containing only the Weinberg mixture, but they are difficult to find now, 10^{10} years after the Big Bang.

In this sense, the solar data in Table 1 represent the modern normality. However, there a few, but highly spectacular, deviant stars. Red stars [112] of spectral class S (T close to 3000 K) show absorption bands of the diatomic molecule ZrO (the far more frequent class M at the same T shows TiO absorption bands) and may also show spectral lines of atoms of technetium ($Z = 43$), of which no isotope is known to have $t_{1/2}$ above 2.6 million years, and a quite conspicuous [46, 58] group is A_p stars (peculiar stars of class A), constituting about 5 percent of stars with T in a narrow interval close to 10000 K. The first one to be discovered was the star of third magnitude (with the ancient name Cor Caroli after the British king Charles II, a benefactor of the Greenwich Observatory) α^2 Canum Venaticorum, where Baxandall in 1913 detected strong Eu^+ Fraunhofer lines of europium (We are not aware of any reaction to this from the 81-year old Crookes).

Some specific examples of A_p stars are reviewed [4], among which the fifth magnitude star $\varkappa$ Piscium, showing very strong lines [59] of osmium atoms and of U^+ and apparently of Ir and Pt^+ as well, but no Au, Hg, Pb, and Bi, were detected. On the other hand, HR 7775 contains 2 kg of bismuth/ton, twice as much as the concentration of iron in the Sun, as well as much gallium and mercury [60]. The most curious case is perhaps the time-evolution [61] of HR 465, which showed Dy^+ in 1979 but in 1981 was gliding toward Ho^+ and Er^+. At the same time, the strong lines of U^+, Th^+, and Th^{+2} increased their intensities. One of the few common characteristics of these A_p stars is that they specialize in one, or a few, among elements such as phosphorus, manganese, gallium, yttrium, europium, holmium, osmium, iridium, mercury, bismuth, thorium, and uranium (and perhaps transuranium elements?). Can it be technological management?

It is likely that the "metallicity" of at least the larger galaxies will increase steadily during the next 10^{10} years because of the productivity of supernovae but nevertheless remaining far removed from the favored thermodynamic equilibrium of iron and adjacent elements (cf. Table 1). However, we have had more than the usual ration for a millenium of revolutions in physics and astronomy since 1887. Contrary to the 30 or 100 Lavoisier elements, it is an ancient trend in chemistry [5] to look for unity in matter [a thought that was not alien to Sir Isaac Newton (1642–1727)]. Before 1934 Dirac convinced physicists that all fermions and most bosons (except photons, neutral pions, and a few unusual cases) each have *anti-particles* with the opposite charge (if any). This is not to say that neutral systems cannot have anti-particles; a neutron and an anti-neutron disappear entirely, when meeting, producing 8 times as much energy as one ^{235}U fission, and the bosons ^4He and anti-helium 32 times more. The best known case is that of electrons and positrons.

We must remember that mc^2 of 1 amu corresponds to 931.50 MeV or 0.9315 GeV (giga-electronvolt) in the units familiar to high-energy physicists. Hence a positron and an electron can annihilate and emit two or three photons with a total energy of 1.022 MeV. Conversely, a high-energy photon passing an abrupt field gradient (e.g. close to a nucleus) can "create" an electron and a positron. This means that the vacuum has become an indefinite storehouse [41], and, in one sense, the "permanent Universe" with particles a kind of epiphenomenon. It became convenient to count only "elementary particles" without including the corresponding anti-particles. In this sense, the distinction by Powell in 1947 between muons and pions (the early work was concentrated mainly on the components of the cosmic radiation detected since 1912) brought the number of particles to seven: the electron, the neutrino, the negative (or positive anti-) muon (0.1134 amu), the positive (or negative) pion (0.1498 amu), the (short-lived) neutral pion (0.1449 amu), the proton, and the neutron. One could live comfortably with 7 essential constituents of matter (i.e., having rest-mass), and there were historical precedents; all theoreticians (having obtained their diplomas) knew until about 1750 that there were, of necessity, exactly 7 metals. However, the large-scale, postwar, high-energy laboratories [62, 63] made elementary particle physics one of the least elementary areas of physics. In 1966 the number of "elementary particles" was 40; by 1980 it had passed 100 (and the number of chemically known Z values).

6 Quarks and Leptons

A novel classification was introduced in 1964 by Murray Gell-Mann [64] (and independently, in a preprint, by George Zweig). "Elementary" particles are either *leptons* [65], all being fermions, such as electrons, muons, and their distinct kind [104] of neutrinos [the Greek word lepton meaning, "not heavy", is now extended to tauons with 1.915 amu and $t_{1/2} = 3 \times 10^{-13}$ s, but otherwise very electron-like] or *hadrons* (i.e., undergoing the strong interactions studied in nuclei). Hadrons come in two categories: *mesons* (which are all bosons, and have $t_{1/2}$ in the 2×10^{-8} s to 10^{-24} s class) and *baryons* (all being fermions), among which the two lightest, the proton and neutron, carry the title *nucleons* (a nucleus having A nucleons). According to Gell-Mann, all mesons consist of a *quark* and an *anti-quark*, and all the baryons consist of *three* quarks. Originally, there were three types ("flavors") of quarks, *u*(up), with the charge $+2/3$ in protonic units; *d*(down); and *s*(strange), both $-1/3$. In 1974 clear-cut experimental evidence was obtained for c(charm) with charge $+2/3$ (predicted in 1970 by Glashow, Iliopoulos, and Maiani) and in 1979 for b(beauty), $-1/3$.

The positive pion has the quark configuration $u\bar{d}$ (where the roof lines indicate anti-quarks), and the neutral pion, which is a kind of Goldstone boson (like the photon), is a mixture of $u\bar{u}$ and $d\bar{d}$. The proton is u^2d, the neutron is ud^2, and the third-lightest baryon, the neutral lambda (1.20 amu) is uds (as usual in atomic electron configurations, the superscript 1 is omitted). A major triumph for the quark model was the fact that the eight baryons with spin $I = 1/2$ (and to a reasonable approximation, their magnetic moments) were immediately classified, and a tenth baryon with $I = 3/2$ (in addition to the nine already known, such as the doubly charged delta u^3 with 1.32 amu) was predicted (s^3) and then discovered (negative omega, 1.795 amu). The impact of the quark model was strongly enhanced by a review [66] in 1974 as well as by extensive studies of high-energy collisions between protons [67], which showed clear evidence for very small, hard "partons", much as Rutherford [39] discovered atomic nuclei by the scattering α-particles.

One is tempted to introduce the neologism that the quark model is "out-prouting" Prout. Here we have 3 to 5 quark flavors explaining almost everything (although there is a recent paradox [68] involving the scattering of polarized muons on protons), but the quarks refuse to manifest their most exceptional property, their electric charge in units of *a-third* protonic charge (although it can be evaluated from parton scattering cross-sections). It must be realized that the blackboards (and wastebaskets) of theoreticians are full of thousands of competitive hypotheses, and the published literature contains hundreds of them. There has been a tendency to elaborate a rigid paradigm, QCD (quantum chromodynamics) in a somewhat remote analogy to the highly satisfactory quantum electrodynamics. According to this treatment, any attempt to separate two quarks induces a kind of dielectric polarization of the vacuum, producing a string of mesons but no fractionally charged end products.

More recently, it is generally believed that sufficiently high T (of the order of 1.5 to 2 terakelvin), to some extent assisted by exceptionally high density and pressure, induce a phase change to a *quark plasma* [69]. This unconfined collective state, which does not show strong interactions in clusters of 3 quarks at a time, much as sufficient heating of CO_2 at constant volume abolishes the individuality of the molecules, may probably be obtained in high-energy collisions between heavy nuclei [70, 71] with the kinetic energy of 0.2 to 200 times the rest-mass mc^2. The major problem for the physicist is that he has only the debris to look at, when the system has normalized ("hadronized") itself after 10^{-22} s. This line of thought [72] has added a prologue to [57] "The Three First Minutes" when the Universe went through a dew-point some 10 to 20 microseconds after the Big Bang. Already because of the exothermic quark trimerization, the transition from a roughly ideal quark gas (of mixed flavors) to coagulated baryons required some time, with two phases coexisting. If the temperature was defined in this situation, it remained invariant, as in a freezing liquid. The mechanism by which the anti-quarks and the anti-baryons were just slightly short of the majority is still hotly debated. It may have happened less than 10^{-10} s after the Big Bang [105].

7 Nonbaryonic Matter

The astrophysicists, who are the most eager consumers of such theories, have acquired a recent craving for *Dark Matter*. A more sophisticated demand is for WIMP, weakly interacting massive particles. The latter concept [73, 74] goes back to as far as 1981 and was imagined as technicolored hadrons, but actually, any species X^- would do, if it is protected by a stringent conservation rule of the same kind as the baryon number B (synonymous with the integer A for nuclei, although it is $-A$ for anti-nuclei and zero for leptons, including muons). It may have any atomic weight, say 10^2 to 10^8 amu. The X^- zooms down on the closest available nucleus (as a negative muon does) and stays there, at least below 20 megakelvin, except for adducts with $Z = 1$ or 2. The strongly bound adducts XZ behave as an isotope of the preceding element $(Z - 1)$ so that, chemically speaking, XC, XTi, XRu, XSm, XTh, and XU are heavy boron, scandium, technetium, promethium, actinium, and protactinium isotopes.

Cahn and Glashow [73] argued that precocious nucleosynthesis may have been triggered in the Population III stars by the smallest addition of such heavy elements (by-passing the bottleneck [74] of instantaneously dissociating ^{8}Be because the lithium isotope X^{8}Be is stable), and the half-life of α-radioactive X^{244}Pu and X^{247}Cm may have been prolonged beyond the geochemical time-scale. Remarkably small amounts of XZ and other WIMP may modify stellar interiors perceptibly and, for instance, decrease the operating temperature at the solar center from 15 to 14 megakelvin, thus solving the "neutrino problem".

However, a real large-scale problem is Ω (the ratio between the average density of the Universe and the critical value 1.88×10^{-29} g/cm^3 times the square of the Hubble constant [56] in the units of 100 km/(s · Mpc)). At $\Omega = 1$, the Universe expands more and more slowly and achieves an asymptotic state in the far future. If Ω was well above 1, we would already perceive [105] the beginning collapse as a prelude to a "Big Crunch". We know the concentration of luminous stellar material quite well, and including a reasonable estimate of interstellar gas and dust, Ω is only 0.02. The density of a few times 10^{-31} g/cm^3 appears very low to chemists (one hydrogen atom per 5 m^3), but we should remember that light is readily detected from galaxies 10^9 light-years or 10^{22} km away.

The first difficulty lies in the fact that the kinematic behavior of many adjacent galaxies shows a rotation strongly suggesting an extended halo of very dim matter stretching as an ellipsoid around the central, very luminous plane full of stars. This plane is not the exclusive residence of bright stars (in analogy to planets in the ecliptic plane). Several categories [106] of straggling stars (types O, B, A, and F with high luminosity) are observed (with velocities of the order $10^{-3}c$) in the halo. The general consensus of such studies and of intergalactic attractions (a very difficult subject) within the closest 50 million light-years suggests that $\Omega = 0.15$ to 0.2. If this matter is baryonic, it might perhaps be Jupiter-like objects without adjacent stars.

Black holes (heavier than 10^{12} kg; otherwise, they would have already Hawking-evaporated) are not strictly baryonic; they do not conserve B because they do not remember whether they swallowed matter or anti-matter. However, any encounter with gas or dust produces a large quantity of X-rays, probably more than is observed. Hence there is a great stimulus [75, 76] to find plausible candidates for dark, i.e., nonluminous, matter. This stimulus has intensified because most cosmologists strongly prefer Ω to be very close to 1, compatible with modern scenarios of "inflation" (well below a picosecond after the Big Bang): Primack [77] compiled an excellent catalog of a dozen, highly differing, dark-matter candidates. Some of them are now declining, such as the electronic type of neutrino (if its rest-mass is below 5 eV, in any case, it is of no use), but we cannot exclude the possibility that the muonic or tauonic neutrino [104] are long-lived (preferably 10^{14} years) or supersymmetric neutral fermions, "photinos", having the desirable rest-mass 30 to 40 eV (4×10^{-8} amu). It is a narrow escape: a long-lived 80 eV neutrino makes the Universe collapse. A quite appealing candidate is *uds*-matter, discussed below.

The ESO (European Southern Observatory) and CERN collaborate in organizing symposia on astronomy, cosmology, and fundamental physics. The second of these was held at Garching-bei-München (Bavaria) in March 1986 and the third [78–80] at Bologna in May 1988. Michael Turner gave an excellent survey of dark-matter (the world looks quite different from 1986), and a newcomer is the almost undetectable axion. If it exists, it can only have a tiny rest-mass between 10^{-5} and 10^{-3} eV. Another early line of thought [81, 82] is that there may exist a highly dense variety (e.g. 10^{15} g/cm^3) of nuclei, which may be either metastable or more stable than conventional (Z, N) nuclei, a rather frightening idea. Chemists are protected by the electrons surrounding even quite moderate Z values; a laboratory does not blow up, although protons (in solvents) have an enormous affinity for nuclei with A above 6. Such nuclei with exceptionally high density attracted general interest [83–85] in 1984 and are now assumed to contain comparable amounts of u-, d-, and s-quarks. If the total number of quarks is $3A$ (the atomic weight may be quite different from A), the charge Z is far smaller than A. It is not known whether the neutral lambda *uds* is able to dimerize to $u^2d^2s^2$, but it seems likely [86] that the neutral $u^6d^6s^6$ and $u^{16}d^{16}s^{16}$ are quite stable.

The "pulsar" is a well-known astrophysical curiosity; it has generally been identified as a neutron star somewhat heavier than the Sun, but at least some pulsars and heavier objects [87, 88] may rather consist of *uds*-matter. This question has recently been further elaborated [107–111]. When the dust settles in a few years [49, 52, 89], it will be interesting to see what the central remainder is in the 1987 supernova in the Large Magellanic Cloud. In high-energy physics "hyperonic" nuclei containing one s-quark are 0.2 amu heavier than corresponding nuclei with a d-quark. The reason why high-density species are stabilized by containing nearly as many s as u and d quarks is the strongly increased kinetic energy of fermions confined in a very small volume (as related to Pauli's exclusion principle). It is the lesser evil to distribute this destabilization on three (rather than on two) quark flavors. On the other hand, c quarks adding 1.5 amu to the rest-mass are unlikely to play a rôle here, although they may become significant when a uds-star collapses to a black hole. At this point, possible attractive "fifth" and repulsive "sixth" force components of gravitational interactions at distances below, e.g. 1 m, may decide the behavior of a collapsing object (mass M kg) crossing its Schwarzschild radius (M times 1.485×10^{-27} m).

Seen from chemist's point of view [4, 90], uds-matter may represent the best opportunity for studying *fractionally charged species* having their chemical properties determined by a charge $(Z_0 + \frac{1}{3})$ or $(Z_0 + \frac{2}{3})$ times that of a proton on their tiny, central volume. The textbook arguments against "free quarks" begin with the fact that our accelerators cannot smash a baryon. It is not perfectly certain that a proton passing very close to a black hole cannot lose one or two quarks because of extreme gravitational attraction. However, the QCD paradigm is probably correct in that the vacuum [41, 91] immediately "dresses" a single quark with a meson cloud, containing a fluctuating or constant number of anti-quarks and an equal number of additional quarks. What "quark hunters" can hope for [92] is a system of *unsaturated quarks*. Several aspects of diquarks [91] suggest that $(3A + 2)$ quarks carrying the charge $(Z_0 + \frac{1}{3})$ are less difficult to find than the more indeterminate $(3A + 1)$ systems $(Z_0 + \frac{2}{3})$ tending to add 2, 5, 8, 11, etc. mesons supplied by the vacuum. This does not prevent the experimental evidence clearly indicating that, for a given x, a set of (3x) diquarks is less stable than (2x) nucleons. The atomic weight of a $(3A + 2)$ system may easily be 100 to 10^6 amu above A, and within a certain interval, it may *decrease* when A increases (because of a kind of color polarization effect destabilizing larger nuclei to a smaller extent), opening the possibility [81, 91, 93] that the vacuum encourages spontaneous growth by supplying A_0 nuclei and anti-nuclei against a fee of A_0 (1.86 GeV).

All of these features assisting the formation of fractionally charged systems are expected to be more pronounced in uds-matter. Rather than the conventional $u^{2Z+N}d^{Z+2N}$ (equivalent to $u^{A+Z}d^{2A-Z}$) nuclei, a superdense nucleus $u^{A+\delta-\varepsilon}d^{A+\varepsilon}s^{A-\delta-1}$ with the moderate $Z = (\delta - \varepsilon + \frac{1}{3})$ may be the most plausible candidate. It has been argued [2, 4] that rare-earth minerals should be particularly apt to contain traces of such species. The rationalization of their chemical properties cannot be based on Mulliken electronegativities [94], but, in particular for the lanthanides, it can be predicted [90] from available spectroscopic parameters.

One may ask whether fractionally charged systems have been observed. There is a case [62] of a 10^{16} eV cosmic-ray primary, but the most intensely studied systems (since 1977 at Stanford University) are 10^{-4} g niobium spheres [95] levitated in a magnetic field in their cooled, superconducting state and consistently showing one fractional charge per 10^{18} niobium nuclei or one per 10^{20} amu (i.e., 6000 per gram). Since there is no particular reason why $Z_0 = 41$ (the chemistry of Z values close to 73, 91, 105, 143, etc. is expected to be rather similar to that of niobium) and since the propensity toward fission should be much smaller [93] in the presence of unsaturated quarks, higher Z values than Z_0 of conventional nuclei are conceivable. There have been about ten negative reports [96] in the literature but usually of the order of, at most, one fractionally charged system per 10^{18} to 10^{20} amu. Some of these estimates have been severely criticized [92].

The next best technique (after one-atom detection by resonance ionization spectroscopy, as suggested by W. Fairbank, Jr., which will only become feasible when the spectral lines are known with at least the precision of 10^{-5} in wave-number) is probably electrostatic deflection of droplet jets [97] giving, as higher limits, one $(Z_0 + \frac{1}{3})$ per 2×10^{19} amu of the liquid, tetraethylene glycol, and one $(Z_0 + \frac{2}{3})$ per 4×10^{19} amu. Such measurements would be highly interesting at slightly higher sensitivity if finely divided compounds (of two-digit Z values) or powdered minerals could be dispersed in the droplets, as has already been done in the technology of pigment droplet printing. Various proposals of preconcentration have been discussed [90], including systematic screening for atomic weights above 300.

It may seem frustrating that a candidate for dark matter (and perhaps 90 percent of the rest-mass in the actual Universe) is so elusive, being fractionally charged or not. The chemical properties predicted [90] for $(Z_0 + \frac{1}{3})$ and $(Z_0 + \frac{2}{3})$ systems would not encourage volatility, and for many $Z = 1.33$; 1.66; 9.33; and 9.66 (anions) or 2.33; 2.66; and 10.33 (cations) neither incorporation in the core of stainless steel (although the smaller ionic charges in many cases yield more pronounced effects of polarizability). Neutral systems as small as nuclei fall (after a collision typically every 20 cm of free fall) to the center of the Earth [4] if their atomic weights are above 10^8 amu (in which case the Brownian motion is too slow to prevent the vertical motion). However, if accompanied by an electronic cloud with a radius of at least 0.3 Å, they should remain suspended in water for a long time, if lighter than 10^{12} or 10^{13} amu, and they may finally concentrate in manganese nodules on the sea floor. Comparison with samples from comets, asteroids, or Jupiter and its moons might still be interesting because the process of formation of an inner planet (like the Earth) may have removed most instances of unsaturated quarks to an inaccessible place (such as the hot center).

As he has done every two years, Dimitri V. Nanopoulos [78, 80] held the latest version of the lecture "Beyond the Standard Model of Particle Physics", including the unique World Sheet theory possibly superseding the supersymmetric string theory. Although there is still a small probability for a fourth neutrino, it now seems almost certain that the other leptons and the quarks ($+2/3$ and $-1/3$) occur in exactly 3 generations (needing a sixth t-(top) quark, at least heavier than 70 amu or twenty times the b-quark). All the participants are eagerly awaiting the 0.6 TeV (600 GeV or 650 amu equivalent) collisions between electrons and positrons in the LEP ring (circumference 27 km) that should be operating in CERN by the summer of 1989, and in the planned 20 TeV US Supercollider hopefully completed by 1996. The concept of three generations [41] of leptons and quarks is quite similar to the Periodic Table, and the expectations of a sixth quark may be analogous to the elements predicted by Mendeleev although they may also be an illusion, like nebulium and coronium [4] proposed by Johannes Robert Rydberg (1854–1919).

8 Constituents Seen from the Point of View of Chemists

Chemists have strongly modified their concept of "constituents" since the time of Lavoisier. Originally, the physical analogy to clocks and machinery [28] involved the tacit, but obvious, assumption of additive masses. The *ratio* R between the binding energy of Z electrons (to one nucleus) and their rest-mass Zm_0c^2 is approximately the ratio [98] between $Z^{2.4}$ rydberg (this unit $= \frac{1}{2}$ hartree $= 13.605$ eV) and 2×137^2 times Z rydberg is $R = 0.015$ for a thorium atom ($Z = 90$) where the mass decrease of 0.0007 amu (3 ppm) would be within the reach of analytical balances if the experiment was feasible. Since 1915 it has been realized [30] and later rationalized [11] by Carl Friedrich von Weizsäcker that the ratio R between the binding energy of A nucleons (Z protons and N neutrons) and the rest-mass of the nucleus formed is close to 0.008 for nearly all stable nuclei having A above 12. What is more astonishing is that the analogous ratio R for $3A$ quarks (dressed with meson clouds [91] by the vacuum) and the atomic weight of the nucleus may be exceedingly large and must almost certainly be above 100, if the data of high-energy physics are considered. Is a weird entity a constituent, if R is 1000 or 100? Another aspect is that just as Moses Gomberg (1866–1947) succeeded in detecting $C(C_6H_5)_3$ in mobile equilibrium with hexaphenylethane in unreactive solvents, we may have R as low as 0.2 in the most favorable cases of unsaturated quarks [e.g. $A = 1000$ with 3001 (quarks minus anti-quarks) and with 3002 quarks, both having the atomic weight of 1200 amu as a numerical example, perhaps inspired by wishful thinking] if they are allowed to react and split into many smaller nuclei (no longer unsaturated). Today there are fewer QCD enthusiasts willing to bet their house that quarks are unconditionally confined [99].

Once we begin discussing "constituents", it becomes irresistible to look for composite leptons and quarks. An early theory of Harari [100, 101] can be nicely represented [41] as an almost inevitable consequence of a beautiful symmetry. One of the chapters in Brock's book [5] on Prout is entitled "The Bottomless Pit", citing Heaviside for our propensity for endless regression. Nature may not make a clear-cut decision. If the ratio R between the binding energy of three rishons [100] in a lepton or a quark and the rest-mass of the product is 10^{14}, we may be in the presence of a pit $f(x) = -2^{16x^2}$ times 1 amu, where the x values 0, 1, 2, and 3 have $-f(x)$ equal to 1; 2^{16}; 2^{64}; and 2^{144}, making the next step (x going from 2 to 3) actually impossible, rather than merely inconceivable. A (probably accidental) result is that the rishon (x = 2) would be comparable to the Planck mass (22 micrograms) and (x = 4) to the observable Universe, as said in another context by Arthur S. Eddington (1882–1944). After all, x may itself have to be a power of two.

It is a very profound question in quantum mechanics [11] (and Niels Bohr said that a "profound truth" is characterized by the opposite proposition also being a profound truth) as to whether "constituents" are *induced by exciting* the system sufficiently. For instance, is an α-particle spherical or tetrahedral? The ground state is a boson and has a vanishing spin quantum number *I*. No experiment on the ground state can show if it has any quadrupole moment (needing *I* at least 1) or octupole moment (as well-behaved tetrahedra have). The very high, lowest-lying excited states lose a proton or a neutron spontaneously (as is true for the ground state of ^{5}He, ^{5}Li, ^{10}Li, etc.). Nevertheless, being a boson, it is an appealing candidate for subnuclear structure (although quark clusters u^6d^6 may be less [43] important than once believed), and an astrophysicist familiar with stellar interiors and moderate temperatures below 25 megakelvin may certainly believe that the difficulty of smashing ^{4}He into four nucleons is, at least conceptually, quite similar to the task of smashing a proton into three quarks. The major difference is the high ^{1}H concentration in stars. On the other hand, the ground state of ^{16}O is also a boson with *I* zero. It is spherically symmetrical *ipso facto*, but it has rather high, but not dissociating states, such as *I* = 3, which *can* be interpreted [11] as rotation of a regular tetrahedron (e.g. consisting of four ^{4}He constituents). Is the ^{16}O ground state then tetrahedral? (as formerly inscribed in trams in Copenhagen "Speaking to the driver is not permitted"). Partons in violent collisions [63, 67] may certainly be quarks provoked by the excitation. But in any case, what are constituents nowadays?

We have been speaking of the serious, open-minded, but cautious Swiss chemist Marignac and the English polyhistor and chief editor of several scientific journals Crookes. We still wonder *why* Crookes had such an astonishing intuition in 1887. We are not arguing that bright intuitions of a chemist are essentially different from the intuitions he or she may enjoy in daily life or in exceptional situations. However, the point is that Crookes did not make a mild prevision of the future (like Jules Verne's inventions or Tsiolkowsky's space-rockets). Sixty years later, in 1948, nuclear physics used exclusively (Z, N) models, nuclear technology was rapidly growing on the same basis, and the 7 elementary particles were half-forgotten.

As a student, one of us [CKJ] remembers Crookes characterized as one of the myriad who always make mistakes. Why does it appear so different in 1988, in view of the last decades? Everyone always saw that Crookes was not pusillanimous, but this is not a sufficient reason for expressing a good feeling for what atomic physics is going to be like a century later. One may be Crookes' involvement with psychic research [102]. He was not a professor or lecturer at a university but a highly respected "private" scientist (as was frequent during this period). He encountered spectacular problems around 1875, when his friend, the gifted illusionist Florence Cook, was performing as a medium and was caught in an obvious fraud. There is another side of this activity (forbidden, but not denied, by many religious denominations) that one may consider, jocularly or seriously, for a moment. Were there any intelligent entities (whether being fallen angels, irreversibly retired professors, or the Committee at Higher Places) who injected their knowledge (for some purpose known only to them) and did not prevent him from expressing it in contemporary words (an admittedly difficult task)?

9 References

1. Weeks ME (1968) The discovery of the elements, J. Chem. Educ., Easton PA
2. Jørgensen CK (1987) Inorg. Chim. Acta 139: 1
3. Szabadváry F (1988) In: Gschneidner KA, Eyring L (eds) Handbook on the physics and chemistry of rare earths, North-Holland, Amsterdam, vol 11, pp 197-2
4. Jørgensen CK (1988) In: Gschneidner KA, Eyring L (eds) Handbook on the physics and chemistry of rare earths, North-Holland, Amsterdam, vol 11, pp 33-80
5. Brock WH (1985) From protyle to proton, William Prout and the nature of matter 1785-1985, Adam Hilger, Bristol
6. Jørgensen CK (1969) Oxidation numbers and oxidation states, Springer, Berlin Heidelberg New York
7. Kauffman GB (1969) J. Chem. Educ. 46: 128
8. van Spronsen JW (1969) The periodic system of the chemical elements, Elsevier, Amsterdam
9. Marignac JCG (1860) Bibl. Univ. Archives (Geneva) vol 9, pp 97-107
10. Marignac JCG: Oeuvres Complètes, 2 vols, C. Eggimann, Geneva (can be obtained from Dr. J. Deferne, Muséum d'histoire naturelle, 1 route de Malagnou, CH 1211 Geneva 4)
11. Jørgensen CK (1981) Structure and Bonding 43: 1
12. Jørgensen CK (1979) Centenaire de l'Ecole de Chimie de l'Université de Genève 1879-1979, Section de Chimie, CH 1211 Geneva 4
13. Strutt RJ (1901) Philos. Mag. 1: 311
14. Crookes W (1888) J. Chem. Soc. 53: 487
15. Crookes W (1895) Die Genesis der Elemente (2nd German edn Preyer W), Friedrich Vieweg, Braunschweig
16. Reisfeld R, Jørgensen CK (1977) Lasers and excited states of rare earths, Springer, Berlin Heidelberg New York
17. Reisfeld R, Jørgensen CK (1987) In: Gschneidner KA, Eyring L (eds) Handbook on the physics and chemistry of rare earths, North-Holland, Amsterdam, vol 9, pp 1-90
18. Crookes W: Proc. Roy. Soc. (1883) 35: 262; (1885) 38: 414; (1886) 40: 502
19. Crookes W (1905) Proc. Roy. Soc. (London) A76: 411
20. Urbain G (1909) Ann. Chim. Phys. (Paris) 18: 289
21. Palilla FC, Levine AK, Rinkevics M (1965) J. Electrochem. Soc. 112: 776
22. Brecher H, Samelson H, Lempicki A, Riley R, Peters T (1967) Phys. Rev. 155: 178
23. Sovers OJ, Yoshioka T (1969) J. Chem. Phys. 51: 5330
24. Crookes W (1886) Nature 34: 423
25. Jørgensen CK, Frenking G (this volume)
26. Wolfenden JH (1969) J. Chem. Educ. 46: 569
27. Urbain G (1925) Chem. Rev. 1: 143
28. Wilson DB (1987) Kelvin and Stokes – A comparative study in Victorian physics, Adam Hilger, Bristol

29. Ador E (1894) Archives des Sciences (Genève) 32: 5 and 183
30. Harkins WD, Wilson ED (1915) J. Am. Chem. Soc. 37: 1367
31. Thomson JJ (1897) Philos. Mag. 44: 293
32. Pais A (1977) Rev. Mod. Phys. 49: 925
33. Rutherford E (1900) Philos. Mag. 49: 1 and 161
34. Crookes W (1900) Proc. Roy. Soc (London) 66: 409
35. Rutherford E, Soddy F (1902) Philos. Mag. 4: 370 and 569
36. Rutherford E (1907) Philos. Mag. 14: 740
37. Rutherford E, Hahn O (1906) Philos. Mag. 12: 371
38. Battelli A, Occhialini A, Chella S (1910) Die Radioaktivität, Johann Ambrosius Barth, Leipzig
39. Rutherford E (1911) Philos. Mag. 21: 669
40. Curie M, Gleditsch E (1908) Compt. Rend. (Paris) 147: 345
41. Jørgensen CK (1982) Naturwissenschaften 69: 420
42. Rose HJ, Jones GA (1984) Nature 307: 245
43. Blendowske R, Walliser H (1988) Phys. Rev. Lett. 61: 1930
44. Richards TW (1925) Chem. Rev. 1: 1
45. Kauffman GB (1982) J. Chem Educ. 59: 3 and 119
46. Trimble V (1975) Rev. Mod. Phys. 47: 877
47. Kuroda PK (1982) The origin of the chemical elements, Springer, Berlin Heidelberg New York
48. Trimble V: Rev. Mod. Phys. (1982) 54: 1183; (1983) 55: 511
49. Pinto PA, Woosley SE (1988) Nature 333: 534
50. Burrows A, Lattimer JM (1988) Phys. Reports 163: 51
51. Applegate JH (1988) Phys. Reports 163: 141
52. Gehrz RD (1988) Nature 333: 705
53. Rudorf G (1900) The periodical classification and the problem of chemical evolution, Whittaker, London
54. Rowe MW (1986) J. Chem. Educ. 63: 300
55. Schidlowski M (1988) Nature 333: 313
56. Tully RB (1988) Nature 334: 209
57. Weinberg S (1979) The first three minutes, Bantam, New York
58. Maeder A, Renzini A (1984) Observational tests of the stellar evolution theory, IAU Symposium no. 105, Reidel, Dordrecht
59. Galeotti P, Lovera E (1974) Nature 249: 130
60. Jacobs JM, Dworetsky MM (1982) Nature 299: 535
61. Cowley CR, Rice JB (1981) Nature 294: 636
62. McCusker B (1983) The quest for quarks, Cambridge University Press, Cambridge
63. Pickering A (1984) Constructing quarks – A sociological history of particle physics, Edinburgh University Press, Edinburgh
64. Gell-Mann M (1964) Phys. Lett. 8: 214
65. Walker DC (1985) Accounts Chem. Res. 18: 167
66. Feynman RP (1974) Science 183: 601
67. Close FE (1979) Introduction to quarks and partons, Academic, London
68. Sachrajda CT (1988) Nature 333: 703
69. Satz H (1986) Nature 324: 116
70. McLerran L (1986) Rev. Mod. Phys. 58: 1021
71. Fraser G: CERN Courier (April 1988) p 5
72. Suhonen E (1982) Phys. Lett. 119B: 81
73. Cahn RN, Glashow SL (1981) Science 213: 607
74. Jørgensen CK (1981) Nature 292: 41
75. Scherrer RJ, Turner MS (1986) Phys. Rev. D33: 1585
76. Peebles PJE (1986) Nature 321: 27
77. Primack JR (1987) In: Cabibbo N (ed) Proceed. Internat. School of Physics Enrico Fermi, Course 92, pp 137–241, North-Holland, Amsterdam
78. Fraser G: CERN Courier (July/August 1988) p 1
79. Lindley D (1988) Nature 333: 391
80. Caffo M, Fanti R, Giacomelli G, Renzini A (eds) (1989) Proceedings of the Third ESO-CERN Symposium "Astronomy, Cosmology and Fundamental Physics", Bologna, May 1988, Kluwer, Dordrecht
81. Bodmer AR (1971) Phys. Rev. D4: 1601

82. Lee TD (1975) Rev. Mod. Phys. 47: 267
83. Witten E (1984) Phys. Rev. D30: 272
84. DeRújula A, Glashow SL (1984) Nature 312: 734
85. Farhi E, Jaffe RL (1984) Phys. Rev. D30: 2379
86. Michel FC (1988) Phys. Rev. Lett. 60: 677
87. Madsen J, Heiselberg H, Riisager K (1986) Phys. Rev. D34: 2947
88. Alcock C, Farhi E, Olinto A (1986) Phys. Rev. Lett. 57: 2088
89. Brown GE, Kubodera K, Page D, Pizzochero P (1988) Phys. Rev. D37: 2042
90. Jørgensen CK (1989) From atoms to polymers: Isoelectronic analogies, chapter "Chemistry of systems containing unsaturated quarks", VCH, New York, (Molecular Structure and Energetics) pp 257–277
91. Jørgensen CK (1984) Naturwissenschaften 71: 151 and 418
92. Lackner KS, Zweig G (1982) AIP Conf. Proc., American Institute of Physics, New York, vol 93, p 1
93. Jørgensen CK (1983) Nature 305: 787
94. Lackner KS, Zweig G Phys. Rev. (1983) D28: 1671; (1987) D36: 1562
95. LaRue GS, Phillips JD, Fairbank WM (1981) Phys. Rev. Lett. 46: 967
96. Milner RG, Cooper BH, Chang KH, Wilson K, Labrenz J, McKeown RD (1987) Phys. Rev. D36: 37
97. Van Polen J, Hagstrom RT, Hirsch G (1987) Phys. Rev. D36: 1983
98. Jørgensen CK (1988) Chimia 42: 21
99. DeRújula A, Giles RC, Jaffe RL (1978) Phys. Rev. D17: 285; (1980) D22: 227
100. Harari H (1979) Phys. Lett. 86B: 83
101. Harari H, Seiberg N (1981) Phys. Lett. 98B: 269; 100B: 41; 102B: 263
102. Oppenheim J (May 1986) Physics Today p 62
103. Wolke RL (1988) J. Chem. Educ. 65: 561
104. Boehm F, Vogel P (1987) Physics of massive neutrinos, Cambridge University Press, Cambridge
105. Rozental IL (1988) Big bang, big bounce – How particles and fields drive cosmic evolution, Springer, Berlin Heidelberg New York
106. Trimble V (1988) Nature 336: 111
107. Horvath JE, Benvenuto OG (1988) Phys. Lett. 213B: 516
108. Brown GE (1988) Nature 336: 519
109. Eichler D (1988) Nature 336: 557
110. Friedman JL, Imamura JN, Durisen RH, Parker L (1988) Nature 336: 560
111. Madsen J (1988) Phys. Rev. Lett. 61: 2909
112. Landolt-Börnstein Tables (New Series) (1982), Group VI, vol 2 (Astronomy and Astrophysics), subvol 2b: Stars and Star Clusters, Springer, Berlin Heidelberg New York

Author Index Volumes 1–73